Purification with

ACTIVATED CARBON:

Industrial
Commercial
Environmental

by

JOHN W. HASSLER

Specialist in Activated Carbon
Research, Manufacture, Marketing
since 1915

Developed original process to
manufacture activated carbon in
America.

1974
CHEMICAL PUBLISHING CO., INC.
NEW YORK, N.Y.

PREFACE TO THE THIRD EDITION

Much of the text has been re-written to incorporate major changes that have developed during the past decade. The title has been altered to capture the attention of those entering fields in which the purification properties of activated carbon could be useful.

Activated carbon enters many diverse applications, and the human life span is too brief to gain first hand knowledge of all facets. Much information can be borrowed from the scientific literature, and some workers are in positions to open lines of communication with authorities in related fields. Through those avenues, much has been received to aid in the preparation of this text. Among the many individuals, who have contributed varied forms of aid, are the following:

F. M. Middleton and Jesse Cohen of Taft Laboratory

L. F. Gleysteen, J. R. Conlisk, R. W. Behrens of Atlas Chemical Industries

F. Schwartz of North American Carbon

P. Wiley, A. Y. Hyndshaw, B. Jensen, J. Lienhardt, J. Fillicky, S. Smith, J. Drudy, J.M. Wafer of Westvaco

H. Todd, Jr., of Connor Engineering Corp.

H. L. Barnebey of Barnebey-Cheney

J. C. Cooper, R. S. Joyce, P. Walker, F. M. Williams, D. G. Chalmers, E. J. Cunningham of Calgon Corp.

Don Clendenin of U.S. Steel

Charles M. Staffer of Witco Chemical Company

R. Hagberg of Sucrest Corp.

Y. Hara and T. Oda of Takeda Chemical Industries

Y. Eguchi of Hokkaido University

H. Yanai of Muoran Institute Technology

T. Mc Donald of ACS at Ohio State University

W. W. Hassler of Indiana University of Pennsylvania

J. C. Enneking of Union Carbide Corp.

S. B. Dhungat of Londa, India

Y. Kishimoto, Charcoal Consultant, Tokyo

W. C. Bokhoven, C. van der Meijden of N. V. Norit-Vereening Vekoop Centrale

A. Kleinmayer of Carbomafra, Brazil.

W. A. Welch, Lancaster, Pa

ABRIDGED PREFACE TO JAPANESE EDITION OF ACTIVATED CARBON.

Although technical writing is a solitary and monetarily unrewarding task, there are compensations, chief of which are the enriching personal contacts. During the writing of *ACTIVATED CARBON*, I had the priviledge of discussing many aspects with numerous authorities.

Following publication, I made many new friends as a result of a mutual interest in activated carbon. A memorable instance occurred in 1965 when I received a most friendly letter from Takashi Oda of Takeda Chemical Industries in which he expressed interest in preparing a Japanese translation of *ACTIVATED CARBON*. This I deemed a high honor and I transmitted the message to the Chemical Publishing Company of New York, N.Y. who are the copyright owners. An agreement was reached with the Kyoritsu Shuppan Company of Japan for the publication of a Japanese translation.

The translation has been made by Yoshitomo Eguchi. During a visit to the United States, he together with Yujiro Hara came to my home to discuss the many varied aspects of activated carbon. It was a delightful and memorable experience, and I will always treasure a fond memory of those two gentlemen who so truly represent the courteous and friendly people of Japan.

PREFACE TO SECOND EDITION

During the decade since the publication of the first edition of this text, entitled *Active Carbon,* new vistas have unfolded for the industrial user of activated carbon. There has been a growing awareness of the many diverse forms of carbon usefulness; new processing techniques have been developed; and granular decolorizing carbons are now available commercially.

Such forward steps have increased the spheres of activity in the industrial use of activated carbon. They often introduce problems as to the right path to follow, however, and in many ventures the potential user is denied the guidance that could be available. To explain, let us review the earlier history of adsorptive carbon in industry. We find that markets were first established in processes already using adsorbents, such as bone char and fullers earth, for the purification of sugar, fats, glycerol, etc. Within those processes there are but few trade secrets, and the success of the operation depends mainly on efficient methods of manufacture and marketing. Consequently the users welcomed any and all aid they could obtain from suppliers of carbon. In general that attitude still holds in established applications and through such cooperation future growth can be assured in those markets.

Much growth, however, also can develop through participation in new products and in new processes. Unfortunately the opportunities for the supplier to contribute know-how in such ventures are frequently restricted because of the secrecy that so often surrounds the development. The reason for secrecy is understandable: Research and development costs are high, and these costs must be recouped in the relatively brief interval between the date a new product is placed on the market and the time for competition to catch up. Understandably, profit-minded industrialists, aware of ubiquitous competition, are unwilling to make premature disclosures even to those that could provide assistance.

An unfortunate consequence of this situation is that in many ventures that could become more efficient through the use of activated carbon, it may not be used; or it may be used ineffectively because of lack of know-how. In such situations a suitable written text can be useful because it asks no questions and carries away no data. This potential has guided the preparation of this revision. No attempt is made to furnish a definitive treatise; instead, the text is an introduction to basic principles and practices that should be considered in the industrial use of activated carbon. To that end the organization of subject matter has been altered and additional material is included.

Much of the added material is drawn from experiences during nearly fifty years' work with activated carbon. In 1915, I started with the West Virginia Pulp and Paper Company in its endeavor to pioneer the development of decolorizing carbons in this country. That endeavor culminated in the manufacture of Filtchar, the first commercial decolorizing carbon produced in America. From then till 1958, I participated in the research, manufacture, and marketing of other activated carbons including Nuchar and Suchar. On reaching retirement, I engaged in free-lance consulting. I am now associated with Barnebey-Cheney, a producer of activated carbon. I am also associated with the AMERICAN SOCIETY FOR TESTING AND MATERIALS in a research study to develop standard testing methods and nomenclature for activated carbon.

The experience of a single individual or group cannot cover an adequate understanding of all aspects of this many-sided subject. Therefore to provide a more comprehensive know-how I have sought the cooperation of others. In this I have had the assistance of sales, service, and research groups who have read sections of the manuscript, and supplied many valued suggestions. For making much of this help available, I am especially indebted to the following:

H. E. Pennington and H. B. Allport, of National Carbon Company; Jonathan C. Cooper, of Pittsburgh Chemical Company; R. W. Behrens, of Atlas Chemical Industries Inc;. Joseph M. Wafer, of West Virginia Pulp and Paper Company; H. L. Barnebey, of Barnebey-Cheney; W. C. Bokhoven and Chr. van der Meijden, of N.V. Norit-Vereeniging Verkoop Centrale.

I am also grateful to many who have contributed much in varied ways: Some supplied information needed for presenting specialized topics; some assisted in the preparation and editing of the manuscript; some drew attention to important items that otherwise would have been overlooked; still others corrected errors in the presentation. The information has been gathered for a number of years and limitations of space and memory preclude naming all who contributed. A limited list includes:

Ted Barnebey, J. D. Clendenin, Joseph E. Drudy, Marjorie Halstrick, T. J. Hassler, W. F. Heneghan, Donald K. Luke, Jr., F. M. Middleton, John J. Schanz, Jr., F. R. Schwartz, G. H. Scheffler, Evan A. Sigworth, J. H. Steen, Philip L. Walker, Jr., and Frank M. Williams.

Permission to reprint tables and figures has been acknowledged with the respective items. Some tables and figures for which permission was requested and granted have not been used because of subsequent changes in the manuscript.

I have made extensive use of the Deitz *Bibliography of Solid Adsorbents.* As it is in the form of a collection of abstracts, this publication offers the reader a better view of the content of each article than does a bare reference to the original source. Similar use has been made of *Chemical Abstracts.*

A word of explanation is in order for the change in title. The designation *Active Carbon,* used in the earlier book, is favored in a number of scientific circles, whereas most industrial workers prefer the term *Activated Carbon.* Inasmuch as this text is focused primarily on industrial aspects, it seems fitting to call this work *Activated Carbon.*

The preface to the first edition is included to establish continuity, and also to give recognition to the many persons who assisted in the preparation of the original text.

Finally, I am happy to express my perennial gratitude to my wife, Clara—my partner in all undertakings.

JOHN W. HASSLER

June, 1963.

PREFACE TO FIRST EDITION

This text has been prepared for operators and research workers in industry. I have attempted to survey principles and practices involved in the use of active carbon with the belief that an insight into the underlying features may suggest possible improvements in existing applications and also stimulate a search for new procedures.

The subject matter is grouped into three principal parts. Since many readers will be interested in only certain aspects, each part has been prepared so that it can be read independently of the others. Following an outline of the history and methods of manufacture, the next ten chapters discuss fundamental factors involved in adsorption by active carbon.

Applications are described in Chapters 13–30. A general outline is given of a number of applications, together with the specific objectives that are sought in each case. An operator will often find more helpful information from a description of how things are done in other processes than by copying what is being done elsewhere in his own field. With this in mind, some rather impractical procedures are described because they present novel approaches which could be useful in other applications.

The last four chapters cover experimental methods that have been found helpful in developing industrial applications.

Part of the information in this text has been drawn from personal experiences during thirty-five years' association with the manufacture and marketing of active carbon and this includes information received directly from other workers. To an even larger extent, the discussion is based on information in the literature. Some published data are obscure and other data are in dispute, so that it becomes the responsibility of the author to try and sift the facts. Consequently, the reader should be aware of the fact that the text contains not only facts, but also beliefs and opinions. The inquiring reader may find statements that he will question. This is as it should be. Questions stimulate independent thinking and this is necessary to integrate new facts with older knowledge. New facts become useful when they enlarge our vision and provide a new approach to problems.

I wish to express appreciation to those authors and publishers who granted permission to reproduce data. I have been fortunate in having received suggestions, criticisms, and information from many workers who have specialized knowledge. A number of authors checked references to their work and others reviewed portions of the manuscript. Professors Elroy J. Miller, Harold J. Cassidy, and F. E. Bartell gave helpful suggestions for portions of the text dealing with fundamental aspects of adsorptive behavior. Dr. Homer Adkins reviewed the chapter on catalysis, and Dr. H. L. Riley the section on the structure of carbon. Dr. Arthur Grollman read the chapters dealing with biochemicals; solvent recovery was reviewed by Dr. A. B. Ray, and air purification by Mr. Hugh Porter. The chapters on the removal of toxic gases and that on laboratory methods of gas adsorption were reviewed by personnel of the Chemical Corps Technical Command. Mr. Robert H. Buckie reviewed the entire manuscript and assisted with a number of translations. Much information on industrial applications was furnished by my associates throughout the West Virginia Pulp and Paper Company, and I regret that space does not permit detailed mention of the many helpful individual contributions. Finally, I wish to express my deep appreciation to Mr. Joseph Wafer, for without his help and encouragement, the text would never have been started and carried through to completion.

February, 1951   JOHN W. HASSLER

TABLE OF CONTENTS

PART I

INTRODUCTION

1

History and Market Review

"For there is nothing good or bad but thinking makes it so"
Shakespeare Hamlet

1. INTRODUCTION

Purity is a subjective concept. Salt is desirable on a breakfast egg, but not in a glass of drinking water. Pesticides benefit a growing crop, and detergents help in the laundry, but both are considered contamination when discharged into waterways.

No substance—of itself and by itself—is an impurity. Conversely, any substance can become contamination, if on entering a system, it damages desirable properties and characteristics.

Some types of contamination can be converted into acceptable forms, for example, the bleaching of an unwanted color with chlorine. But in general purification is accomplished by providing effective means of separation. The separation may be accomplished by removing the desired constituent from the system, as when pure water is distilled from salt brines; and when white sugar is crystallized from a syrup. Alternatively, it may be the contamination that is removed, as when dissolved mercury salts are precipitated and separated from the solution by filtration; or when acid gases are removed from air by contact with an alkaline solution. Some separations are accomplished through the ability of fluid molecules to adhere to the surfaces of solids—a phenomena known as *adsorption*. For practicable use, certain requirements must be met. A large surface area is essential because only a very small weight of molecules adheres to each square meter of surface.

1

The needed surface exists on porous solids known as *adsorbents*. Activated carbon adsorbents contain a myriad of micropores, the walls of which have surface areas that range from 400 to over 1800 square meters per gram in the various commercial brands. But more than large surface area is needed. To provide a means of separation, it is obvious that adsorption must be selective; certain species of molecules should be adsorbed in preference to others. To provide adequate purification, an adsorbent should be able to to take up and hold molecules of the substance to be removed without disturbing other constituents in the system. Moreover no single type of adsorbent surface will be appropriate for all diverse forms of contamination; therefore a variety of adsorbent surfaces should be available to handle the different needs.

It so happens that activated carbons can meet many of the diverse needs.[1,2,3] The adsorptive properties, which exist in primitive form in ordinary wood charcoal, can be developed in various forms by appropriate changes in manufacturing processes. Consequently, brands of commercial activated carbons made by dissimilar processes differ in adsorptive characteristics. Some excell for gas masks, others are superior for sugar refining, still others are best for water purification, and so on. Hence, we can quite properly consider that the term *activated carbon* comprises a family of adsorbents. That aspect extends and widens the potential utility because if one brand is tried and found wanting, possibilities remain that another brand will be suitable. Also to be considered is that the performance of carbon can often be guided into diverse channels by appropriate conditions. Consider the addition of activated carbon to a mixture of aniline and phenol in dilute aqueous solution: at pH 7, both will be equally adsorbed; at pH 10, aniline will be preferentially adsorbed; at pH 3, phenol will be preferentially adsorbed.

Features to be considered in the selection of carbon for use are considered in later chapters, but at this time we should mention the separate spheres of application of powdered and granular carbon. Powdered carbons are applied in a so-called *batch-contact* treatment: in this measured amounts of carbon and substance to be treated are mixed and subsequently separated by filtration. With granular carbons, the gas or liquid to be purified is passed continuously through a bed of carbon. For many years granular carbons (except bone char) were used primarily in vapor phase systems because the early commercial brands lacked adsorptive characteristics needed for most liquid phase purifications. After World War II, new granular brands were developed having a broad spectrum of adsorptive powers, and today are in use in many liquid phase appplications. They have opened new markets that offer promise of important future growth.

The adsorptive properties of carbon were well known long before the terms *active* and *activated* had been coined. In early literature data on the adsorptive properties appear under many varied names: bone char, blood char, coconut char, and others. More recently the information appears under *decolorizing carbon* and also under individual commercial brand names.

2. HISTORY

Early history[1-2] The use of charcoal for purposes other than as a fuel and in metallurgy is very old; the use in medicine being mentioned in an Egyptian papyrus from 1550 B.C. In the time of Hippocrates wood chars were used to treat various ailments. Kehl in 1793 discussed the use of char for removal of odors from gangrenous ulcers.

The earliest date at which adsorptive powers were definitely recognized was 1773 when Scheele[1] described experiments with gases. In 1785, Lowitz[4] called attention to decolorizing effects of charcoal on solutions. A few years later, wood char was employed to purify cane sugar, and in 1808 was applied to the then infant beet-sugar industry. Figuers[1] discovery in 1811 of greater decolorizing power of bone char led to its almost immediate adoption by the sugar refiners. At first, pulverized bone char was applied on a single use and discard basis, but limited supplies made regeneration necessary. A method of regenerating granular bone char was developed and that process is still in general use in refining cane sugar.

During the 19th century, many studies were made to develop decolorizing carbons from other source materials. Bussy[5], in 1822, heated blood with potash and produced a carbon with with 20 to 50 times the decolorizing power of bone char. Blood char so produced was used for years in many laboratory studies. Hunter[6], in 1865, reported the the gas-adsorbing power of cocoanut char. Stenhouse,[7] in 1865, prepared a decolorizing char by heating a mixture of flour, tar, and magnesium carbonate. Winser and Swindells[8] heated paper mill waste with phosphates. Many of the disclosures are similar to processes now in industrial use, and it is natural to inquire why they were not then developed on a commercial scale.

One answer is found in the manufacturing problems that are involved. Although activated carbon can be prepared with relative ease in the laboratory, the industrial production is attended by engineering difficulties. The corrosive action of many activation conditions requires special structural materials not then available. Moreover, successful industrial production depends on the skill of controlling manufacturing conditions within narrow limits, and suitable instrumentation for such control is relatively recent. The development was also retarded by the absence of apparent needs for a more adsorptive carbon than bone char. However, it is of interest to note that some early suggestions have today become important markets for activated carbon. Thus, Lipscombe[9] in 1862 prepared a carbon to purify potable water; and Stenhouse in 1854 described the forerunner of the modern gas mask.

Recent History

Work by Ostrejko[10] patented in 1900 and 1901 led to the development of modern commercial activated carbons. In one process, metallic chlorides were incorporated with carbonaceous materials before carbonization; another basic patent described selective oxidation with carbon dioxide at elevated temperatures.

Early interest in the development of activated carbon was stimulated by a belief that a large potential market existed in the cane sugar industry. Some early studies seemed to indicate that powdered activated carbon could be applied directly to the raw cane juice and thereby eliminate the need for an intermediate state in which raw sugar is first produced, which then must be redissolved and recrystallized. However, this hope was not realized; factory operation disclosed that the intermediate stage of raw sugar was still necessary.

Powdered activated carbon was found to be an effective substitute for bone char in established operations but required the installation of new specialized equipment. Inasmuch as the existing equipment for bone char represented a large capital investment, very few refiners were induced to make the change.

Sales opportunities were found for the purification of other products that require decolorization but the total extent of such markets fell far short of fulfilling the hopes of those that had pioneered in the development of what were then known as decolorizing carbons.

Such was the situation prevailing in 1915 when an event which was to shock the world brought fame to active carbon. Following the initial stage of rapid offensive operations, World War I had become a defensive action, with the opposing armies landlocked in trenches. Seeking to break the impasse, the Germans, on April 22, 1915, released chlorine from cylinders on a four-mile front near Ypres and a breeze carried the yellowish-green vapor over no-man's land into the Allied trenches. Against this weapon, the Allied soldiers had no protection; the agonized men choking and gasping could only flee, leaving a breach in the line. Had the Germans conducted the gas attack over a wider front and provided effective reserves to follow through the breach, they might have won a decisive victory. But the German soldiers had no protective armour against the gas and could not follow through. Soon after, another attack was made with similar results. The German command turned a study of other methods of gas warfare and the interval enabled the allies to provide protection in the form of gas masks containing activated carbon. Had it not been for that development, history might have followed a different course. Although powdered activated carbon had been manufactured for some years the powdered form was not suitable for gas masks. Consequently granular grades had to be developed. The rapid progress in developing effective methods of production is a brilliant episode in the annals of industrial chemistry.

3. MARKETS

The publicity given activated carbon in World War I stimulated a search for new peacetime markets, and the years since have witnessed steady growth.

Vapor and Gas Phase Applications

At the conclusion of the war, research found uses for granular carbon in various industrial operations. Processes were developed to recover gasoline from natural gas, and to extract benzene from manufactured gas.

Granular carbons are employed to purify various industrial gases. Sulfur compounds are removed from hydrogen and from acetylene; pyridine is removed from ammonia. Carbon dioxide for dry ice and carbonated beverages is deodorized.

A large market developed for recovery of vapors of organic solvents used in diverse industrial operations: as processing mediums in the manufacture of plastics and explosives; and as agents to apply a product to its intended use as in painting and printing. Many such solvents are volatile and the escape of the vapors into work rooms creates hazards to health, and from fire and explosion. Moreover, the amount of vapor thus vaporized constitutes an appreciable cost and recovery becomes important for economic reasons.

The need to protect the environment makes it necessary to end the discharge of alien vapors and gases to the atmosphere. Much research is now devoted to the removal of sulfur dioxide from stack gases.

The gas mask principle is employed to provide freshness to air in work rooms and living quarters. Purity of indoor air can be maintained by continuous recirculation through a bed of activated carbon. The incorporation of such use of carbon into an air conditioning system avoids the cost of ventilation with outdoor air during hot summer days. In areas where smog prevails outdoor air should be passed through carbon before being admitted to living quarters. Individuals living in smog areas can then go indoors to get a breath of fresh air. In submarines, activated carbon preserves the freshness of air that is reused very many times.

Liquid Phase Applications

The publicity given to activated carbon in World War I renewed the search for additional liquid phase markets. Although the prewar endeavor had fallen far short of the hopes of the entrepreneurs, it resulted in commercial production of carbons having great adsorptive power, and the knowledge and skills acquired were put to use. In addition, experience brought an awareness of the practical value of certain purification properties that had received very little prior attention. As we reflect back on the early use of powdered carbon we realize the dominant objective was color removal, and the term then generally used, *decolorizing carbon,* mirrors the thinking. However, those who used carbon for decolorizing found other benefits of practical value such as: improved flavor of food products; longer shelf life of packaged goods. In some applications, the use of carbon provides benefits to the operation of various stages of the manufacturing process. The presence of traces of adsorbable impurities can create difficulties listed in Table 1-1. Pretreatment of the process liquor with a suitable adsorbent will often avoid those difficulties.

TABLE 1:1

SOME OPERATIONAL DIFFICULTIES WHICH CAN OFTEN BE CORRECTED BY ACTIVATED CARBON

Slow and/or difficult filtration
Retarded and incomplete crystallization
Emulsion formation
Foaming during concentation and/or distillation

The liquid phase applications are reviewed in Chapter 7. Purification of most liquid systems by carbon is accomplished by adsorbing the contamination, the desired purified product remaining in the liquid. A different procedure is employed when a valuable adsorbable substance is present in minute concentrations. Then the desired substance is removed by selective adsorption and subsequently extracted in a purified and more concentrated condition. The operation known as desorption* is effected by specific changes in the environment; thus many vapors can be desorbed by an elevation of temperature. The desorption of substances taken up from liquids is more involved, and the details are examined in Chapter 5. The general features are illustrated by production of penicillin during World War II. In that process, activated carbon was added to an inoculated broth containing about 30 ppm penicillin. After removal from the broth, elution of the carbon yielded a relatively con-

*The overall process is known as adsorption-desorption.

centrated penicillin solution. At the time, adsorption-desorption provided the only immediately available method for large scale production of penicillin. Since then, experiences with adsorption-desorption in liquid systems have provided many disappointments, and the hopes once held for large markets remain unfulfilled. In contrast adsorption-desorption processes have been very successful in many vapor phase applications

When increasing amounts of carbon are added to a system, additional molecules become bonded to the carbon surface. Finally, practically all the adsorbable molecules are anchored at the surface and are immobilized. This feature is utilized to prevent the migration of color bodies in plastic solids as applied in the manufacture of white-sidewall tires. These are formed by placing a layer of white rubber over a carcass of dark rubber. Color molecules in the carcass rubber can migrate and stain the white-sidewall. This staining is prevented by incorporating activated carbon in the carcass rubber. Then, color molecules will migrate only until they reach a particle of carbon, whereupon they are adsorbed and anchored.

The *advanced treatment* (see chapter 8) of domestic and industrial waste waters is a promising future market for activated carbon. Aspects of that market recall the search for ways and means to purify potable water supplies. During the early decades of this century, an acute problem was the presence of objectionable tastes and odors in municipal water supplies. Several factors were responsible. One was the discharge of much greater quantities of domestic and industrial wastes into waterways. Another was the use of chlorine for disinfection.

Intensive research demonstrated that the use of activated carbon was essential to furnish a completely palatable water. Today, the use of activated carbon to purify municipal water supplies extends throughout the entire world. The development encountered many difficulties, and much of the rapid progress can be traced to a spirit of cooperation that prevails within the water works industry.

Most new specialized applications of carbon are initiated by personnel working within the industry because they are in a position to know what needs to be done to improve the operation. However, the progress of a new application can often be expedited by consultation with the carbon supplier.

REFERENCES

1. Deitz, V. R., Bibliography of Solid Adsorbents, United States Cane Sugar Refiners and Bone Char Manufacturers and National Bureau of Standards, Washington, D. C., 1944: History of adsorbent carbon, p. ix ff; abstracts of reviews, histories, and general discussions, pp. 689–696.
2. Bancroft, W. D., J. Phys, Chem., 24: 127, 201, 342 (1920).
3. Zerban, F. W., Louisiana Agric. Exper. Sta. Bulletin, 161 (1918).
4. Lowitz, Crell's Chem. Ann., 1:211 (1786).
5. Bussy, A., J. Pharm. Sci. Accessorires, 8:257 (1822).
6. Hunter, J., J. Chem. Soc., 18:285 (1865).
7. Stenhouse, J., British Patent 1395 (1856)., Chem. News, 3:78 (1861); 25:239 (1872).
8. Winser, F., and Swindells, J., British Patent 835 (1868).
9. Lipscombe, F., British Patent 2887 (1862).
10. Ostrejko, R. von, British Patents 14224 (1900); 18040 (1900); German Patent 136,792 (1901).
11. Parry, Richard H., Thesis, Pennsylvania State University, Dept. Mineral Economics, Dec. 1961.
12. Ratcliff, J. D., Readers Digest, July 1958; Today's Living, June 22, 1958.
13. Taste and Odor Control in Water Purification, Industrial Chemical Sales Division, West Virginia Pulp and Paper Co., New York, 1947; contains 1063 classified references. Hassler, W. W., J. Am. Water Works, Assoc., 33:2124 (1941). Wrench, J., Eng. Contract Record, 50:610 (1935).

2

Elementary Aspects of Adsorption

1. INTRODUCTION

Atoms and molecules are held together in a solid by cohesive forces that range from strong valence bonds to the relatively weak Van Der Waals forces of attraction. Molecules in the interior of a solid are completely surrounded, consequently their attractive forces are satisfied on all sides. The attractive forces do not cease abruptly at the surface. Instead they extend outward and can capture wandering fluid molecules—a phenomenon that is known as *adsorption*. The terms *adsorbent* and *adsorbate* describe the solid and its captured molecules.

But the forces radiating from the solid surface do not act alone they join with attractive forces radiating from the fluid molecules, and the combined attraction weaves an adsorption affinity. It follows that substances having great affinity will be adsorbed in preference to those with less affinity *if* other forces do not intervene. Other properties such as condensibility of a vapor or solubility of a solute may sometimes augment a certain adsorption and diminish another.

Adsorption is viewed as a surface phenomenon, and it is well to understand the full significance of that concept. Subconsciously, we think of a surface as the exterior of a material substance. However, a surface is really a boundary, because where the mass of one body ends, the mass of another begins. Consider a solid immersed in a liquid: there, the surface of the solid faces a corresponding surface of the liquid; and the region enclosed by those two surfaces is an *interface*. It is within that interfacial region that adsorption occurs.

In view of this, it is perhaps unfortunate that we generally refer to adsorption as ocurring on a surface, but it would be pedantic to change the custom. However, we can and should keep in mind that the solid surface is just a part of an interfacial region, and realize that the masses on both sides of the interface participate. The adsorption-affinity which pulls molecules to the interface can be assisted or opposed by forces on the other side of the interface. To take a single example: the dissolving power of the solvent will pull molecules away from the surface and will have a stronger pull for those that are more soluble. Our knowledge of the interplay of all the varied forces operating at the interface is dim and indistinct. However experience has taught us ways and means whereby we can often improve the effectivness of an adsorption process.

2. CHARACTERISTICS OF ADSORPTION FROM GAS/VAPOR PHASE[1,2]

Molecules remain in the gas/vapor state because of the energy of separation and translation contained in the moving molecules. In order to adhere to the interface, that energy must be displaced to bring the molecules to a state of rest. When activated carbon is present in the system, we can visualize that the adsorption-affinity squeezes away the energy, which is released in the form of *heat of adsorption*.

If an insufficient amount of carbon is present, some gas/vapor molecules will not be adsorbed but will remain in the gas/vapor side of the interface. These can be adsorbed by adding more carbon; or alternatively by lowering the temperature of the system. When the temperature is lowered, energy is removed from the moving molecules, and less attractive force is needed to displace the residual energy from each molecule. The force thus saved is available to bring more gas/vapor molecues to a state of rest.

Many features of our present knowledge of adsorption of gas

and vapor are outlined in an investigation of de Saussure[3] published in 1814. He found that porous substances such as charcoal will take up many gases but in different amounts, and that the more condensible gases are adsorbed in greater quantities. Many subsequent studies have confirmed that when adsorptions are conducted at an identical temperature, e.g., 20°C, a correlation exists between the adsorbability and the boiling point, also to the critical temperature. Or as we may otherwise express it, vapors that readily condense to a liquid state will be more readily adsorbed than gases that liquefy only if and when cooled to a low temperature. Such interdependence is to be expected inasmuch as molecules which can more readily slow down to the comparative rest or the liquid state should by the same token be more easily captured by a solid surface.

Early investigators described adsorption as surface condensation; and today it is generally accepted that the forces that bind molecules or a liquid (van der Waals) are a large factor in physical* adsorption at solid surfaces. A strong attraction is contributed by the solid surfaces-this is demonstrated by the fact that vapors will condense on a solid surface under conditions that are inadequate to cause normal condensation to a liquid. A familiar illustration is the way in which dust particles enable a supersaturated vapor to condense to a liquid; or the way dew forms on solid surfaces when the humidity is not sufficient to precipitate as a fog. In presenting the contribution of solid surfaces we must not lose sight of the fact that the cooperation of vapor molecules is essential, and a greater measure of assistance can be furnished by molecules that more readily condense to a liquid.

Gases having a low boiling point and critical temperature are not appreciably adsorbed at 0°C. However, they can be adsorbed in large quantities at a sufficiently low temperature. By appropriate adjustment of the temperature it is often possible to separate such gases by adsorption. The proper temperature depends on the gases to be separated. Yagi[4] points out that carbon dioxide is effectively separated from air between –90°C and –110°C; at higher temperatures carbon dioxide is incompletely adsorbed, and at lower temperature adsorption of nitrogen and oxygen occurs.

*The roles of physical and chemical forces in adsorption phenomena are reviewed in Chapter 13.

Although condensibility is a major factor, the amount of gas or vapor adsorbed is not always in exact correlation with the boiling point or critical temperature. Deviations appear in Table 2:1, and Bancroft[5] and Mantell[6] call attention ot others. The deviations suggest that we lack full knowledge of all forces that are involved. In many studies, the adsorption of different vapors are measured at identical temperatures, a condition that augments the influence of the condensibility factor. This can be illustrated by a comparison of methane (boiling point, −160°C) and methanol (boiling point, 64°C). If both are adsorbed at 25°C it is apparent that the adsorption of methanol will be far greater because of the relative ease or condensation at the temperature.

TABLE 2:1

ADSORPTION OF GASES AT 0° C ON CHARCOAL*

Gas	Boiling point	Critical temperature	Gas adsorbed cc/g charcoal	
			Wood, 10 cm pressure	Coconut, 76 cm pressure
	° C	° C		
Helium	−268	−267	—	2
Hydrogen	−253	−241	0.3	4
Nitrogen	−196	−149	1.5	15
Carbon monoxide	−190	−136	—	21
Oxygen	−183	−119	2	18
Argon	−186	−117	—	12
Carbon dioxide	− 78	+131	20	—
Ammonia	− 33.5	+130	50	—

* From A. B. Lamb, R. E. Wilson, and N. K. Chaney, *Ind. Eng. Chem.* 11:420 (1919), by permission of the American Chemical Society.

Various ways have been searched to equalize the condensibility factor. One approach is to measure the adsorption of vapors at equal ratios of partial pressure to saturation pressure.

Pearce[7] sought to equalize the influence of condensibility by conducting the adsorption of each vapor at a temperature having the same definite ratio to its boiling point. Pearce found a relation to exist between molecular structure and the amount adsorbed, but the relation depends on the pressure at which the adsorption is measured.

At low pressures the adsorption increases with molecular size in an homologous series. The adsorption also increases when an atom of oxygen, chlorine, or nitrogen is introduced into an organic chemical molecule-an effect that has been interpreted as indicating a direct attraction to carbon for such substituent atoms.

The foregoing behavior occurs only at low pressures. Increasing the pressure promotes the adsorption of smaller molecules, and eventually reverses the order found at low pressures. The reversal has been ascribed to a crowding effect that appears sooner for large molecules which have greater covering power and take more room.

In evaluating the relative adsorbability of vapors, the interpretation of data can depend on whether the amount adsorbed is calculated as mols or grams. Thus carbon N adsorbed approximately the same number of millimoles of chloroform as of carbon tetrachloride, and on that basis, both vapors would be viewed as equally adsorbable. But on a gram basis the carbon tetrachloride would be judged more adsorbable because of a greater molecular weight.[8]

Mixtures of Gases and Vapors

From a mixture of two gases or vapors, the one more strongly adsorbed from the pure state will usually be preferentially adsorbed from a mixture; but each may be less adsorbed from a mixture than when present separately at the same partial pressure in a pure state. In most industrial applications, the adsorptions of some components are so small as to be negligible, and the adsorbable component can be viewed as though it were present alone. An example is the adsorption of organic vapors from air.

Effects of High Pressure

A large increase in pressure has little effect on the adsorption of the more adsorbable vapors, many of which approach saturated adsorption at normal pressures. The less adsorbable gases, however, often show a marked increase at high pressures. The relative amounts adsorbed from mixtures can vary with changes in pressure, Lorenz and Wiedbrauk,[9] in a study of mixtures of ethylene and carbon dioxide, found the adsorption of ethylene to be greater at low pressures; whereas this was reversed at high pressures. Richardson and Woodhouse[10] observed that at 2870 millimeters pressure, carbon dioxide and nitrous oxide are adsorbed in almost equal volumes; whereas at 72 millimeters pressure, two-thirds of the adsorbed gas is nitrous oxide.

3. CHARACTERISTICS OF ADSORPTION FROM THE LIQUID PHASE

The manifold forces acting at the solid liquid interface provide a complex phenomena that is formidable to analyze. The algerbraic sum of all the forces (Table 2:2) is measured by the quantity of substance adsorbed by a given weight of carbon. We lack quantitative knowledge of *separate strengths* of the individual factors, namely: the solvent the brand of carbon and the solute to be adsorbed. However, experience has developed guide lines that aid in the selection or appropriate operating conditions.

TABLE 2:2

SOME FACTORS INFLUENCING ADSORPTION AT CARBON/LIQUID
INTERFACE

a) Attraction of carbon for solute
b) Attraction of carbon for solvent
c) Solubilizing power of solvent for solute
d) Association
e) Ionization
f) Effect of solvent on orientation at interface
g) Competition for interface in presence of multiple solutes
h) Interactions of multiple solutes
i) Coadsorption
j) Molecular size of molecules in the system
k) Pore size distribution in carbon
l) Surface area of carbon
m) Concentration of constituents.

Adsorbable Solutes

Inorganic compounds reveal a wide range of adsorbability: At one extreme are dissociated salts such as potassium chloride and sodium sulfate which (for all practical purposes) are viewed as non-adsorbable by carbon. At another extreme we find iodine which is one of the most adsorbable substances known. Between those extremes are all degrees of adsorbability. The adsorption of some substances is accompanied by chemical changes.

The published data, on adsorption of organic compounds from solution, provide ample evidence that the architecture of a molecule is an important factor in adsorption phenomena. Some reported trends are listed in Tables 2:3.

TABLE   2:3

INFLUENCE OF MOLECULAR ARCHITECTURE ON ADSORBABILITY

a)   Aromatic compounds are in general more adsorbable than aliphatic compounds of similar molecular size.
b)   Branched chains are usually more adsorbable than straight chains.
c)   Influence of substituent group is modified by position occupied, *e.g.*, ortho, meta, para.
d)   Stereoisomers show inconsistent pattern; thus fumaric acid (*trans*) is more adsorbable than maleic (*cis*), but the *trans* form of hydrobenzoin is less adsorbed than the *cis* form.
e)   Optical isomers, *dextro* and *levo*, appear to be equally adsorbed.

In comparing the data of different investigations, one should be aware that quantitative data (i.e., precise quantities of substance adsorbed) are reproducible only when the studies are conducted under identical operating conditions. A different temperature or concentration, or the use of a different brand of carbon can alter the amount adsorbed; and on occasion may even reverse the apparent relative adsorbability of some compounds.

Much published data are based on studies of adsorption from solutions containing only a single solute. Such data can seldom be directly related to industrial processes in which the impurities to be removed are generally a mixture of substances of unknown identity. However, studies using a single solute are essential in research seeking to trace the influence of the separate factors in Table 2:2. On this, our information is far from complete. As of now, we see only the probable direction and the possible strength of each influence, the latter only in a very qualitative way. Let us examine portions of what we now know.

Solubility

An increase in solubility reflects a greater affinity between the solvent and the solute, and acts to oppose the attraction to the carbon. Consequently, any change that increases solubility may be accompanied by reduced adsorbability. Thus, polar groups (characterized by an affinity for water) usually diminish adsorption from aqueous solutions. Conversely, the greater adsorption of the higher aliphatic acids and alcohols is attributed in part to their relatively smaller solubility in aqueous solution.

Solubility has been established as a factor in adsorption, but it is important to recognize the true nature of its influence which is to serve as a restraining force-a brake against the pull of the carbon. Even great solubility does not prevent the adsorption of substance that is strongly attracted to the carbon surface, *e.g.*, the very soluble chloracetic acid is well adsorbed. Conversely, a slightly soluble substance will be adsorbed only if and when an attraction exists to the carbon surface.

Ionization

Ionization is usually adverse to adsorption by carbon. Ions of inorganic salts are not appreciably adsorbed. Undissociated molecules of organic compounds are more adsorbable than the ions of dissociated molecules.

Although as a general rule, ions are not well adsorbed by carbon, an exception is found in case of hydrogen ions, which are appreciably adsorbed under some conditions. Because of this, other negative ions become more adsorbable when associated with hydrogen ions as in an acid. Even mineral acids such as sulfuric are appreciably adsorbed from relatively strongconcentrations.

In many systems the acidity or basicity of a solution is an important factor. A low *p*H promotes the adsorption of organic acids; whereas a high *p*H is favorable for adsorption of organic bases. The optimum *p*H is specific for each solute.[11]

The adsorption of some non-electrolytes can be affected by changes in *p*H,[12] and curiously the *p*H of the carbon used can influence adsorption from some non-aqueous liquids, *e.g.,* the decolorization of an edible oil.

Multiple Solutes

Purifications in industrial processes usually involve the removal of mixtures of solutes, consequently, it is helpful to have some knowledge of influences exerted by multiple solutes on adsorptive phenomena. The effects are specific depending as they do upon the relationships existing between properties of the separate solutes.

The adsorption of a solute is usually diminished by the presence of other types of adsorbable solutes, although an adsorbable increase is sometimes observed. Compounds that have greater relative adsorbability, as measured from pure* solutions are often preferentially adsorbed from mixtures, but there are frequent exceptions. The quantity adsorbed from a pure solution does not necessarily reflect the strength of the attachment.

When several species of solutes are simultaneously adsorbed, the total amount can depend on surface area relationships. When and where the different species are attached to separate areas of the surface are factors that might increase opportunities for a larger overall amount of adsorbed substances. Frequently, the different species compete for the same sites and then the adsorption of each is less than from a pure solution.

Another major factor evolves from the ability of some solutes to alter solution status of other specific solutes. Thus, potassium iodide increases the solubulity of iodine and thereby reduces the amount of iodine that is adsorbed. Conversely, the adsorption of fatty acids is increased when the solubility is reduced by the addition of sodium chloride to the aqueous solution. The effect on fatty acids is in sharp contrast to that observed with mercuric chloride. The adsorption of mercuric chloride may be diminished when sodium chloride is present[13]: an action ascribed to the formation of a weakly adsorbable complex, $Na_2 Hg Cl_4$.

Co-adsorption

Instances are reported of solutes that are able to enhance the adsorption of certain other specific solutes. Eisler[14] observed that cholesterol and saponin mutually increase each others adsorption. Rossi and Bassini[15] found that char after adsorbing congo red had greater power to adsorb indigo; but had less power to adsorb methylene blue. Miller[16] found that a char impregnated with a water insoluble acid dye was able to adsorb alkali from an aqueous solution, a property not shown by the untreated char. This cooperative action of adsorbates - a phenomenon known as co-adsorption is considered further in Chapter 5.

---

*The expression pure state or pure solution as used here refers to adsorption conducted from a solution in which only a single solute is present.

Solvent Influence

Water is the solvent most generally used in industrial processes that employ carbon for purification. The influence exerted by the liquid side of an interface is clearly revealed when another solvent is substituted (Table 2:4). The adsorption of most organic compounds is less from an organic solvent than from an aqueous solution. One cause is the greater solubility of organic compounds in organic solvents. Another factor is solvent adsorption. In contrast to water which is weakly adsorbed by carbon, organic solvents are strongly adsorbed, and consequently leave less free available surface space for solute molecules.

TABLE   2:4

INFLUENCE OF SOLVENT ON ADSORPTION OF DYE*

Carbon Code	Methylene Blue[a]		Malachite Green[a]		Alizarin Red[a]	
	In water	In ethanol	In water	In ethanol	In water	In ethano
A	0.84	0.26	1.07	0.113	1.25	0.35
C	0.70	0.15	0.74	0.029	1.00	0.43
D	0.37	0.07	0.45	0.017	0.62	0.14
E	0.30	0.08	0.34	0.008	0.45	0.17
F	0.44	0.12	0.19	0.018	0.39	0.12
H	0.73	0.07	1.10	0.050	0.95	0.24

[a] Millimoles per gram of carbon at a concentration of 0.1 millimoles per liter in equilibrium solution.

* Data from *Ind. Eng. Chem.* 37 : 645 (1945); reprinted by permission of the American Chemical Society.

The amount of the reduction varies: the use of ethanol in place of water as a solvent made a greater reduction in the adsorption of malachite green than of methylene blue. One must also reckon with specific effects of the type of carbon used. The ratio of dye adsorbed from different solvents can depend on the carbon used.

Mixed Solvents

Studies of adsorption from mixed solvents reveal diverse effects. In some cases, such as a mixture of toluene and benzene, the adsorption is directly related to the percentage of each solvent in the mixture. With some mixtures, one of the components may have a major influence even when present in minor quantities. With mixtures of some solvents, maximum adsorption has been observed at an intermediate mixture of the solvents. Other instances of points of minimum adsorption have been reported, a behavior that recalls maximum and minimums in vapor pressure observed in mixtures of some solvents.

Temperature

Except when a chemical reaction is involved, an elevation in temperature increases the escaping tendency of a vapor or gas from the interface and thereby diminishes the adsorption. The same influence functions at the carbon liquid interface, but here the escaping tendency is often outweighed by the effect of temperature on the other factors listed in Table 2:2. The resulting action can be quite specific. Instances are known in which an elevation in temperature increased the adsorption of one solute and simultaneously decreased the adsorption of other solutes present in the same solution.

Adsorption involves specific relationships between the properties of the carbon and those of the solute; consequently the quantitative effects of temperature are not alike with all carbons.

4.   RATE OF ADSORPTION

When an adsorbable vapor or solute molecule makes contact with a suitable unoccupied space on the carbon surface, the molecule will adhere almost instantly. Therefore the time required for an adsorption process depends on the rate at which vapor, gas, or solute molecules can diffuse to the carbon surface-plus the time to find an unoccupied space. In systems where a gas, vapor, or liquid solution is passed through a bed of granular carbon, the molecules need travel only a short distance to reach the carbon surface. Having established contact the adsorbable molecules must now find a suitable unoccupied site. Initially many sites are available on the exterior portions of the surface, but when these are filled, molecules arriving later must find sites in the interior and this can require a search through tortuous channels. This does not appear to present difficulties for gases or vapors as is evident from the ability of a thin layer of activated carbon in a gas mask to provide complete protection.

With liquids, the experience is somewhat different. After a brief rapid initial rate, the subsequent rate becomes much slower especially for large molecules. Granular carbons (12 × 30) may require many hours to utilize an appreciable portion of the total potential adsorptive capacity. The rate of adsorption from liquids is accelerated by pulverizing to expose more of the interior surface. With powdered carbon, if mixing is adequate, the major part of the adsorption is accomplished within an hour. Then adsorption may continue at a diminishing rate for days and longer.

REFERENCES

1. Adam, N. K., *Physics and Chemistry of Surfaces,* Oxford University Press, London, 1941.
    Adamson, A. W., *Physical Chemistry of Surfaces,* Interscience Publishers, Inc., New York, 1960.
    Alexander, J., *Colloid Chemistry,* 6 volumes; Chemical Catalog Co. and Reinhold Publishing Corporation, New York, 1926–1946. Brunauer, S., *The Adsorption of Gases and Vapors,* Physical Adsorption, Princeton University Press, Princeton, 1943. Buzagh, A., von, *Colloid Systems,* Technical Press Ltd., London, 1937.
    Freundlich, H., *Colloid and Capillary Chemistry,* E. P. Dutton and Co., New York, 1922.
    McBain, J. W., The *Sorption of Gases and Vapors* by Solids, G. Routledge Sons Ltd., London, 1932.
    Weiser, H. B., *Colloid Chemistry,* John Wiley, New York, 1949.
2. Deitz, V. R., *Bibliography of Solid Adsorbents,* United States Cane Sugar Refiners and Bone Char Manufacturers and the National Bureau of Standards, Washington, D. C., 1944; ab-

stracts of scientific literature on theories of adsorption for the years 1900 to 1942 inclusive.

Deitz, V. R., *Bibliography of Solid Adsorbents 1943-1953* National Bureau of Standards, Circular 566, Washington, D.C., 1956.

Witco Chemical Company Inc. Abstracts from 1946 to 1969.

3. McBain, J. W., *The Sorption of Gases and Vapors by Solids,* G. Routledge and Sons, Ltd., London 1932, p. 2.

4. Yagi, S., *J. Soc. Chem. Ind. Japan,* 34 (suppl. binding): 203 (1931).

5. Bancroft, W. D., *Applied Colloid Chemistry,* 3rd Edition, McGraw-Hill Book Company, New York, 1932, p. 5.

6. Mantell, C. L., *Industrial Carbon,* D. Van Nostrand Company, New York, 1946.

7. Pearce, J. N., and Eversole, J. F., *J. Phys. Chem.* 38:383 (1934).

    Pearce, J. N., and Hanson, A. C., *J. Phys. Chem.,* 39:679 (1935).

    Pearce, J. N., and Johnstone, H. F., *J. Phys. Chem.,* 34:1260 (1930).

    Pearce, J. N., and Peters, P. E., *J. Phys. Chem.,* 42:229 (1938).

    Pearce, J. N., and Taylor, A. L., *J. Phys. Chem.,* 35:1091 (1931).

8. Pennington, H. E., Personal communication.

9. Lorenz, R., and Wiedbrauck, E., *Z. anorg. allgem. Chem.,* 143: 268(1925).

10. Richardson, L. B., and Woodhouse, J. C., *J. Am. Chem. Soc.,* 45:2638(1923).

11. Phelps, H. J., *J. Chem. Soc.,* 1929, 1724.

12. Kniaseff, V., *J. Phys. Chem.,* 36:1191(1932).

13. Lawande, Y. V., and Karve, D. D., Proc. Indian Acad. Sci., 21A:41(1945)

14. Eisler, M., Biochem. Z., 172:154(1926).

15. Rossi, G., and Basini, A., Ann. chim. applicata, 16:306(1926).

16. Miller, E. J., J. Phys, Chem., 36:2967 (1932).

PART II

APPLICATION TO INDUSTRIAL AND ENVIRONMENTAL LIQUID SYSTEMS

3

BASIC ASPECTS and CONCEPTS

1. INTRODUCTION

This chapter will review ways and means to explore the potential of activated carbon for a new venture—the present discussion being limited to liquid systems.

An appraisal of the suitability of activated carbon for an untried operation involves two basic considerations:

a) Will activated carbon accomplish the desired objective?

b) If so, how does the efficiency compare with that provided by other means?

Crystallization, distillation, precipitation, chemical agents, are among the other methods to be considered. No one method of separation is universally applicable; the usefulness of each is related to the system to be treated and on the task to be done. With some operations there is little or no difficulty in deciding which separation method to use, but frequently we encounter twilight situations in which any of several methods could function. Further, there are certain systems that require a suitable combination of several separate methods.

A first step in a study of purification is to delineate the characteristics of the system under study and the objective sought, after which a tentative selection can be made of the method or methods that appear to warrant further investigation.

The purpose of this chapter is to furnish an understanding of ways and means to evaluate the suitability of activated carbon for a new application. No attempt will be made to furnish the know-how in the form of specific procedures; it would not be feasible to cover the many divergent situations to be encountered.

At one extreme we may find a simple project that seeks only to make a modest improvement in an established operation, and which may require nothing more than the selection of a suitable type of carbon. From such a simple task we can pass, by degrees of complexity and novelty, to projects that envision the creation of new kinds of usefulness. The understanding needed to conduct this research can be described variously as a pragmatic or working philosophy, a habit of mind, or a way of thinking things through. Its contour will become clarified as the discussion proceeds. For now we may say in brief that is is a philosophy that contains a recognition that:

a) The solution for most situations can be sought by *cut and try*;
b) We must have an awareness of what to try;
c) Having made a try, we must have a sort of sixth sense to discern whether we are headed in the right direction.

Experience may contain the awareness of what to try. It may be an item from some particular experience, or it may be a hunch that arises intuitively from a general background of experience. Although experience often furnishes a direct clue to the solution of a new problem, it is seldom that an established operation can serve as an exact pattern for conducting a new and untried operation—not even when the same product is involved. Consider the processing of sugar from cane and from beets: In both juices the main ingredient is sucrose and it certainly influences the adsorptive behavior of carbon. Even more, however, may depend on the properties of the impurities to be removed. As the impurities in cane differ from those in beets with respect to chemical identity and concentration, the method of treatment can and does differ. Still further, we find that various details of the treatment given to cane juice must be modified according to the species of cane processed, and to the region in which it was grown. Beyond this is the fact that some conditions of treatment provided at the beginning of a sugar-grinding season often require some change as the season approaches an end.

If such a long-established operation as sugar refining cannot be placed in a fixed groove, one might well expect perplexities to arise in developing proper operating conditions for a new venture, and especially for one that departs from traditional forms. The difficulty is usually less than might be anticipated; most untried operations are new primarily in that they are a new permutation of steps taken from operations done before and elsewhere; consequently a thoughtful study of diverse operations and experiences will often furnish outlines from which a resourceful imagination can carry on.

To illustrate what we mean by a thoughtful study, consider Table 3:1. Little or nothing will be gained by memorizing isolated items; thus the fact that 2% of Carbon A at 80°C removed 80% of color from a sample of phosphoric acid is an observation unlikely to be of great use in some future situation. However it is of some value to observe that the decolorization of phosphoric acid increased at a higher temperature. We gain more by observing that whereas a higher temperature improved the decolorization of one batch of hydrol, a change in temperature had little or no effect on the decolorization of another batch of hydrol obtained from a different factory. And we gain still more understanding when we discern that a lower temperature is more effective for decolorizing

TABLE 3:1

INFLUENCE OF TEMPERATURE ON DECOLORIZATION

Substance	Carbon		Color removed at indicated Temperature %		
	Type	Dose, %	25° C	50° C	80° C
Phosphoric acid	A	2	60	75	80
Hydrol (batch #1)	F	1	70	85	95
Hydrol (batch #2)	F	1	60	65	65
Lactic acid	A	1	70	60	55

a particular sample of lactic acid. Separately each trend has some significance, but primarily they are meaningful in that they all lead to a salient point, namely: The direction and extent of the influence of temperature depends on the specific system under study; and only by cut-and-try can we learn what the effect will be on an untried system.

It is in this fashion that experimental data and operating experiences are searched to learn of other influencing factors and the trends they can impose. Thereby we gain a kit of tools with which to explore the suitability of activated carbon for a new venture. In this chapter we seek to give a panoramic view of lessons learned over a period of years devoted to the application of activated carbon in industrial operations. Contributions are from varied sources: personal communications;[1] published literature;[2,3] and direct experience. Many items are based on long established plant-scale operations; others are from laboratory notes covering projects that never reached the production stage. Some observations are abstracted from very recent developments; others are recollections of operations discontinued years ago, such as the recovery of iodine from salt waters. Throughout, the selection is based on the criterion of sound possibilities for extending the usefulness of activated carbon into untried fields of endeavor.

Inasmuch as the intent of this text is not to recommend specific types of carbon; but instead to aid the reader to develop his judgment in this field, data relating to commercial brands of carbon will be quoted under code. The use of a code is warranted also for other reasons. Most name brands of commercial carbons are offered in a number of grades, each having distinctive standards of quality; consequently data quoted under brand names can be misleading unless the exact grade is known and specified. Still further, quality standards change and develop, hence data obtained in earlier years may not apply to production today. Data relating to proprietary and trade marked items will also be reported under a code.

Experimental Aspects

Batch type experimental techniques employed to search out a new venture in industry are essentially similar to those employed in academic studies, but the flexibility is less. In academic studies, the purpose of which is to increase our store of general knowledge, the investigator works with substances of known identity and concentration, moreover he has much freedom in the choice of experimental conditions.

In contrast, materials for an industrial study must be handled on an *as received* basis. Knowledge is often lacking as to the identities and concentrations of the contaminating substances, and they are the ingredients to be removed, Moreover, in the selection of experimental conditions, one must consider possible adverse effects on the prime ingredient—the product. Thus although a sugar syrup may be effectively decolorized at pH 3, such a low pH cannot be used because of damage to the sucrose by inversion. One must also be guided by practical requirements of plant operation; e.g., it may be necessary to use an elevated temperature in order to provide rapid filtration.

It is also necessary to consider the possibility that the product itself may be very adsorbable, and if so take steps to minimize loss of product through this cause.* Losses also occur as a consequence of the liquid being held as in a sponge; up to three parts of liquid may be retained by each part of carbon used (for some carbons the retention is even higher), all of which limits the dosage of carbon. The allowable maximum will vary from one industrial operation to another depending on such affiliated factors as the cost of product lost by retention, the enhancement of the quality of the product by the purification, and operational considerations such as filter capacity. Although carbon dosages of 5 per cent and more are known, for most applications, the carbon dosage is less than 2 per cent.

As such small quantities of carbon cannot adsorb large gravimetric amounts of contaminants, consequently one may wonder how carbon is able to accomplish the abundant benefits so generally reported. When we seek for the answer by examining established operations, we find that the benefits are measured, not by weighed amounts of contaminating substances removed, but by the correction or elimination of unwanted characteristics such as color, odor, foam, and other properties that impair the salability of the product or impede processing operations. In most applications for which activated carbon has been used successfully, the unwanted characteristics are caused by small gravimetric concentrations of parent substances with very potent properties. Consequently we can expect the use of carbon to be practicable when a large nuisance is created by a small amount of parent substance. It is in such areas that fertile markets for activated carbon are to be found, and for this ample evidence exists. Perhaps the best documented field is that of water purification where even the more severe conditions of taste are caused by contaminants present in parts per million or less. And in this

*Here we are considering the more traditional type of application in which unwanted ingredients are adsorbed leaving the purified product in the liquid phase. Processes in which the desired substance is itself adsorbed and subsequently eluted are coverd in Chapter 5.

field literally thousands of problems of taste have been cured by the use of activated carbon.

But not all problems of taste, color, and other unwanted characteristics are caused by small concentrations of parent substances, and when unwanted properties are associated with large gravimetric concentrations other means of purification must be sought.** Therefore a first step in determining the suitability of carbon for a new application is to ascertain the adsorbability and the concentration of the offending parent substances. Such information is seldom directly available because the contamination is usually a mixture of diverse compounds whose individual identities are often unknown because of analytical difficulties, especially during the development stage of a process. The lack of knowledge that sometimes exists as to the nature of the contamination is evidenced by instances in which studies have been conducted to remove a color or odor that was subsequently found to be a characteristic property of the product itself.

2. PRELIMINARY SURVEY

When, as is generally the case, analytical data are lacking on the contamination, it becomes necessary to utilize the behavior of the unwanted properties to diagnose a situation. As any measurable property can serve to describe the *modus operandi* we shall consider a hypothetical situation in which an unwanted color is to be removed. The first step to be taken is to add a suitable dose of appropriate activated carbons to portions of the liquid to be decolorized, and maintain contact under conditions that could be employed in an industrial operation. If no decolorization results (assuming that the color is not a property of the product itself), this indicates that the colored bodies are not adsorbable and so we should turn to other means.

If, however, the liquid is decolorized sufficiently, this indicates that the contamination is adsorbable and is present in small

---

**There are exceptions; e. g., a product that can be marketed with much color and therefore may require only partial decolorization to become acceptable in the market.

concentrations in which event a brief further study may suffice to learn optimum operating conditions.

Another possibility is that the preliminary survey may show appreciable decolorization, but not sufficient to meet standards of quality for the product. Such a finding can be interpreted as showing the presence of:

a) Colored bodies that are adsorbable but present in concentrations too large to be satisfactorily adsorbed by the carbon dosage used;

b) A small concentration of a weakly adsorbable colored body.

Either conclusion would suggest the advisability of searching other methods of separation to be used, either in conjunction with activated carbon, or as a complete substitute for it. Before turning to other methods, however, it is well to explore the possibility that more suitable experimental conditions might supply the additional decolorization needed. Such further study is especially indicated for borderline situations.

Thus far we have considered situations that tacitly assume the presence of only a single contaminant, whereas actual situations usually involve several impurities. Moreover the separate impurities can have different adsorbabilities. This may cause a behavior in which the decolorization proceeds smoothly until a certain amount of color is removed, beyond which further additions of carbon provide little or no additional decolorization. In such a situation, carbon may be considered for the easily adsorbable color, and other means sought for the less adsorbable portion of color.

Industrial applications often seek to accomplish more than one objective; for example, the treatment of corn sugars and syrups may aim to remove color, bitter flavor, iron, hydroxymethylfurfural, and foam-formers. Inasmuch as the operating environment favorable for improving one property (*e.g.*, color) may not be best for another (*e.g.*, foaming), an experimental study of a new venture should include a simultaneous examination of all properties to be corrected. To do this, the filtrate obtained from each variation in experimental conditions is examined to learn the effect a particular set of operating conditions may have on each of the properties to be improved. This brings us to a discussion of useful functions of activated carbon in industrial applications.

3. DECOLORIZATION

More data are available on decolorization than on other liquid-phase functions of activated carbon, and this is understandable in view of the comparative ease of measuring color. Long before the introduction of modern instrumentation, there were satisfactory visual methods of measuring color which continued in use until recently. Photoelectric colorimeters which are now in general use avoid the personal equation and expedite evaluations. Spectro-photometers offer further advantages, especially when it is desirable to measure the separate adsorption of several different colored bodies.

Data on decolorizing are presented in varied forms and in diverse units of measurement, depending on the terminology in vogue for the substance or product being treated, and on the instrumentation in use. Consequently, sets of data taken from different sources are not always readily correlated if left in their original form. Inasmuch as our present purpose is to learn relationships and trends, this goal can be attained more easily by expressing data in a common language. Whenever feasible, data will be expressed in terms of the relative capacity of a carbon to provide a desired purification.

Decolorizing Carbons

Probably the greatest single factor contributing to effective decolorization is the selection of an appropriate carbon. No one carbon is best for all applications. Thus, as shown in Figure 3:1, carbon E which is above average on several products, is inferior for the decolorization of the vinegar.

All the carbons, **A, B, C, D, E,** in Figure 3:1 are marketed as decolorizing carbons, but to put an identical label on these carbons would be misleading. For as can be seen, the ability of a carbon to decolorize depends on the specific sample of product being treated; consequently the survey of a new venture should include varied types of activated carbon; preferably as many different brands as available time and facilities permit. In the selection, consideration should be given to aspects such as filtration and density which can influence performance in a plant operation. The choice of carbons will be governed by whether the treatment will be conducted in

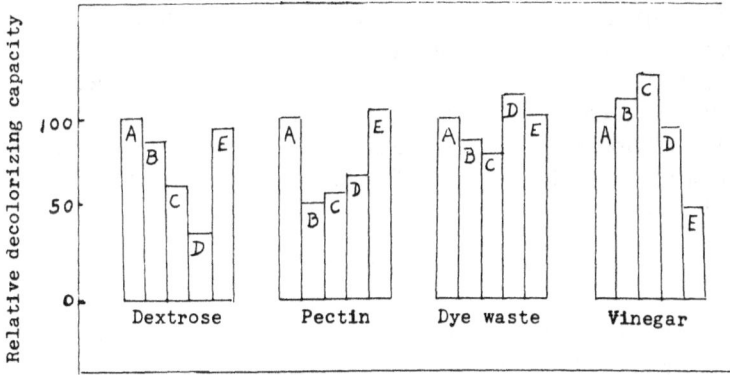

Fig. 3:1   Relative decolorizing power of five different grades of carbon.

batches with powdered carbon, or by percolation through a bed of granular carbon. Properties such as ash and pH may need examination; a low pH cannot be used for cane sugar, and a high soluble ash is not acceptable for glycerine.

An expensive carbon may provide additional benefits in some operations, but a price tag must not be taken as an index of value for all purposes. A carbon may be expensive because of a higher cost required to impart special adsorptive powers or to ensure freedom from soluble ash. Except where such properties are needed, the purchase of such a carbon can be an extravagance; in fact, in some applications a suitable low-cost carbon can be more effective on a pound-for-pound comparison.

From the carbons selected, appropriate dosages of each are added to measured portions of the liquid to be treated under conditions deemed suitable for plant-scale operation, after which the carbon is separated by filtration. Those brands of carbon that show reasonable promise should be included in further subsequent tests involving variations in experimental conditions. It is to be pointed out that all types of carbon do not respond similarly to changes in operating conditions, consequently the carbon that shows the most promise in the initial test may not prove to be best under the operating conditions finally adopted.

Variations in experimental conditions should be kept within

limits that can be tolerated by the product, and should be such as can be carried through on a plant scale. Experimental changes in pH, temperature, and other conditions should be made in small steps inasmuch as the optimum benefit is often found within a narrow range. Small benefits are not to be ignored because the cumulative effect of several beneficial conditions can convert a marginal operation into full success. Even attainment of a marginal operation may be justified if it outdistances other methods of purification.

Time

Granular carbons take much longer periods of time to reach equilibrium. Therefore, when granular and powdered carbons are being evaluated simultaneously in batch-contact tests, granular carbons are usually pulverized and tested in powdered form.

Because of over-all economic and technological considerations, industrial operations are seldom conducted for the full period of time needed to utilize the total ultimate adsorptive capacity of a carbon. Instead, it is usual practice to add enough carbon to give the desired decolorization within a relatively brief period, e.g., 10 minutes to an hour. (Table 3:2). Much more time may be required for special situations, and to accomplish special tasks. The time factor for a granular carbon is often expressed in terms of rate of flow through a column.

TABLE   3:2

EFFECT OF TIME ON DECOLORIZATION

Material	Carbon		Color removed, %			
	Grade	Dose g/100 ml	10 minutes	20 minutes	40 minutes	80 minutes
Corn syrup	B	1.0	60	65	68	68
Vinegar	F	0.5	70	75	80	80
Nitrobenzene	F	1.0	40	50	55	60
Sorghum syrup	A	1.0	50	60	65	67

Temperature

An increase in temperature accelerates the velocity of adsorption by decreasing the viscosity of the liquid, stepping up the diffusion of solute molecules, and hastening the displacement of pre-adsorbed gases from the pores of the carbon.

Temperature can alter the amount of decolorization, but the effect is seldom large. Depending on characteristics of the system involved, decolorization may increase, decrease, or show no change when the temperature is elevated (see Table 3:1). Not infrequently, maximum decolorization is obtained at an intermediate temperature. Unexpected behavior is sometimes encountered. In an experience with a long-chain nitrile no appreciable decolorization occurred until 250°C was reached, but then it increased rapidly and attained the maximum decolorization at 275°C. As can be observed from Table 3:3, the influence of temperature is not the same for all types of carbon.

TABLE 3:3

INFLUENCE OF TEMPERATURE ON
DIFFERENT CARBONS

Material	Carbon		Color removed, %	
	Grade	Dose g/100 ml	25° C	90 °C
Molasses solution	J	1.0	70	72
Molasses solution	K	1.0	75	95

In the treatment of some products, decolorization improves with higher temperature up to a certain point, above which the liquid darkens. Such an effect can usually be traced to ingredients susceptible to decomposition or oxidation. Running a blank control will indicate the extent to which any such reaction is being catalyzed by the carbon. Systems susceptible to such adverse oxidation may require treatment in an inert atmosphere or maintenance of vacuum conditions. These add to the operational cost, and when the oxidation is minor in degree; it may be possible to minimize ill effects by: reducing the temperature; shortening the time of contact; and avoiding excessive agitation.

Influence of pH

The colorants in agricultural source materials are natural dyes often acidic in type, but basic types are sometimes present. Consequently, and as would be expected, the adsorption of color can be influenced by changes in *p*H. A *p*H above neutral is seldom favorable and in many applications it has been found detrimental. In contrast, a *p*H below 7 frequently aids decolorization, the optimum *p*H depending on the particular system under study.

In many operations, however, it is not feasible to alter the natural *p*H of a product because of ill effects on other properties of the system. For example, many products are damaged in quality at the *p*H most favorable to decolorization. Even when a product can tolerate a change in *p*H, other difficulties can arise. Thus iron will be extracted from many types of carbon at a low *p*H, and this would involve the extra cost of either purchasing a carbon low in iron, or providing means for subsequently removing dissolved iron from the treated filtrate. Moreover, the optimum *p*H for decolorization may not be acceptable for marketing the finished product.

When the optimum *p*H for treatment by carbon is not acceptable for the finished product, it may be practicable to adjust the *p*H of the system temporarily to the optimum level for decolorization and subsequently re-adjust to the natural *p*H for the product. The addition of reagents for *p*H adjustment can be objectionable in some process liquors. The practice is more apt to be feasible for processes that include a subsequent crystallization or distillation. Another possible opportunity for utilizing the optimum *p*H for adsorption is afforded when a product exists in different forms at separate stages of processing; as, for example, in the manufacture of an organic acid in which the formation of a salt precedes the formation of the free acid. Salts of organic acids usually give an alkaline reaction unfavorable to decolorization; but on the other hand the acid stage has the unfavorable effect of dissolving iron from the carbon. Both disadvantages are avoided by applying the carbon at a favorable intermediate *p*H during the transition from the salt to the free acid.

In evaluating the effect of *p*H, it is always well to examine for possible indicator action which could be misinterpreted as an adsorption effect. Many colors in agricultural and industrial products behave as acid-base indicators, in which case a change in *p*H may alter the hue or change the intensity of the color, *e.g.*, the color of a molasses solution diminished by one-third in going from *p*H 9 to 4. When it is desired to compare decolorizations conducted at different *p*H levels, all filtrates should be adjusted to an identical *p*H when reading the residual color intensities.

Such an adjustment of *p*H in the filtrate should also be made when comparing brands of carbon having dissimilar *p*H. Some carbons contain sufficient acidity or alkalinity to alter the *p*H of unbuffered solutions. Incidentally, the *p*H of a carbon can influence the decolorization of non-aqueous liquids (Table 3:4).

TABLE   3:4

EFFECT OF *p*H OF CARBON ON DECOLORIZATION OF
COCONUT OIL

pH of Carbon	*Color removed, %*
8.5	55
7.0	55
6.0	60
4.5	62
4.0	64
2.5	66

Conditions: Used 0.25 % carbon H at *p*H indicated; contact with carbon for 20 minutes at 80° C.

Influence of Concentration

In processes that include a concentration stage, a study should be made to determine the proper point to introduce the carbon inasmuch as the decolorizing efficiency in a dilute solution often differs from that obtained in a more concentrated solution. Thus for a batch of raw cane sugar syrup, the decolorization with carbon was more effective in a concentrated solution; whereas a sample of sodium lactate was more effectively decolorized in a dilute solution. In a study of glycerine, activated carbon removed proteins effectively from concentrated solutions, but the decolorizing effect was greater with dilute solutions.

4.   TASTES AND ODORS

Problems that involve taste and odor can embrace diverse objectives. The conditions that are most amenable to correction by activated carbon are those in which a serious nuisance in taste or odor is created by relatively small concentrations of parent substances. In such situations, practically any and all adverse conditions can be corrected—a claim that is so well documented in the purification of water. Generally, the only problem is the relatively minor one of selecting the grade of carbon that will do the job at the least cost. Any difficulties on this score arise from the fact that tastes and odors are measured by subjective sensory means, and opinions as to the best grade of carbon and dosage to use may vary from one individual to another. Reasonably good agreement is usually to be had when complete freedom from taste and odor is desired. Somewhat greater difficulty in reaching a decision can be presented by products, *e.g.*, gelatin, that have an inherent natural flavor and odor. Here it can be difficult—at least for the novice—to know just when all foreign odor has been removed; and especially so, for a foreign odor that changes in quality as it becomes more dilute. That successful decisions are often reached, however, is evidenced by the widespread and extensive use of activated carbons for improving the salability of products that have an inherent natural flavor and odor.

A much more difficult type of application is one in which a foreign odor is to be removed from honey, maple syrup, or other materials that have a bouquet or natural flavor composed of constituents present in minute quantities. Such applications require what may be termed *fractionating adsorption*. Unfortunately an undesired fractionation often results, and the removal of desired components of the bouquet leaves the product with an unbalanced flat flavor; however, success has at times been achieved. A method occasionally found helpful is to allow the liquid to remain in contact with minute quantities of carbon; the time of contact may range from a day to a month or more. Off-flavored brandies, whiskies, and other alcoholic beverages have been so treated. One explanation of the improvement is that the desired type of fractionation takes place during the longer contact; another interpretation is that activated carbon catalyzes beneficial chemical changes.

Catalytic action is known to solve certain odor problems. An important example is the removal of free chlorine from potable water supplies. Free chlorine will gradually disappear from water as a consequence of a chemical reaction with the water. The reaction is greatly accelerated by the catalytic effect of activated carbon. When potable water containing free chlorine is percolated through a suitable bed of activated carbon, the chlorine is removed almost instantaneously.

When the correction of an odor requires the removal of large gravimetric quantities of parent substances, other methods of separation must usually be employed, although in many cases activated carbon may still fill a need for the removal of residual traces. In this connection it may be of interest to relate an experience in which the roles were reversed: A small dose of activated carbon effectively removed a very pungent odor from a fungicide but left a mild unpleasant odor that was subsequently corrected by a masking agent.

When a situation requires the removal of both color and odor, the conditions found favorable for decolorization will usually be suitable for removing tastes and odors. This is not an invariable rule; however, e.g., a low pH that is favorable for decolorizing glycerine can result in creating a rancid odor. Such an effect can often be traced to a side reaction; thus, natural glycerine may contain esters which at a low pH form fatty acids with a rancid odor. A similar explanation applies to the adverse effect of an activated carbon of low pH on coconut and cottonseed oils (Table 3:5). In general, an increase in odor should always be regarded as a warning signal of improper treatment conditions.

TABLE   3:5

EFFECT OF pH OF CARBON ON ODOR OF
EDIBLE OILS

pH of Carbon	Odor of oil after treatment	
	Coconut oil	Cottonseed oil
8.0	Excellent	Excellent
7.0	Excellent	Excellent
6.0	Excellent	Excellent
4.5	Fair	Good
4.0	Bad	Bad
2.5	Very bad	Very bad

Conditions: Used 0.25% carbon H at pH indicated; contact with carbon, 20 minutes at 80° C.

Considerable latitude can be exercised in the selection of the carbon to be used for the correction of tastes and odors. Unlike decolorization, the removal of tastes and odors seldom requires a specific type of carbon for a given situation. A carbon found satisfactory for one odor situation will often be acceptable for others. When both odor and color are to be removed, the choice of a carbon can usually be based on the grade suitable for the decolorization. When decolorization is not necessary, however, money can be saved by purchasing one of the so-called *water-grade* carbons; such carbons are particularly preferable for applications in which the color is to remain unchanged. Both granular and powdered water-grade carbons are available. For industrial applications, the powdered form should be furnished in a particle size suitable to give adequate rate of filtration in industrial filters.

5. SHELF LIFE; STABILITY OF PRODUCT

After treatment with activated carbon, a product may in time develop defects of quality such as color and odor reversion. Or again, defects may appear when a product is put to its intended use, e.g., corn sugar in candy manufacture, or linseed oil in the varnish kettle.

Most such difficulties fall into two general classifications. In one group we find colorless ingredients with unwanted characteristics that are hidden from immediate view; or the objectionable characteristics may develop in consequence of a slow chemical change during storage. Also in this latter group are ingredients that act as catalysts to accelerate changes that impair the quality of the prime ingredient—the product.

Whatever the nature of the contamination, the cure is usually to be sought in more effective or complete removal. This may be accomplished: by a greater dosage of carbon; by changes in operating conditions; and by the selection of a brand of carbon that has greater adsorptive power for the hidden offenders. The search is simplified if and when the chemical identities of the precursors are known, in which case appropriate control tests can be instituted. But this short cut is seldom at hand and the selection of appropriate conditions of treatment is usually based on actual experience with shelf life.

Mention should be made of a quite different type of influence on shelf life that is found when the source materials contain adsorbable ingredients that give stability to the product. An illustration is supplied by vegetable oils containing natural anti-oxidants that retard the development of rancidity. These natural anti-oxidants are adsorbed by activated carbon; therefore to compensate for their loss, suitable synthetic anti-oxidants are added to an oil after treatment with carbon.

Chemical changes that occur during storage may form compounds that accelerate further destructive changes—a process commonly known as *audo-catalysis*. Milone[4] found that esters of chloromaleic acid are stabilized by the presence of 2% carbon. Englehardt[5] controlled the formation of gum in organic liquids by storage in the presence of 1% activated carbon; and a similar procedure has prevented rancidity from developing in lard.

6. OTHER FORMS OF USEFULNESS

Many other forms of usefulness for activated carbons are known[1,2,3] (Tables 3:6 and 3:7). Although some of these have considerable utility in diverse applications, few have captured the

TABLE 3:6

REPORTED USES OF CARBON FOR PARTIAL OR COMPLETE
REMOVAL OF VARIOUS SUBSTANCES
(Abridged List)

Oil from boilers water
Aldehydes from fermented liquids
Mercury compounds from hyposulfites
DDT from water supplies
Carcinogens
Fluorescence from certain oils
Soaps from alkali-refined oils
Peroxides from rancid oils
Conradson residue from used crank-case oil
Sulfonates from mineral oils
Tetraethyl lead from gasoline
Hydroxymethylfurfural from sugar syrups
Unsaturates from mineral oils
Bloom from vegetable oils
Staining impurities from dye intermediates
Radioactive substances
Separation of antimony from copper sulfate

TABLE 3:7

LESSER KNOWN FORMS OF USEFULNESS FOR CARBON
(Abridged List)

Prevention of gum in certain organic liquids
Storage of toxic volatile laboratory reagents
Removal of certain catalyst residues
Removal of substances that shorten effective life of ion exchangers
Scavenger for irradiated meats
Reduce formation of sludge in heat-transfer oils
Assist chemical reactions by adsorbing inhibitors
Safen insecticides and fungicides for sensitive foliage
Stabilize esters

general interest or attention accorded decolorization and deodorization. This somewhat negative attitude is a consequence of
several factors: Some of these other forms of usefulness apply to
specialized situations and do not have general utility; others represent properties and characteristics that are not readily measured
so it can be difficult to appraise their net worth. Then when
there is a time interval between the treatment with carbon and the
appearance of the improvement, we can fail to recognize that the
betterment is caused by the carbon treatment, e.g., a higher melting
point of a finished crystalline product, or a longer shelf life. Such
delayed benefits are in contrast to a deodorization in which an
improvement is immediately apparent.

We must not overlook the fact that most industrial applications
have their inception in laboratory studies, and in such studies,
changes in properties other than color and odor can easily escape
notice unless attention is specifically directed to them. As a matter
of fact, many forms of usefulness have been discovered, often
quite by accident, in plant-scale operations in which activated
carbon had initially been employed for decolorization or deodorization. This is particularly true in applications in which the
use of carbon has led to betterment in a subsequent distillation,
crystallization, or other processing operations. Such benefits can
be very real to the operating personnel even though one may not
readily place an exact monetary value on them.

The contribution of activated carbon to improved quality of
product is not always fully recognized; characteristics such as
freedom from haze, absence of foam, and other less visible benefits
are often overlooked. It is worth noting that many of these less

evident benefits are often a by-product incidental to the use of carbon for color and odor removal—and hence can often be obtained for little if any extra cost. For the most part, all that is ordinarily required is a selection of appropriate operating conditions.

7. ORGANIC COLLOIDS

Many improvements provided by activated carbon can be correlated with effects caused by organic colloids (see Table 3:8). Colloidal phenomena are considered in Chapter 7; here we shall give but a brief simplified outline using familiar experiences with detergents for purpose of illustration.

TABLE 3:8

IMPROVEMENTS ASSOCIATED WITH THE USE OF CARBON FOR
REMOVAL OF ORGANIC COLLOIDS

Elimination of turbidity, haze, emulsions
Improved filterability of process liquor and finished product
Reduced tendency to foam during evaporation and distillation
Faster rate of settling during clarification
Improved performance at crystallization station: greater yield of purer
 crystals having better form and higher melting point
Greater stability of product, longer shelf life

If air is passed through a column of water, bubbles will form, but they will break almost immediately. Shaking an oil-water mixture will cause the oil to form droplets in the water but the two liquids separate when the shaking ceases. Powdered sand will remain suspended in water only briefly.

However, if the experiments are repeated, but with a detergent solution: then air bubbles will remain as a foam; the oil will remain emulsified; and the fine sand will stay suspended in the liquid. The explanation is that the detergent forms a protective envelope around each air-oil-sand particle, shielding them from one another and preventing them from coalescing. And as the detergent is very soluble in the water, this keeps the enclosed particles anchored and suspended in the liquid space.

This phenomenon, which is known as *protective action*, can be provided by many organic colloids—gums, pectins, proteins, soluble starch, and many synthetic substances. The colloids are described as protective colloids; and the protected particles are

TABLE 3:9

REMEDIAL MEASURES FOR PEPTIZATION

Increase the dosage of carbon
Select brand of carbon having greater adsorptive power for the organic colloids present
Use a hard granular carbon that can withstand peptizing action
Apply carbon in an admixture with another adsorbent such as magnesium silicate, bentonite, activated clay. In selecting an additional adsorbent it is necessary to consider whether the pH and other characteristics are compatible with the system being treated
Conduct the adsorption in separate stages:
    1) Treat the original solution with carbon;
    2) Filter;
    3) Treat filtrate with a type of carbon appropriate for the final stage
Conduct two- or three-stage countercurrent adsorption
Employ a different solvent, one that impairs the colloidal properties; or alternatively, dilute with suitable miscible solvent.
Use good technique for filtration:
    1) Proper formation of cake;
    2) Slow initial rate of flow;
    3) Retentive diatomaceous filter aid;
    4) Avoid interruptions and severe pulsation in pumping
Study changes in treatment, e.g., time, temperature, pH

said to be solubilized or *peptized*. Peptization can be viewed as adsorption in which the more familiar role is reversed; in the orthodox version we think of adsorption as an attachment whereby the solid pulls a solute out of solution, but in peptization the solid is pulled into the liquid space.

Peptization will occur only if and when the concentration of colloid is sufficient to form a blanket around the particles to be solubilized. This is known as the *critical concentration level.*

When activated carbon is added to a solution containing organic colloids they will be adsorbed, and if this adsorption is great enough to cause the colloidal concentration to fall below the critical level any enclosed liquid or solid particles will be set free and swept out of solution during the subsequent filtration. In some systems, however, the initial concentration of colloid is so large that it cannot be reduced below the critical level by practicable dosages of carbon; and then a colloidal suspension, emulsion, or foam will remain as is. In some instances, in fact, the colloidal concentration is high enough to peptize the carbon, which cannot then be separated from the solution in the subsequent filtration. Such peptization has prevented the use of activated carbon in a number of potential applications. Some such situations can be moderated or even cured by various techniques developed out of past experience (Table 3:9).

8. HAZE

An emulsion after having been cleared by carbon will not spontaneously form anew, but this cannot be said of a haze. This may be due to any one of the many factors that can cause a haze to form (Table 3:10).

TABLE   3:10

CONDITIONS CONTRIBUTING TO FORMATION OF HAZE IN
PRESENCE OF ORGANIC COLLOIDS

Supersaturated solution of difficultly soluble substances
Substances introduced during processing that can react with ingredients already present to form an insoluble compound; e.g., when traces of calcium salts are added to a neutral solution containing oxalates
Change in pH during processing that renders certain ingredients less soluble
Presence of salts that can hydrolyze to form insoluble hydroxides
Denaturing of proteins
Fermentation, or other biological changes
Retarded chemical changes

The mechanism involved can be illustrated by citing a common producer of haze, namely a supersaturated solution of a slightly soluble substance. When organic colloids are present in such a system, precipitation is retarded or even prevented; but if and when it does occur, the precipitation is seldom complete and some degree of supersaturation still remains in solution.

Consequently when carbon is applied to remove organic colloids, it will simultaneously remove any existing haze, but it may not immediately affect supersaturated molecules of solute. As a consequence of removing protective colloids, however, any residual supersaturation is better able to precipitate and form a new haze. The precipitation is not always immediate; it may be

TABLE   3:11

ELIMINATION OF HAZE

Case histories selected to show that optimum conditions can be specific to the
system being treated

Case No.	Optimum time and temperature of contact with carbon	Other required conditions
10	20 minutes at 20 C	
11	20 minutes at 90 C	Finely pulverized carbon
12	60 minutes at 90 C	
13	2 hours at 10 C	
14	1 hour at 10 C	pH of solution preadjusted to 4.5
15	24 hours at 25 C	pH of solution preadjusted to 6.0
16	72 hours at 15 C	pH of liquid preadjusted to 7.0
17	1 hour at 60 C	pH of solution preadjusted to 8.0; used finely pulverized carbon
18	1 hour at 25 C	Solution pre-diluted with 10% methanol
19	20 minutes at 25 C	Solution pretreated with enzymes

delayed and in some cases it will develop only under certain circumstances and conditions, *e.g.*, storage at low temperatures.

In order to prevent such regenesis of haze, it may become necessary to do more than reduce the organic colloid concentration
below the protective level; in addition we must also provide conditions that aid in the removal of residual haze-formers. Procedures
that have been so employed, often with success, are listed in Table
3:11. We can only guess as to the molecular mechanics involved
in removing haze-formers. It may be that, under appropriate conditions, haze-formers are adsorbed while still dissolved in a supersaturated state; or it could be that during longer contact, precipitation is initiated on the carbon surface which captures and
holds the freshly formed solid precipitate.

The conditions favorable for the removal of residual haze-
formers may differ from the conditions required for the adsorption
of organic colloids, and in this event it becomes necessary to re-
adjust conditions during the carbon treatment as illustrated by
experiences listed in Table 3:12.

Some source materials that originally are free of haze will
develop a haze after being decolorized with activated carbon. In
such situations it is natural enough to blame the haze on the
treatment with carbon. The charge is perhaps true but not in the

TABLE 3:12

ELIMINATION OF HAZE

Typical case histories in which conditions were readjusted during treatment

Case No.	Conditions of treatment
21	Carbon in contact with solution initially at 90° C for 20 minutes to adsorb organic colloids; then cooled to 20° C and maintained at that temperature for 4 hours to remove residual haze-formers.
22	Treatment conducted in separate stages. In first stage, carbon in contact with solution 1 hour at 80° C, then filtered. Second stage filtrate treated with virgin carbon 4 hours at 10° C.
23	Haze-forming ingredients here were less soluble at pH 4.5 than at the normal pH of 6.0 for the substance being treated. Therefore the adsorption stage was conducted at pH 4.5 in order that the maximum precipitate would form and be removed from the system when the carbon was filtered off. The filtrate was readjusted to pH 6.0 to minimize precipitation of residual haze-formers.
24	Solution treated with carbon to remove protective colloids; then overconcentrated by evaporation. After allowing sufficient time for insolubles to precipitate, the solution was filtered. The filtrate was diluted to normal concentration.

usual sense, namely: that the carbon adds haze-formers to a liquid. This is a false idea that can be unfortunate in delaying the search for the true cause and finding a cure.

All this is illustrated in the experience of a corn sugar operation that was confronted with a problem of haze in the finished syrup. The crystals forming the haze were minute and distorted in shape, thus preventing identification by crystallographic methods. The haze was first thought to be due to calcium sulfate formed by sulfates in the syrup reacting with calcium extracted from the carbon. The next step was to procure and use an activated carbon free of calcium, but this failed to solve the haze problem. An analysis of haze-free syrups from other operations disclosed that many batches of haze-free syrups contained more total ash than samples of hazy syrups. Following this finding, a portion of hazy syrup was ashed but when the ash was dissolved in a satisfactory syrup no haze developed.

Eventually, the cure was found in adjusting the pH when it was discovered that haze developed in syrups produced from thin juices containing sulfates if and when such juices were concentrated to a thick syrup at a pH below 4.5. When a hazy syrup was diluted and reconcentrated above pH 4.8, the haze did not reappear. On the other hand, when a satisfactory syrup was diluted and reconcentrated at pH 4.0 a haze appeared.

Removal of Iron and Ash

Colloidal protection can enter into the use of carbon for removing inorganic substances—a familiar problem being the removal of iron salts from corn sugar syrups. In many batches of syrup, iron salts are present in concentrations that would normally hydrolyze and precipitate as iron hydroxide at the pH of 5 which often prevails in the syrup; but this precipitation does not occur because of the protective action of the organic colloids present. If such a sample of corn syrup is treated with a suitable activated carbon at pH 5, a test of the filtrate will show that most of the iron has been removed—but this removal is not a direct consequence of orthodox adsorption. This deduction is supported by the fact that when the syrup is treated with the carbon at pH 3, no iron is removed from the solution, but the concentration of colloids is reduced below the protective level. The removal of protective colloids can be demonstrated by elevating the pH of the filtered syrup to pH 5 and then the iron will slowly precipitate as a hydroxide. To exclude the possibility that the effect might be a consequence of readjusting pH, the entire procedure was repeated but without carbon. When the final readjustment was made to pH 5 no iron precipitate formed, and an analysis showed that the concentration of iron in the final syrup was the same as originally present.

9. USE OF ACTIVATED CARBON IN CONJUNCTION WITH OTHER SEPARATIONS

Thus far we have discussed activated carbon acting as an agent of and by itself. More often it functions as a partner in a process, allied and associated with one or more other means of separation.[1,2,3] In such combinations, the characteristics of the system involved and the objectives to be accomplished will govern the role of each cooperating means of separation. Each method may do that part for which it is best qualified; several partners may operate collectively to attain a degree of purification that could not be accomplished by any single member acting alone; and, in still other cases, one method may precondition a system so that another member of the team can become more effective. The reader should be cautioned that in many applications the true function of each stage may not be so clearly defined as might be inferred from the foregoing remarks; often we seem to stumble on a right combination.

The assistance furnished to or by associated treatments ranges in significance and magnitude. On occasion much benefit may be derived from a very simple auxiliary treatment; *e.g.*, many operations have been aided by pre-filtration through diatomaceous earth to remove slimes that would otherwise coat the surface of the carbon and diminish the adsorptive capacity. In contrast are the specialized and elaborate techniques employed in some biological separations. Let us briefly review types of auxiliary aid and associated procedures.

Solvents

In many operations the material to be purified is received as a liquid or in solution and is processed as is. Modification may be needed as with a high viscosity liquid that may require dilution with a miscible solvent.

Fats, waxes, and similar solids are usually treated at a temperature sufficiently high to convert them to a fluid state, although on occasion a fluid state is effected by dissolving the substance in a suitable solvent. This is done when it is desirable to minimize losses of a costly fat or wax by retention; also when it is desired to conduct an adsorption at temperatures below the melting point of the solid material being purified.

Many solid materials, such as sugar, gelatine, and tartaric acid, must be dissolved to bring them to a liquid state for treatment with carbon. Generally, water is the preferred solvent because apart from advantages of availability, low cost, freedom from fire hazards, adsorption with activated carbon is almost always more effective from an aqueous solution. Situations arise, however, when it becomes necessary to use an organic solvent, as in the case of water-insoluble organic substances. On occasion an organic solvent is desirable for substances that form a colloidal solution in water. This latter is illustrated by an experience with a synthetic lacquer which could not be decolorized in an aqueous solution because the carbon became peptized. The peptization was avoided by using a solvent consisting of equal parts of water and methanol.

When an organic solvent is to be used, one should be mindful that the best solvent will depend on the over-all characteristics of the system under study; and usually the specific solvent properties needed for a particular application can best be learned by trial and error.

Chemical Agents

Pretreatment with carbon of a material to be purified often enables a chemical agent to function more effectively (Table 3: 13).[1,2,3] In a process to reclaim iodides from petroleum brine, the initial step consisted of oxidizing the iodides to elemental iodine, and this reaction had been prevented by impurities in the brine that acted as inhibitors of oxidation. It became necessary to remove these inhibitors by pretreating the brine with a special type of activated carbon, after which the oxidation proceeded without difficulty. A similar type of behavior probably explains why hydrogen peroxide is a more effective bleaching agent for certain waxes after they have been given prior treatment with activated carbon.

TABLE   3:13

PREPARATORY USE OF ACTIVATED CARBON TO AID
SUBSEQUENT PROCESSING WITH CHEMICAL AGENTS

Case No.	Material	Subsequent treatment
101	Petroleum brines	Oxidation of iodides to iodine
102	Waxes	Bleaching with peroxide
103	Waxes	Bleaching with chlorine
104	Glutamic acid	Bleaching with peroxide
105	Vegetable oils	Hydrogenation
106	Crude animal fats	Caustic refining
107	Crude vegetable oil	Caustic refining
108	Pyrethrins	Solvent-extraction

In other applications the roles may be reversed with the stage of chemical reaction preceding the use of carbon (Table 3:14).

In sugar refining, the syrup is pretreated first with lime and then with phosphoric acid in order to precipitate and sweep out much contamination. This leaves a relatively smaller amount of adsorbable impurities to be removed by activated carbon. Oxidizing agents, such as chlorine, chlorine dioxide, and hydrogen peroxide, are on occasion successfully employed to convert impurities into a more adsorbable form. Such action is not always effective, however, and the prior use of an oxidizing agent is sometimes disappointing; in fact, in some applications the impurities are less adsorbable after oxidation. Moreover, the chemical treatment has been known to confer objectionable characteristics such as chlorinous odor.

TABLE 3:14

PRETREATMENT WITH CHEMICAL AGENTS TO
IMPROVE ADSORPTION BY ACTIVATED CARBON

Case No.	Material	Pretreatment
150	Sugar	Chlorine
151	Medicinal oil	Chlorine
152	Carnauba wax	Peroxide
153	Naphthenic acid	Sulfuric acid
154	Lactose	Mineral acid to precipitate proteins
155	Sugar	Lime added to pH 9, followed by phosphoric to pH 6.5
156	Phthalate ester	Permanganate
157	Nitrile	Solvent-extraction
158	Bio-chemical	Carbon impregnated with thiosulfate to minimize catalytic oxidation

Mention should be made of a novel approach for using activated carbon as an adjunct to purification by chemical agents. In this approach, the chemical reaction is relied upon to effect the entire removal of color, odor, or other undesired characteristics, after which activated carbon is added to mop up residues of the chemical reactants.

An unusual use of chemical reaction as an adjunct was employed by Hoare for purifying isatin. In his process, isatin is converted to sodium isatin bisulfite in which state it is treated with carbon, and after purification is reconverted to isatin.

Crystallization

Pretreatment of a process liquor with carbon is often helpful to processes that include a crystallization; and in many operations such pretreatment is a *must*. In addition to providing a rapid rate of crystallization and a greater yield of crystals, solutions treated with carbon furnish crystals of truer shape and higher melting point. Processes that include a final crystallization seldom require complete purification at the solution stage, inasmuch as very pure crystals can generally be obtained from partially purified solutions.

When very large concentrations of adsorbable impurities are present in a raw liquor, carbon adsorption acting alone may not provide an adequate degree of partial purification. Then supplemental measures are employed. Thus, in production of white cane sugar, an initial crystallization separates the sugar from the major portion of impurities; after which carbon can remove sufficient residual impurities to enable pure white sugar to form in a second stage of crystallization.

When a solution is purified with carbon prior to a crystallization stage, it is essential to keep the solute concentration below the saturation point. If carbon is added to a super-saturated solution, the adsorption and separation of protective colloids can result in spontaneous crystallization before the carbon can be separated from the system.

Activated carbon is employed to purify mother liquors so that they can be re-used a greater number of times as a recrystallization medium. The ability to re-use a mother liquor is especially worthwhile for the recrystallization of very soluble substances.

Distillation

A combination with distillation is illustrated by the purification of glycerine. Ordinarily, it is difficult to decolorize crude glycerine to water-whiteness, but the task is easier after one distillation, and even more so after redistillation. In industrial practice, virgin carbon is applied to the twice-distilled glycerine, and the spent carbon from this stage is added to the crude glycerine to remove foam-formers that hinder distillation.

Adsorbents

In operations that do not include a stage of distillation or crystallization, ion-exchange resins have been employed to remove inorganic compounds, alkalinity, or acidity not adsorbable by activated carbon. In general, the use of the ion-exchanger should follow the treatment with carbon, especially if the carbon contains any appreciable amount of soluble inorganic compounds.

Magnesium silicate and bentonite, when used with activated carbon in aqueous solution, are reported to aid filtration and settling; and also to minimize peptization of the carbon.

Activated clays and fuller's earth are used extensively for decolorizing fats and oils—animal, vegetable, and mineral. In a number of applications, they are used alone: but for some fats and oils an admixture with activated carbon is required in order to attain a desired decolorization. Moreover, carbon will remove off-odors and tastes. whereas clays frequently impart a so-called earthy flavor and odor. Consequently, even when not needed for removal of color, carbon is often admixed with clay to ensure freedom from any earthy flavor.

Occasionally, two different grades of carbon are employed when a single grade of carbon does not have all the properties needed for purification. Usually each grade is used as a separate stage; for example, a granular carbon at one stage, and a powdered carbon used at another.

REFERENCES

1. Suppliers of activated carbon can furnish much helpful information.
2. Deitz, V. R., *Bibliography of Solid Adsorbents,* United States Cane Sugar Refiners and Bone Char Manufacturers and the National Bureau of Standards, Washington, D. C., 1944. Supplementary volume issued in 1956 for years 1943 to 1953.
3. The original sources of information on specific items can usually be located in Chemical Abstracts in which they are indexed under *Carbon, active* and also under *Charcoal, active.*
4. Milone, C. R., U.S. Patent 2,385,018.
5. Engelhardt, A., U.S. Patent 2,306,870.
6. Hoare, R. C., U.S. Patent 2,086,805.

4

Interpretation and Evaluation
of Adsorption Data

1. INTRODUCTION

When a dye solution is placed in contact with activated carbon,
part of the color will leave the solution and become attached to
the carbon surface. But the transfer will be only partial because
the dissolving power of the liquid exerts a pull to keep some dye
in solution. An equilibrium is reached in the distribution of the
dye between the carbon and the liquid phase. After equilibrium
is reached, if we add a more concentrated dye solution to the
carbon—additional dye will transfer to the carbon until a new
equilibrium is reached. The transfer process can be continued by
introducing stronger and stronger dye solutions.

The customary laboratory procedure for determining the distribution between the quantity adsorbed and the concentration
remaining in solution is as follows:*

  a) Different weighed amounts of carbon are added to separate
      flasks, all containing equal volumes of the solution being
      studied.
  b) The contents of the flasks are stirred at uniform temperature
      until adsorption equilibrium is reached.
  c) The carbon is separated from the system by filtration and
      the concentration remaining in the filtrate is measured.

*More detailed description in Chapter 17.

The data are coded: C_o original concentration

$\quad\quad\quad\quad\quad\quad\quad$ C   final concentration

$\quad\quad\quad\quad\quad\quad$ amount adsorbed: C_o minus C

$\quad\quad\quad\quad\quad\quad$ M   weight carbon used

$\quad\quad\quad\quad\quad\quad$ X/M amount adsorbed per unit weight of
carbon

Typical data appear in Table 4:1. It should be emphasized that adsorption data are specific for each system. Thus, the data in Table 4:1 would be different if an acid other than benzoic had been adsorbed, or if a different brand of carbon had been used. Consequently an experimental study of a new application should employ components identical with those projected for the operation.

Knowledge of the relationship between the amount of substance adsorbed and the concentration remaining unadsorbed is of practical utility. Much research has shown that the concentration on the adsorbent is usually proportional to a power or a fractional power of the concentration that remains unadsorbed. Expressed mathematically:

$$X/M \text{ is proportional to } C^{1/n}.$$

This is converted to an equation by introducing a constant K which forms the classical Freundlich equation

$$X/M = K \, C^{1/n}$$

For data collected at identical temperature, an isotherm is formed when X/M is plotted against C. If the plotting is done on logarithmic paper, the isotherm will generally be a straight line over a wide range of concentrations.

The adsorption data should be expressed in the same units of mass. In Table 4:1 all data are expressed as grams (or fractional parts of a gram), and as such can be plotted directly. However, in many industrial operations, the identity and gravimetric concentration of the impurity are not known. In such cases it becomes necessary to convert some property of the impurity into units that can express the *relative* gravimetric concentration. Methods for establishing such units are described in Chapter 17.

Inasmuch as decolorization is a frequent industrial application, and as color units offer a convenient way to present adsorption data, we will use them frequently in further discussion in this chapter.

TABLE 4:1

BENZOIC ACID ADSORPTION*

(Tests of 50 milliliter sample. $C_0{}^a = 0.1500$ gram)

m grams carbon	C Residual benzoic acid (grams/50 milliliter	x Acid adsorbed C_0-C (expressed in grams)	x/m Grams acid adsorbed per gram of carbon
0.0	0.1500	—	—
0.2	0.0900	0.0600	0.3000
0.4	0.0560	0.0940	0.2350
0.7	0.0256	0.1244	0.1791
1.0	0.0136	0.1364	0.1364
1.5	0.0070	0.1430	0.0953
2.0	0.0040	0.1460	0.0730

a C_0 is the weight of benzoic acid in 50 ml of the original solution.

* From data of W. A. Helbig, in *Colloid Chemistry*. Edited by J. Alexander, Reinhold Publishing Corporation, New York, 1945, Vol. 6, p. 816; reprinted by permission of the copyright owners.

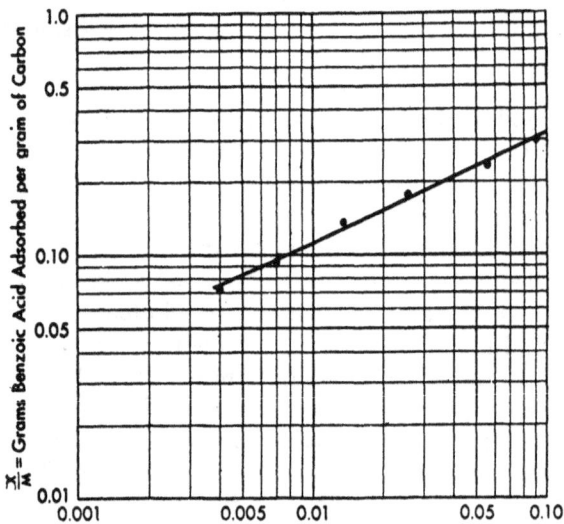

Fig. 4:1 Benzoic acid adsorption isotherem. From W. A. Helbig, in *Colloid Chemistry*, (Edited by J. Alexander), 6:814 (1946); reprinted by permission of Reinhold Publishing Corporation, New York.

The exponent 1/n in the Freundlich equation is of practical utility because it discloses the adsorption pattern when different quantities of adsorbate are removed from a solution. The dimensions of the exponent can be determined by measuring the ratio of the rise to the run of the isotherm. Thus in Figure 4:2 the isotherm, in moving one inch to the right, rises one-half inch vertically. Hence 1/n is 1/2, or expressed decimally 1/n is 0.5. This means that the concentration on the carbon (X/M) is proportional to the square root of the concentration remaining in solution.

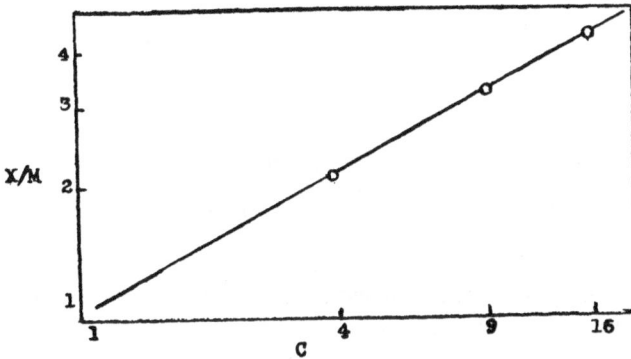

Fig. 4:2   Adsorption isotherm (1/n equals 0.5)
            C, concentration dye remaining in solution (color units/liter)
            X/M, amount of dye adsorbed (color units per gram carbon)

TABLE 4:2

"DATA OF AN ADSORPTION IN WHICH 1/N IS EQUAL TO 0.5"

color units adsorbed per gram carbon	color units unadsorbed per liter solution
4	16
3	9
2	4

When an isotherm has a 45° angle slope (*i.e.* 1/n equals 1), the rate of change in adsorption concentration is identical with that occurring in solution. Thus, in Figure 4:3, the reduction in dissolved color units from 90 to 10 parallels the reduction of X/M from 90 to 10.

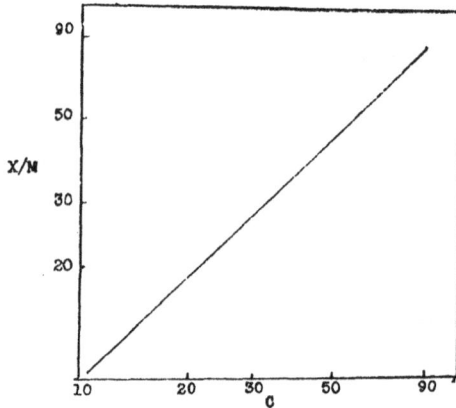

Fig. 4:3   Adsorption isotherm (1/n equal to 1)
           C, concentration dye remaining in solution (mg/liter)
           X/M, amount dye adsorbed mg/g carbon

When the isotherm slope is steeper than 45 (*i.e.*, 1/n is larger than 1.0) changes in the adsorbed concentrations outweigh changes in the solution concentrations.

An isotherm can be viewed as a map of the way in which an adsorbable solute distributes itself between the adsorbent and the solvent. Isotherms can convey an overall picture of many data more clearly than could be derived directly from numbers.

Let us examine some isotherms keeping in mind that the higher the level of an isotherm, or any portion of it, the greater is the relative distribution of the adsorbable solute on the adsorbent— *i.e.*, the carbon. Conversely, isotherms at low levels signify proportionately greater concentration in the solvent.

Thus when an isotherm is at a high level and has only a slight slope, as shown at *A* in Figure 4:4 this means that the adsorption is large throughout the entire range of concentration studied; conversely, a similar slope isotherm at a lower level, as at *B* in Figure 4:4 indicates proportionately less adsorption. Next, consider an isotherm having a steep slope, as shown in Figure 4:5; this typifies an adsorption that is large at strong concentrations, but is much less at dilute concentrations.

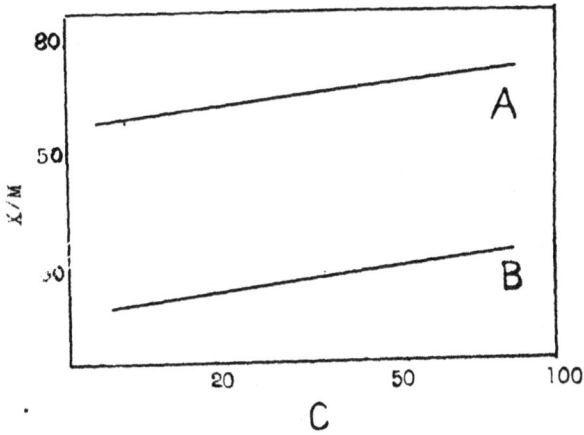

Fig. 4:4 Adsorption isotherms (plotted logarithmically):
C, concentration dye remaining in solution (mg/liter)
X/M, amount of dye adsorbed (mg/g carbon)

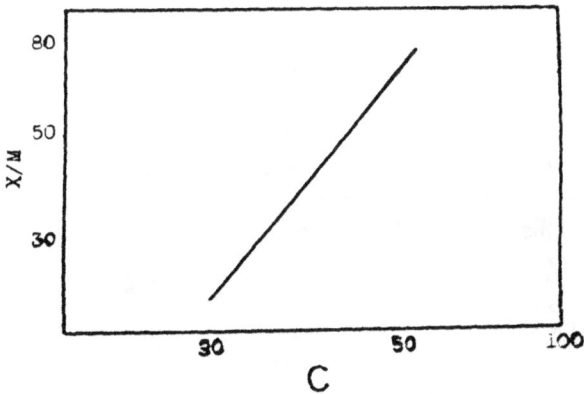

Fig. 4:5 Adsorption isotherm (plotted logarithmically)
C, concentration of dye remaining in solution (mg/liter)
X/M, amount of dye adsorbed (mg/g carbon)

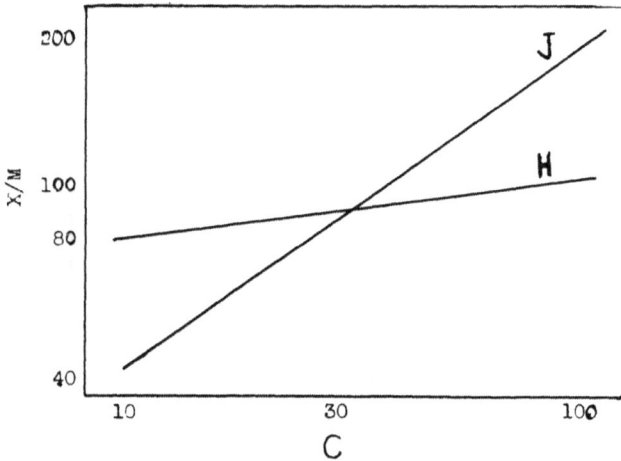

Fig. 4:6   Adsorption isotherm (plotted logarithmically)
            C, concentration in ppm remaining in solution
            X/M, ppm adsorbed per g of carbon

The relative efficiency of different carbons is in direct proportion to the X/M values measured at an identical concentration remaining in solution; and so, the isotherms afford a full view of the relative efficiency of carbons over the entire range of concentration studied. Thus, in Figure 4:6 carbon H has twice the relative efficiency of carbon J at a residual solution concentration of 10 ppm; at a concentration of 30 ppm both carbons are equal: and at a residual solution concentration of 100 ppm, carbon J has double the efficiency of Carbon H.

2.   CHANGES IN OPERATING CONDITIONS

Isotherms provide an overall view of the influence of operating conditions. Thus Figure 4:7, shows the decolorization of a process liquor, was greater at a temperature of 90°C than at 30°C. The effect of the higher temperature increases with greater decolorization of that particular liquor.

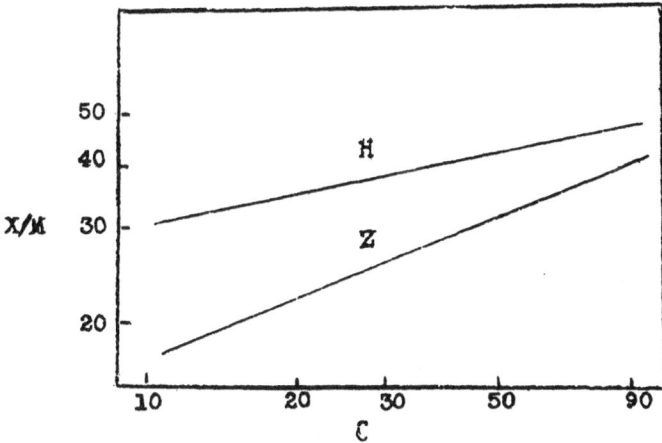

Fig. 4:7 Adsorption isotherms decolorization lactic acid
Z decolorized at 30°C, and H at 90°
C, concentration color remaining in solution (color units/L)
X/M, amount color adsorbed (color units per gram carbon)

A similar procedure can be employed to study effects of changes in other operating conditions, *e.g.*, time, pH.

3. COUNTERCURRENT ADSORPTION

Many industrial applications can use carbon only if and when it can:

(1) accomplish complete purification
(2) do so with a small quantity of carbon, *i.e.*, each gram of carbon must carry away a relatively large amount of impurity

The isotherms show that those two objectives are counteractive, because when additional amounts of carbon are employed to obtain the required purification there is a simultaneous decrease in the amount of impurity carried away by each gram of carbon. The adverse effect is especially great with steep isotherms.

In many applications such adverse effects can be offset by countercurrent use of carbon. The underlying principle of countercurrent application can be found in experiments such as the one described at the beginning of this chapter. It will be recalled that a carbon in equilibrium with a dilute dye solution was able to remove additional dye color from more concentrated solutions.

The principle is applied to an industrial application as follows: A batch of raw liquor is treated with sufficient carbon to provide complete purification. The carbon is separated by filtration, and reused to bring another batch of raw liquor to an intermediate purification. Less virgin carbon is then needed to bring the liquor from the intermediate stage to complete purification. The effectiveness depends on the additional reserve adsorptive capacity at the higher solution concentration, and this is revealed by the isotherm.

Let us consider a process liquor containing 100 color-units and which is to be decolorized to a finished color of 10 units (Figure 4:8). When sufficient carbon is used to produce this decolorization in one step, each gram of carbon will hold 20 color units. If this carbon is then filtered off and re-used on another portion of the untreated liquor, this portion will be partially decolorized to an

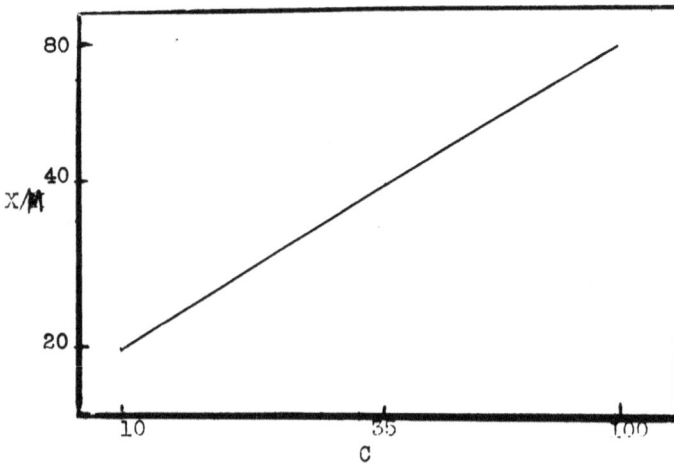

Fig. 4:8   Adsorption isotherm: Decolorization Process Liquor
C, color units/liter remaining in solution
X/M, color units adsorbed/gram carbon

intermediate color, e.g., 35 color-units. An inspection of the isotherm shows that the amount of color held by each gram of carbon increases to 40 color-units per gram—the increase being the extra reserve capacity at the stronger solution concentration.

The increase can be even greater when the liquid is percolated through a bed of carbon, and if the percolation process is continued until the carbon is in equilibrium with the original color concentration. Thus, if the isotherm (Figure 4:8) is extrapolated to the color concentration of the original solution, namely 100 color units, we find each gram of carbon then holds 80 color-units.

The amount of reserve capacity at higher solution concentrations is correlated to the slope of the isotherm. This is illustrated in Figure 4:6 where the extra reserve capacity is large for carbon J, but is neglibible for carbon H. In general, the advantage of either countercurrent batch or column percolation over single stage batch is more evident when the isotherm is steep and when more complete purification is required.

4. LIMITATIONS

Although extremely useful in any serious adsorption study, isotherms are not essential to the successful use of activated carbon in a new application. Situations arise in which the properties of the impurities are such that it is not feasible to translate the concentrations into gravimetric equivilents. Then it is necessary to use other approaches as mentioned in Chapter 17.

REFERENCES

1. Adam, N. K., *Physics and Chemistry of Surfaces,* Oxford University Press, London, 1941.
   Alexander, J., *Colloid Chemistry,* 6 volumes; Chemical

Catalog Co. and Reinhold Publishing Corporation, New York, 1926–1946.

Freundlich. H., *Colloid and Capillary Chemistry,* E. P. Dutton and Co., New York, 1922.

McBain, J. W., *The Sorption of Gases and Vapors by Solids,* G. Routledge Sons Ltd., London, 1932.

Deitz, V. R., *Bibliography of Solid Adsorbents,* United States Cane Sugar Refiners and Bone Char Manufacturers and the National Bureau of Standards, Washington, D.C., 1944; abstracts of scientific literature on theories of adsorption for the years 1900 to 1942 inclusive.

Deitz, V. R., Bibliography of Solid Adsorbents 1943–1953, National Bureau of Standards, Circular 566, Washington, D.C., 1956.

5

Adsorption Desorption Operations

1. GENERAL ASPECTS

Desorption processes[1] have theoretical and practical importance. From a purely scientific angle, they shed light on the nature of adsorption. They have practical value in that at times they constitute a stage in preparing a carbon for reuse. Desorption is generally an essential stage in the recovery of substances of value such as biological productsand vapors of industrial solvents.

Except in those cases in which the adsorbed molecules are attached irreversibly to the carbon surface, the adsorbed layer is in a dynamic condition. Molecules are continually leaving the surface to be replaced by others arriving from the fluid phase. The action is illustrated in a study by Lindau and Salomon.[2] They placed carbon, containing adsorbed optically active tartaric acid, in a solution containing an equivilent amount of the optical antipode. Within 20 seconds the solution—and also the carbon—held equal parts of each active form of the tartaric acid, indicating the rapidity of the exchange.

From this, one would expect that an adsorbed layer would be stripped from a surface if the molecules were removed from the system as they left the surface, and if no other molecules were present to replace them. In other words, it should be possible to extract a substance adsorbed from an aqueous solution by washing the carbon with plain water. But such elution is seldom of practical utility. At most, the utility is limited to extractions in which the prior adsorption is conducted from very concentrated solutions, and when only partial extraction is desired. Even such partial utility exists only when the adsorption isotherm has a steep slope.

The behavior can be understood when we explore the probable results that would ensue if the carbon in the system Figure 5:1 were washed with plain water. In that system at point S, one gram of carbon containing 200 mg adsorbed aniline is in equilibrium with a solution containing 10 mg per liter. The first liter of wash water will extract less than 10 mg of aniline because as soon as the washing begins the equilibrium moves to the left. Consequently each subsequent portion of wash water removes less and less aniline. By the time that a total of 100 mg adsorbed aniline has been extracted from a gram of carbon, the equilibrium has moved to point T, and there, less than 1 mg adsorbed aniline is eluted by a liter of water. The behavior is analogous to dissolving an imaginary precipitate that becomes more insoluble as the washing is continued.

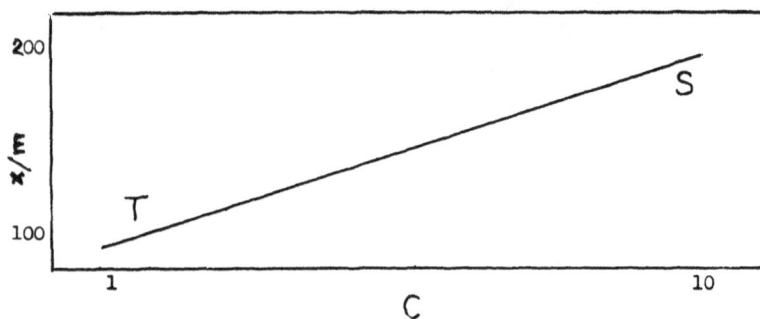

Fig. 5:1   Adsorption isotherm (plotted logarithmically)
            C, concentration of aniline in solution (mg/liter)
            X/M, amount of aniline adsorbed (mg/g carbon)

In view of the foregoing, those beginning a study of activated carbon may be surprised and puzzled on learning that carbon can be and is utilized to adsorb substances from dilute solutions, and subsequently release them in a much stronger concentration. This seeming anomolous phenomenon can be traced to the powerful influence of operating conditions on the course of an adsorption. Certain conditions decrease adsorption, whereas other appropriate conditions can cause an increase.

2. ADSORPTION-DESORPTION

The amount of substance that can be recovered by an adsorption-desorption process may be viewed as the difference between the amount adsorbed under favorable conditions versus the amount adsorbed under less favorable adsorptive conditions. From this, it follows that effective use of adsorption-desorption depends on learning the conditions that provide the greatest amount of adsorption and the least amount of adsorption. The process can be viewed as a two stage series in which the first accomplishes maximum adsorption and the second provides minimum adsorption. To avoid confusion in the discussion we will employ conventional terminology and refer to the initial stage as *adsorption* and the final stage as *desorption*.

Adsorption Stage

Many of the factors to be considered for this stage are reviewed in Chapter 3.

When the adsorbent is *powdered* activated carbon, countercurrent application is advisable. The advantages described in Chapter 4 are supplemented by benefits at the elution stage. Countercurrent adsorption concentrates the adsorbate on a smaller quantity of carbon; the use of less carbon results in less total adsorbate being retained after desorption. Those benefits are also accomplished by percolation through a bed of granular carbon.

Desorption Stage

Temperature control, which is useful in vapor phase application, has only a minor direct influence in liquid phase systems. However, the selection of the appropriate temperature can contribute to the effectiveness of other measures.

There are three basic approaches to desorption, and they may be employed either singly, or in various combinations. They are:

(a) a chemical change of the adsorbate to a less adsorbable form

(b) Use of a fluid phase that has sufficient attraction for the adsorbate to pull it away from the carbon. An example is a liquid that is able to solubilize the adsorbate.

(c) the use of a fluid phase which is itself very adsorbable and can displace previously adsorbed substances.

The desorption of iodine is a classic example of a change in chemical form. Elemental iodine is very adsorbable on carbon whereas iodides are not. When carbon containing adsorbed elemental iodine is contacted with alkaline sodium sulfite, the adsorbed iodine is converted to the non-adsorbable iodide, and is readily leached from the carbon with water.

An adjustment in pH will suffice to elute some substances. Adsorbed organic acids and bases are converted to less adsorbable salt forms by suitable pH adjustment. However the elution is seldom complete because organic salts hydrolyze, and in that form are adsorbable to some extent. Adjustment of pH is often a useful adjunct to other desorption measures.

Adsorbed iodine can be eluted from carbon by potassium iodide solution. This is an example of solvent power acting alone, because iodides are not adsorbable and therefore have no displacing power.

Many of the organic solvents are effective because being very adsorbable they can displace many substances previously adsorbed. To be fully effective they must have solvent power for the displaced adsorbate.

Mixtures of solvents are used when a solvent capable of displacing an adsorbate cannot simultaneously dissolve it. Steenburg[3] found that methylene blue can be displaced by chloroform, but the action takes place only if and when an additional suitable solvent is also present to dissolve the displaced dye.

Hamilton[4] developed a desorption process in which minimum boiling point azeotropic mixtures are employed to elute adsorbates. After desorption is complete, the azeotropic mixture is distilled from the eluted material and is then recycled back into the process. Some mixtures have a further advantage of being compatible with wet carbon from an aqueous application. This eliminates a drying step that is necessary when a water immiscible solvent such as chloroform is used as an eluant.

Mixtures of immiscible solvents are utilized in a rather unique application by Miller.[5]

Adsorbed benzoic acid is eluted from carbon by benzene, but the extraction is not complete. Some benzoic acid remains adsorbed. Miller accomplished complete extraction by incorporating an alkaline aqueous phase (sodium hydroxide solution) in the system. The chain of transfer is:

(a) the carbon is preferentially wet by the benzene
(b) adsorbed benzoic acid dissolves in the benzene
(c) sodium benzoate forms at the interface between the benzene and the alkaline aqueous phase
(d) the sodium benzoate dissolves in the aqueous phase
(e) the chain of transfer continues until all the adsorbed benzoic acid appears as sodium benzoate in the aqueous phase.

It is often beneficial to add an adsorbable solute to eluting solvents. The action of many solutes is specific, therefore cut and try experimental studies are often necessary to discover a suitable solute. A displacing solute is more effective under those conditions that render it more adsorbable; for example, aniline is more effective above pH 7, and phenol below pH 7. The effect is illustrated by the fact that aniline will displace phenol at a high pH; whereas at a low pH, phenol will displace aniline.

Mixed solutes are often very effective. A mixture of methanol, pyridine, and hydrochloric acid is much more effective for eluting vitamine B_2 than any of the ingredients used separately.[6]

The enhanced eluting power of mixtures may be a result of the adsorbed substance being held at different types of surface or in pores of different sizes—each requiring a specific displacing agent. The eluting power of a mixture may also depend on a property to which Bancroft[7] calls attention: namely, the ability of one component to increase the solubilizing power of another. Thus, pyroxylin will dissolve more readily in a mixture of ether and ethanol than in either solvent used separately.

3.   IRREVERSIBLE ADSORPTION

Based on the data in Table 5:1, if a gram of Carbon H, after adsorbing 90 mg dye from an aqueous solution, were placed in ethanol we might expect that much of the adsorbed dye would be released to the ethanol. When such a lab test was conducted, very little dye was extracted. The reluctance of an adsorbed substance to be desorbed is not an uncommon experience. It is termed irreversible adsorption. The behavior can be visualized in a kitchen experiment: when a spoonful of salad oil is placed in a glass of water and stirred, the oil will not adhere to the glass. But the oil will adhere when placed in a dry glass and will be only partially displaced if and when water is added later. Oily patches and globules will be found clinging to the inside of the glass.

TABLE 5:1

ADSORPTION OF RHODAMINE B BY CARBON H

Condition of Adsorption*	mg. dye adsorbed/g. carbon
Aqueous Solution	190
Ethanol Solution	45

*Adsorption conducted so that the residual dye concentration in the ethanol solution was identical with that in the aqueous solution.

Much experience has shown that the extent of irreversible adsorption is often inter-related with the method used for desorption. The possibility often exists that some portion can be overcome by more suitable agents. Miller[5] found that traces of hydrochloric acid which resisted elution could be extracted by electrodialysis.

Diverse factors can contribute to irreversibility. A chemical change may take place that necessitates desorption agents appropriate to the altered form. Polymerized products may form that are insoluble in known solvents. The possibility should not be overlooked that an adsorbate may be desorbed and escape recognition because of the altered chemical form.

Obviously chemical changes occur with some adsorbates, but they are probably less frequent than might be inferred from the foregoing. Penicillin, insulin, and other sensitive biochemicals are stable when adsorbed on activated carbon.

Some changes that do occur may be minimized by altering the adsorption and desorption conditions. Moloney and Findlay[8] report that methylene blue could be desorbed when the prior adsorption was conducted at pH 3.8; but not when the prior adsorption was at pH 2.2. Catalytic oxidation is avoided by operating in an inert atmosphere. The catalytic activity may be controlled in some applications by poisoning the active centers before adsorption.

4. COADSORPTION

Irreversible adsorption is often less extensive when natural substances are processed in their crude state. This behavior has been traced to the presence of constituents in the crude material which have been termed coadsorbates. When the coadsorbates are removed during refining, the finished product suffers extensive irreversible adsorption during adsorption-desorption.

This has led to studies of adding synthetic coadsorbates to a system. Williams[9,10] and coworkers studied the adsorption-desorption of folic acid. They found ammonia elutes the active principle from crude preparations. In contrast, the elution was less successful for folic acid adsorbed from relatively pure solutions. Apparently, certain substances in the crude preparations cause the folic acid to be adsorbed in a mode from which it is readily released by ammonia.

In further studies, they found preadsorption of aniline on carbon prevented the irreversible adsorption. Folic acid adsorbed from a pure solution on aniline-treated carbon was readily eluted with ammonia

Weiss[11] found streptomycin can be eluted more completely from carbon pretreated with oleic acid. Asatoor and Dalgliesh[12] report that pretreatment of carbon with octadecylamine minimizes the irreversible adsorption of some aromatic compounds.

The *modus operandi* of coadsorbates is a bit uncertain as is apparent from the diverse descriptive names applied: *modifier, deactivator, mordant, complexing-agent*. It may well be that coadsorbates operate in diverse ways depending on characteristics of the system involved. In one view, the coadsorbate occupies and deactivates a surface area having strongest adsorptive power.

Then the substance to be reclaimed enters the less active surface from which it is readily dislodged. From another view, primary forces hold the carbon to the coadsorbate, from which secondary forces hold the substance to be reclaimed. This envisions the coadsorbate as having two hands, one of which holds the carbon firmly, while the other loosely holds the substance to be reclaimed. The latter hold can be broken by appropriate conditions.

The extent of irreversibility may be controlled in some systems by the selection of a suitable carbon. Thus carbons A and B may adsorb equal amounts at the first adsorption stage; but under desorption conditions carbon B may hold the adsorbate less tenaciously, and therefore, be more suitable.

In some applications, the carbon after desorption is recycled for re-use. This is seldom practicable when desorption is accomplished by displacement with a strongly adsorbable substance. But some desorption processes enable a carbon to be recycled repeatedly.

An example is provided in a process to recover uranium.[13] Granular carbon is impregnated with a modifier, namely, an organophosphoric ester which can form chelate compounds with uranium. When the adsorption stage is completed, the uranium is desorbed by washing the carbon with sulfuric or phosphoric acid; after which the carbon is re-used. The service life for repeated re-use is limited only by the slow hydrolysis of the organophosphoric ester, up to 80 cycles are reported.

In some applications less irreversible adsorption has been reported when a carbon is re-used. This is attributed to the initial adsorption saturating areas of the surface that cause irreversible adsorption. Kamikubo[14] observed that the repeated use of spent carbon promoted the elution of vitamin B_{12}.

5.   INDUSTRIAL PRODUCT RECLAMATION

Adsorption-desorption systems have long been used with much success for the recovery of organic vapors. Success in liquid systems has been much less evident even though many avenues of approach have developed through research in diverse applications.

In the early 1920's a process, for recovering iodine from petroleum brines, became one of the largest consumers of activated carbon in America. The undertaking prospered for a number of years, but became a casualty of the depression. Interest in adsorption-desorption subsided until World War II when a process with activated carbon provided the only available means for large scale production of penicillin at a time it was urgently needed. The dramatic success gave great impetus to possible application to other biochemicals. An important development was the manufacture of streptomycin. Again, adsorption-desorption processes gave promise of furnishing a major market for carbon, but interest evaporated when producers of penicillin and streptomycin turned to other methods of manufacture.

Much has been published on the application of carbon for reclamation of gold in mining operations, but information on the size of the market is not readily available.

Interest in reclamation of other metals has received increasing attention in recent years.

It is apparent that a better understanding is needed of varied aspects: the causes of irreversibility, the role of coadsorbates, the function of mixtures of displacing agents. New test procedures are required to disclose conditions favorable for the release of an adsorbate.

6. RESERVOIR CONCEPT OF ADSORPTION-DESORPTION

The gradual release of an adsorbed substance from carbon by extraction with water has led to applications in which the carbon serves as a reservoir. Years ago, carbon was used as a carrier of medicinals to be released gradually in the gastro-intestinal tract. More recently carbon has been used to make the action of fungicides and insecticides on sensitive foliage more safe.

The successful use of the reservoir concept depends on the adsorbed substance being eluted in concentrations effective for the required task. Consider a hypothetical situation in which it is necessary to maintain a fungicide concentration of 15 to 5 ppm in the soil surrounding a plant. If carbon containing adsorbed fungicide is placed in the soil, the only portion to be effective is that released in concentrations between 15 and 5 ppm.

In Figure 5:2, carbons A and B initially contain identical amounts of adsorbed fungicide (point 0 in Figure 5:2) .When placed in moist soil, the fungicide will move from the carbon when necessary to maintain an equilibrium in the surrounding soil. As the fungicide in the soil becomes depleted, it will be replenished by additional migration from the carbon. In the case of carbon A, most of the fungicide is transferred from the carbon within the desired range of 15 to 5 ppm. In contrast, only a small portion of the fungicide is released from carbon B within the effective range. It is apparent that the reservoir concept is more apt to have economic utility when the isotherm has a steep slope.

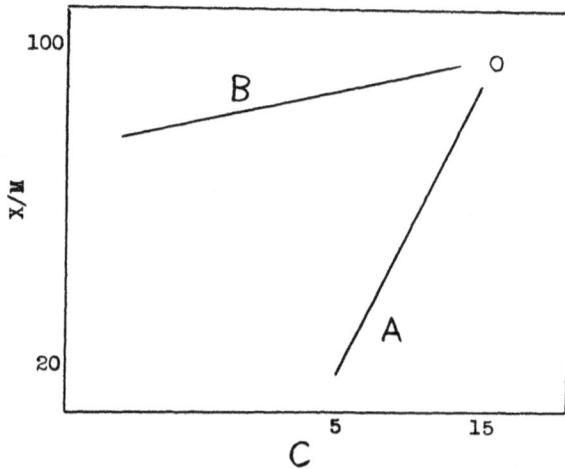

Fig. 5:2   Hypothetical isotherm to illustrate elution of fungicide
(from carbons **A** and **B** that have different isotherm slopes)
C, concentration in ppm of fungicide in soil
X/M, relative amount of adsorbed fungicide held by carbon

REFERENCES

1. Deitz, V. R., *Bibliography of Solid Adsorbents*, United States
    Cane Sugar Refiners and Bone Char Manufacturers, and
    National Bureau of Standards, Washington, D.C., 1944.
    Supplementary volume published 1956.
2. Lindau, G. and Salomon, G., *Ber.* 67 B 1296 (1934).
3. Steenberg, B., *Naturwissenschaften*, 29:79 (1941).
4. Hamilton, C. E., *Chem. Abstracts* 65, 16142d; 60, 13013 C

5. Miller, E. J., et al., *J. Am. Chem. Soc.*, 46:1150 (1924); *Mich. Agr. Expt. Sta. Rept.* for 1930–32,
6. Guha, B. C., and Chakrovorty, P. N., *J. Indian Chem. Soc.*, 11: 295 (1934).
7. Bancroft, W. D., *Applied Colloid Chemistry, 3rd Edition*, P. 208, McGraw-Hill Book Company, New York, 1932.
8. Moloney, P. J., and Findlay, D. M., *J. Phys. Chem.*, 28:402 (1924).
9. Mitchell, H. K., Snell, E. S., and Williams, R. J., *J. Am. Chem. Soc.*, 66:267 (1944).
10. Frieden, E. H., Mitchell, H. K., and Williams, R. J., *J. Am. Chem. Soc.*, 66:269 (1944).
11. Weiss, D. E., *Discussions Faraday Soc.*, 7:142 (1949).
12. Asatoor, R., and Dalgliesh, C. E., *J. Chem. Soc.*, 1956–2291; 1958:1498.
13. McClaine, L. A., Noble, Paul Jr., and Bullwinkel, Edward P., *J. Phys. Chem.*, 62:299 (1958).
14. Kamikubo, T., *Chem. Abstract*, 61, 16350.

6

Unit Operations

1. INTRODUCTION[1,2,3]

Bone char, the forefather of activated carbon has long been employed in sugar refining. The granular char is placed in columns through which the hot syrups percolate until the char is spent as evidenced by the color of percolated syrup.

When activated carbon was first produced, efforts were made to use it as a replacement for bone char. The carbons that were initially made for many years, were too soft and friable to be supplied in granular form; and it became necessary to search for processing techniques suitable for carbon in powdered form. The *batch-contact* unit operation was generally accepted.[3,4,5] In this, predetermined quantities of carbon and the liquid to be purified are mixed under appropriate treatment conditions, after which the carbon is separated from the system.

The batch-contact unit operation was readily adaptable to many varied industrial applications, and for many years was employed in most liquid-phase purifications that required a decolorizing type carbon.

Hard granular activated carbons were produced during World War 1, and were soon available commercially. They have since been employed extensively in gas and vapor systems. However, they lacked the decolorizing properties needed in many liquid phase applications. But shortly after 1950, developments in the manufacturing art resulted in the production of decolorizing carbons in the forms of hard granules and pellets. This has led to extensive expansion in the use of continuous percolation through granular carbon.[6,7,8,9]

Purification with either powdered or granular activated carbon can be conducted with standard forms of equipment available from suppliers listed in buyers guides. Equipment manufacturers will gladly join with carbon suppliers in assisting in the selection of appropriate items and processing design.

Whether powdered or granular carbon should be selected for a contemplated new application will depend on the particular circumstances surrounding each individual situation.

Granular carbons provide an advantage in that when removed from the system they carry away a larger quantity of impurity than can be accomplished by powdered carbon in batch operation. Then too, percolation is viewed by many as easier to operate than batch treatment.

For operations in which the daily carbon requirements are large, granular carbons offer the advantage of established methods of regeneration which greatly reduce the carbon cost. However this gain may not continue if present research develops a satisfactory method of regenerating powdered carbon.

Powdered carbons offer advantages in many current applications. Because of a lower initial cost per pound, powdered carbon is more economical when the daily carbon requirements are not large enough to warrent installation and operation of regeneration equipment. Batch treatment is practicable for processes that operate intermittently in which the product is subject to chemical or biological deterioration. Obviously, batch treatment provides greater flexibility for manufacturing operations in which the same equipment is diverted from time to time for production of different products. Also it is not to be overlooked that some specially required specific adsorptive powers are available only in certain types of powdered carbon. Batch treatment can often be used when it is desired to selectively remove an undesired property (such as an objectionable odor) without simultaneously removing other adsorbable constituents that should remain in solution.

2. CONTACT-BATCH UNIT OPERATION

A typical arrangement of equipment for the contact-batch operation is shown diagrammatically in Figure 6:1, but it is to be emphasized that this is only one of many variations that are in use.

The equipment should be constructed of materials suitable to the liquid being processed. Carbon steel is acceptable for non-corrosive systems. The new organic protective coatings are successful on mild steel for slightly corrosive substances. More corrosive situations require stainless steel, glass-lined steel, rubber-lined steel, or similar resistant materials.[3]

The capacity of the tank will depend on the amount of liquid to be processed and on the time of treatment, which must include time for filling and emptying. The tank should be equipped with an agitating device to enable the carbon particles to meet fresh

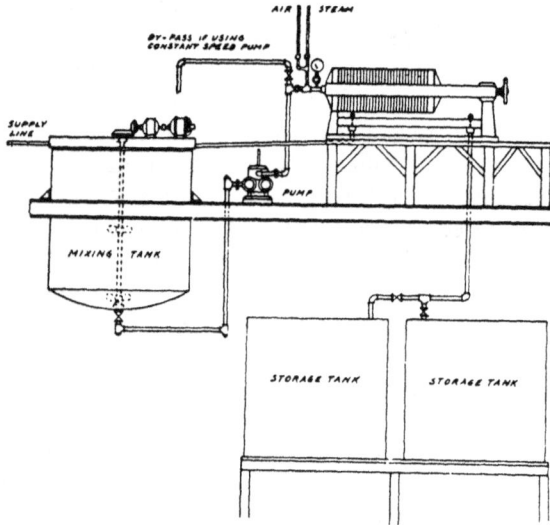

Fig. 6:1 Typical layout for carbon process for purification of liquids
From *Active Carbon the Modern Purifier*, published by West Virginia Pulp and Paper Company, New York; reprinted by courtesy of copyright owner.

portions of the liquid. The action should cause a mild turbulence to accomplish thorough mixing, but it is important to avoid excessive turbulence which would aerate the liquid. When the liquid contains ingredients very sensitive to oxidation (as caused by aeration), the treatment with carbon should be conducted under partial vacuum, or under an inert atmosphere, such as nitrogen or carbon dioxide. Should elevated temperatures be required, heating coils are installed.

A pump is necessary to transfer the liquid-carbon mixture from the treatment tank and force it through the filter. In this, means must be provided to control the rate of liquid flow through the filter.

Filter cloth is commonly used for dressing a filter, although for some applications, fine mesh wire screen or paper is appropriate. The real function of the filter cloth or other membrane is to furnish a support for the real filter medium, namely, the bed of deposited solids that form the filter cake. The ability to maintain maximum flow of clear filtrate depends on furnishing a filter cake of proper permeability.

The conveying pipes may be made of materials compatible with the processed liquid; rubber hose or suitably lined pipes are serviceable for corrosive fluids. Sharp bends are to be avoided.

Plate and frame filters were generally used during the early development of activated carbon, but since then practically every type of filter has been used effectively on one occasion or another. The type to be chosen depends on circumstances surrounding the individual situation. Unless there is a good knowledge of filters within the organization, consultation with filter manufacturers is desirable. It is important to provide adequate filter capacity because this might permit the use of a carbon that has a slow rate of filtration, but has specific adsorptive powers needed in the application. Rates of flow range from 3 gallons to over 50 gallons per square foot per hour.

Laboratory data provide a guide as to the required dosage of carbon, but for reasons yet unknown the dosage required for plant operation is usually appreciably less than quantities indicated by laboratory tests. Experience will establish a factor of the relation between laboratory data and the dosage required in each plant operation.

Special attention must be given to the method of placing the carbon into the liquid to be treated so as to avoid carbon dust from, being blown or otherwise scattered throughout the premises, because dirt and black dust can be a significant deterrent to the use of powdered carbon. That the dust nuisance is avoidable is demonstrated by the successful use of powdered carbon in dry-cleaning establishments and water plants.

A method used with success in many situations, and which in brief consists of applying powdered carbon in the form of a slurry. The slurry is prepared in a separate room to isolate any dust. Either water or a portion of the liquid to be purified is mixed with the carbon in a proportion suitable for mechanical handling; e.g., 1 pound of carbon per gallon of fluid. The slurry is then brought to the treatment station and placed in a corrosion-resistant storage container for use as required.

Laboratory data will indicate the time required in the treatment tank which will usually range from 30 minutes to an hour. Longer time is needed when the liquid is viscous or if very small dosages of carbon are used.

Usually the treatment is conducted at the operating temperature normal for the process except when other considerations enter; for example, viscous liquids may require an elevation in temperature to aid in the subsequent filtration.

When the adsorption is considered complete, the carbon-liquid mixture is pumped to the filter,

As previously mentioned, it is the bed of deposited solids that constitutes the actual filter medium. Therefore, until sufficient solids are deposited to provide a filter cake the filtrate may be turbid; and if so, it should be recirculated to the treatment tank. Should the turbidity continue after the bed begins to form, arrangements should be made in future runs to precoat the filter with $\frac{1}{8} - \frac{1}{16}$ inch of a suitable filter aid.*

In many applications, sufficient filter aid is added with the carbon to the treatment tank to create a more porous structure in the filter cake as it forms.

Satisfactory filtration requires careful regulation of the rate at which carbon-liquid mixture is delivered to the press.

At the start, the pressure should be low to permit a proper flow of clear liquid from the filter. Then as the filtration continues and the filter cake becomes thicker, the pressure should be increased, but the increase should be gradual.

*To precoat the filter: The filter aid (diatomaceous earth, pulp, etc.) is mixed with a portion of the liquid to be treated or with water; the slurry is fed to the suction of the pump to the filter. The filter aid is retained on the filter membrane as a light coating and the effluent liquid is returned to the slurry container. The recirculation is continued until the effluent is entirely clear of solid particles

There are two reasons for careful control of the initial pressure. A high pressure at the start will move the liquid with swift velocity that will carry fine particles into the pores of the filter fabric. The other adverse feature is associated with a liquid that contains floculant precipitates and other non-rigid solids. When a strong pressure is applied, the non-rigid solids are compressed, thus creating a structural condition that offers resistance to the flow.

Once such an adverse condition is established, the flow cannot be increased by greater pressure from the pump; the only course is to clean the filter and make a fresh start.

When the filtration is complete, some process liquor retained in the filter cake can be recovered. With sugar liquors the filter cake is washed with hot water (sweetening-off) to recover sugar. In processing oils, the filter is blown with steam to recover loosely held oil; and in some operations more complete extraction is obtained by a suitable solvent.

If the process aims to recover an adsorbable substance of value, an eluting solvent can be recirculated through the cake while it is still in the filter, but this is seldom satisfactory because the eluting solvent channels through thin portions of the bed. A preferred method is to remove the carbon from the filter, mix it with eluting solvent and then separate it by a filtration.

3.   COUNTERCURRENT APPLICATION

The foregoing procedure—known as *single-stage batch contact* is very effective for many purifications. However, when impurities that are difficult to adsorb are present in large concentrations, then complete removal in a single stage unit operation may require the use of a greater amount of carbon than is practical. To meet this difficulty, M.T. Sanders in 1923 reported the development of *countercurrent* application which can reduce the amount of carbon required to accomplish a purification. Savings up to 50% are not unusual. The benefits not only reduced carbon costs in many existing applications, but also enabled carbon to enter new markets.

The potential benefits can be assessed by an inspection of the adsorption isotherms. The benefits increase as the slope of the isotherm becomes steeper, also when more complete purification is required. Usually two stages are included, occasionally a third stage is added, but the extra benefits seldom warrant the extra handling and filtration station. A *two-stage* countercurrent operation is conducted as follows: A batch of process liquor is partially purified by a *once-used* carbon and after treatment and filtration in the usual manner, the now *twice-used* carbon is discarded, and the filtrate is given a second treatment with sufficient *virgin* carbon to bring the liquor to the desired final purity. It is customary to apply both the semi-spent and virgin carbon at the same stage of the operation, but in some processes they are applied at different stages. Thus, in one corn sugar operation, virgin carbon is applied to the concentrated liquor, and the *once-used* carbon to the thin syrup before concentration.

Nomographs prepared by Sanders[4] and Helbig[5] simplify the selection of the proper carbon dosage for countercurrent application.

4. SPLIT APPLICATION

Countercurrent application is less adaptable to processes that operate intermittently, especialy with products susceptible to deterioration through chemical and/or biological action. In such a situation, *split-application* will require less carbon than single stage, but the savings are less than with countercurrent.

In split application, the carbon dosage is divided into two portions. One portion is added to the raw liquor in the conventional manner and is discarded after filtration. The filtrate is then treated with the other separate portion of carbon to yield a completely purified filtrate. The carbon from this stage is also discarded.

It is to be emphasized that a filtration stage must be provided between the separate additions of carbon.

5. POWDERED CARBON CONTINUOUS TREATMENT

In most water purification operations, the carbon is fed by proportioning equipment to the flowing water, and after mixing, the carbon is separated by flocculation and sedimentation.

In other operations the carbon is separated by filtration. In some types of processing the carbon and liquid are not premixed before filtration; instead a layer of carbon is precoated* on the filter through which the liquor to be purified is passed continuously. This method, *precoat filtration,* is employed in systems that contain a liquid phase which serves to pick up impurities produced by the operation, *e.g.* dry cleaning and electroplating. An accumulation of the impurities is avoided by recycling the liquid phase through the precoat layer

6. CONTINUOUS PERCOLATION THROUGH GRANULAR CARBON[3,6,9,13]

In a percolation system, the raw liquid is passed through a bed of sufficient depth to provide a finished liquor of the required purity. Adsorption in a column is a continuous process, but for convenience of presentation we will envision it as taking place in steps. Thus, we can envision that when a colored solution passes through a bed of carbon, some color is removed from layer *a* at the entrance leaving a less colored solution to contact the next layer *b* which removes more color. The reduction of color continues as the liquid contacts further layers of unused carbon until it finally reaches layer *z* where all the adsorbable color is removed. The layers *a* to *z*, where the transfer of color from the liquid to the carbon takes place, is known as the *adsorption zone* or *mass transfer zone,* or just *MTZ.* As the flow of liquid continues, the *MTZ* advances through the column progressively contacting new layers of unused carbon, and leaving behind layers of carbon saturated with color.

The flow of liquid through a column of granular carbon will range from 1 to 10 GPM per square foot of cross section of the bed. The depth of the bed may be from 10 to 40 feet. The longer bed depths usually are divided into two or more sections,** with each unit section placed in a separate adsorbing column.

---

*The carbon precoat is formed by a procedure similar to that previously described for forming a precoat of filter aid.

**Two units have been reported to provide optimum economic operation.

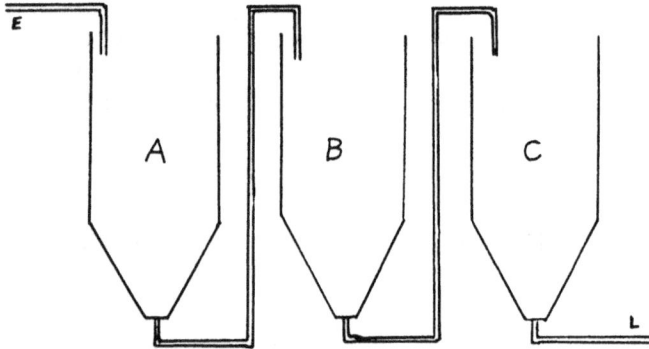

Fig. 6:2   Series arrangement for percolation columns **A, B** and **C** contain granular carbon; liquid enters system at *E* and leaves at *L*.

It is possible to make excellent use of the countercurrent principle by placing the columns in a flexible series arrangement in which each column can take turns at being at each successive stage. The scheme of rotation can be illustrated by reference to Figure 6:2. A large portion of the impurities in the entering liquid is adsorbed by the carbon in column *A*, and much of the remainder is taken out in column *B*. A final cleanup is accomplished in column *C*, which delivers a finished liquid of the desired purity. The flow of liquid continues until the carbon in column *A* becomes saturated with impurities. It is then taken off stream and the entering liquid is diverted to column *B*. Fresh carbon is substituted for the exhausted carbon in column *A* and the column is placed back on stream, but it now becomes the final column in the series and delivers the finished purified liquid. This scheme of rotating the order of flow is repeated each time the carbon becomes exhausted in the column that receives the entering liquid, and it thus provides continuous production of required purity.

A typical column is a vessel having either flat or dished heads, with a carbon-retaining screen and supporting grid installed in the bottom. The retaining screen should be fastened to the upper face of the supporting grid, which can be made of grating, a drilled plate, or bars covered with a heavy screen. The grid should be strong enough to support the weight of carbon and fluid pressure drop across the carbon bed.[9]

Minimum recommended height-to-diameter ratio of a column is 2 to 1, but ratios greater than 2 to 1 will improve the performance.

Cone-botton columns (as in Figure 6:2) are recommended for complete and rapid discharge of the carbon. When the carbon is spent and washed free of the product, a valve at the base of the cone is opened and the slurry is discharged by gravity or by pressure.

Columns of granular carbon are designed for economical and convenient working pressure. Actual operating pressure seldom rises above one pound per foot of bed depth even with a liquid of 10 centipoise viscosity. Good design allows for some excess pressure.

In the operation just described, the raw liquid enters at E (Fig. 6:2) then passes downward through each adsorber, and finally leaves at L.

In an alternative procedure the raw liquid enters at L (Fig. 6:2) then passes by up-flow through each adsorber and finally leaves the system at E.

Parallel Arrangement

This method of operation is used when only partial removal of impurities is required; and is described in Chapter 8.

Moving Bed Systems, Pulse Bed Adsorbers

A moving bed incorporates an effective countercurrent operation within a single column.[6,9] The systems are especially applicable to smaller size operations.

In a moving bed column, the flow of liquid is up through the bed of carbon, while the carbon advances periodically down through the column. At definite time intervals, a small amount of spent carbon is removed from the bottom of the bed to be regenerated, and an equivilent amount of regenerated carbon is added to the top of the column. The required final purity is assured since the liquid passes through fresh carbon before leaving the bed.

Continous Adsorption Process (CAP)

Irregularities in the sizes and shapes of granular carbon particles can result in non-uniform packing of a column. If and when this happens, the resistance to the flow of liquid will not be uniform for all portions of the cross-sectional area. In those areas and regions where the resistance is less than average a greater-than-average quantity of liquid will flow. Carbon particles in the path of this faster-flowing portion of the liquid stream will soon become saturated with color and impurities, and then color will pass through to appear in the effluent. Because of this, the column must be taken off the line before the adsorptive capacity of the entire bed has been fully utilized.

It is reported that this behavior, called *channeling,* is eliminated in the *Continuous Adsorption Provess (CAP).* The operation of this process is in some respects similar to the *Moving Bed System* previously described, but with this important difference: the upward-flowing stream of liquid moves at a velocity sufficient to provide an *expanded state.* In this expanded state, the individual carbon particles are separated very slightly from one another by a moving liquid film. Each carbon particle, however, maintains its position relative to all adjoining particles. and the expanded bed moves downward like a regiment on parade.

The rate of flow required to maintain an expanded bed depends on the viscosity of the liquid. If, for a given viscosity, the rate of flow of the liquid is too slow, the expanded bed collapses and becomes a packed bed. On the other hand, if the flow of liquid is too rapid, the bed becomes fluidized and the particles of carbon move about at random like shoppers throughout a supermarket.

7. STORAGE AND HANDLING[3,11,12,13]

Activated carbon is commonly packaged in bags, drums, and cartons. It is also available from some suppliers in bulk, i.e., in hopper-bottom railroad cars and tank trucks.

If stored in rooms free of chemical or other vapors, most types of activated carbon will retain activity indefinitely. Long storage in the form of an aqueous slurry has been found entirely satisfactory in many applications.

Most types of activated carbon are less readily ignited than many common materials such as paper pulp, starch, and powdered bituminous coal. As they are not subject to spontaneous combustion, the only precautions necessary are those required for storage of any combustible material. Packages should not be piled against steam pipes or electric wires. Hot electric light bulbs such as used in drag cords should not be laid on the bags. Activated carbon, in dry form, should never be mixed with hypochlorites, permanganates, nitric acid, or other strongly oxidizing chemicals.

Should a fire occur, powdered activated carbon, when undisturbed, will merely smolder. The fire can best be smothered with a foam type of extinguisher or by a fine spray of water. Never use a strong stream of water, because this will scatter the burning particles of carbon. As burning activated carbon can cause painful burns, it is well for an attendant to stand at least 15 or 20 feet away from any pile of burning carbon.

Many types of activated carbon have high electrical conductivity; therefore, dustproof motors and enclosed switches should be installed in rooms where carbon dust is present.

In the case of powdered carbon, problems from dust and dirt can be minimized by observing a few simple precautions:

1. When a shipment is received remove loose dust from the packages by brushing or by a vacuum cleaner.
2. When moved into storage, lay the packages in place—never throw or drop them.
3. When feeding powdered carbon, avoid drafts and keep free fall of loose carbon to the minimum.

Special appliances are available for feeding dry carbon into closed vessels under dust-free conditions. As mentioned elsewhere, feeding as a slurry provides an effective means of controlling dust.

Another source of dirt is in the transporting of spent carbon from the filters, but this can often be remedied by proper placing of the treatment tanks and by the choice of appropriate filters.

Guidance on details for carrying through measures of dirt control can be obtained from the supplier.

REFERENCES

1. Deitz, V. R., *Bibliography of Solid Adsorbents,* United States Cane Sugar Refiners and Bone Char Manufacturers, and National Bureau of Standards, Washington, D.C., 1944, pp. 633–641.
   Supplementary volume issued in 1956.
2. Aehnelt, W. R., *Entfarbungs-und Klarmittel,* T. Steinkopff, Dresden und Leipzig, 1943.
   Counter-current Handbook, and *Adsorption from Solution,* Darco Corporation, New York.
   Helbig, W. A., in *Colloid Chemistry,* Volume 6 (J. Alexander, Editor), Reinhold Publishing Corp., New York, 1946, p. 814. Kausch, O., *Die aktive Kohle,* W. Knapp, Halle, 1928, Supplement 1932.
   Mantell, C. L., *Industrial Carbon,* D. Van Nostrand Co., New York, 1946.
3. Much useful information is available from suppliers.
4. Sanders, M. T., *Ind (Eng. Chem.,* 15:784 (1923).
5. Helbig, W. A., *Two-step Divided Application,* Atlas Chemical Industries, Wilmington, Del.
6. Cooper, Jonathan C., Pittsburg Chemical Company; personal communication.
7. Marcy, W., *Inter. Sugar J.,* Nov. 1961, p. 340.
8. U. S. Patent 2,954,305.
9. Literature from Pittsburgh Chemical Company, Pittsburgh, Pa.
10. *Operating with Activated Carbon,* Atlas Chemical Industries, Wilmington, Del.
11. *Taste and Odor Control J.,* 16 No. 3 (1950); 23: No. 9 (1957).
12. Visscher, M. J., *Bulk Transport, Automatical Dosage, Continuous Contact and Filtration of Activated Carbon.* United Norit Sales Corporation, Amsterdam, Holland, 1962.
13. Williams, Franklyn M., personal communication, 1962.

7

Representative Industrial Applications

1. INTRODUCTION

It is well to begin this chapter with a clear statement of purpose. First of all, the brief descriptions of separate applications are not offered as capsule guides of correct methods of using carbon in individual applications. The veteran operator already has far more knowledge of his specific field than could be told here; and the novice needs much more information than can be included in a general text. That brings us to the real purpose of this chapter which is based on a recognition that we often discover new approaches by the learning of practices in enterprises other than our own. To that end, this chapter will endeavor to outline some of the varied ways and means employed to accomplish diverse objectives. In addition to current practices, some items are taken from abandoned processes, and still others from research that never got beyond the laboratory or pilot plant stage. The items selected are those which appear to have potential utility.[1,2]

Activated carbon will provide effective purification at low cost if and when it is employed under appropriate conditions to do tasks for which it is fitted. In some situations carbon by itself can accomplish complete purification; but more often it is joined with other cooperating separation methods.

When several different methods of separation are joined to accomplish total purification, they seldom act independently of one another, and the effectiveness of each step often depends on what proceeds and on what is to follow. Consequently the overall effectiveness can depend on arranging a correct sequence.

The proper sequence may be evident as when carbon is used to adsorb substances that impede a crystallization stage—obviously the carbon treatment should then precede the crystallization. Conversely, when used to remove adsorbable residues left by other operations—then carbon should be used at the end of the line. But all situations cannot be so readily diagnosed. Some purifications require much research to learn the optimum procedure.

2. SUGAR

Cane Sugar[2,3,4,5,6]

Sugar juice is synthesized by the growing plant. Roller mills squeeze the juice from the cane, which after being strained is clarified.

Lime, the basic clarification agent, converts many organic acids into insoluble calcium salts, and also removes much colloidal impurity. The clarified juice is concentrated by evaporation to a syrup (50% to 60% sugar), which is then boiled in vacuum pans to a point where crystals form and separate from the syrup. These crystals are known as *raw sugar*.

Some raw sugar is sold for direct consumption, but the major portion is packaged and shipped to refineries in industrial areas to be converted into white sugar.

The first stage of refining is known as *affination* and consists of washing the crystals in a centrifuge to remove an adhering molasses film. For this, a saturated sugar syrup is used which washes away the molasses film without eroding the crystals. The washed crystals—now about 99° purity are dissolved (*melted*) to form a 50–60% syrup which is further clarified with lime—often in conjunction with phosphoric acid. The combined action causes co-precipitation of many residual impurities, and the filtered syrup is ready for the adsorption station.

Bone-char

In many refineries the hot clarified sugar syrup is percolated through beds of bone-char which adsorbs and removes color and other organic substances together with inorganic substances. The char becomes less effective as the percolation continues. When finally exhausted, the flow of liquor is ended and the char is *sweetened-off* by adding water to displace the remaining syrup. The char is reactivated by burning off the adsorbed impurities and then is returned to repeat the cycle.

Powdered Activated Carbon

Compared to bone-char, powdered activated carbon is a more powerful decolorizer; consequently smaller quantities are employed for refining. Because of the powdered form it cannot be applied in bone-char refineries. Instead, refining with activated carbon is by the batch-contact process conducted as a two-stage counter-current operation.

Granular Activated Carbon

Granular activated carbon has been used for a number of years for refining both cane and beet sugar. Some early applications involved replacement of bone-char in traditional refineries.

In an operation in Europe complete replacement of bone-char extended the use of the char filters to over 26 days as compared to 4 to 6 days with bone-char. The char was satisfactorily reactivated in the char retorts. The granular char was less effective for ash removal from the syrup. The best overall operation was provided by the addition of 30 to 40% granular char to the bone char.

Installations specifically designed for refining with granular activated carbon are in successful operation. The following is a description of a fixed bed refining: The hot clarified sugar syrup is percolated through the granular carbon placed in cone-bottom columns 9 feet in diameter and 20–24 feet in height. Each column is charged with 35,000 lbs. granular carbon together with 2,500 lbs. magnesite to prevent a drop in pH. With light colored Louisiana sugar, the filter columns run continuously for 20–30 days. On dark sugars, the runs are shortened to 16–24 days.

The pulse bed (moving bed)* has also been adapted to sugar refining.

*See chapter 6

Beet Sugar

Sugar from beets can be purified more readily than cane sugar, and refined beet sugar can be produced without a separate raw-sugar stage. In the United States, carbonation is commonly employed for clarification.

Much information has been published on the application of powdered activated carbon to beet sugar juices and liquors.[1] Various points of application are described—first and second carbonators, evaporators, thick liquors, and remelts.

Linsbauer and others found that application of activated carbon to thin juices is very effective for color removal; in addition the removal of nitrogenous substances, calcium, iron, and colloids at this stage is an important advantage. A disadvantage is that a subsequent application of the carbon may be needed when appreciable caramelization occurs during evaporation.

Studies have been made of adding carbon to the juices as they enter the evaporators, but opinions vary as to the benefits.

Modifications of the Carbon Process for Sugar

Powdered carbon generally is applied as a suspension, but in Europe some study has been devoted to *layer filtration*. In this, the sugar liquors are pumped through a layer of activated carbon previously deposited in the filter.

Suggestions have been offered for using various chemical treatments in conjunction with activated carbon. Reichert and Elliot[8] treated sugar solutions with hydrogen peroxide before adding the carbon. Vincent[9] reports benefits from adding sodium and calcium chlorite before the carbon.

Effects of Carbon Treatment

Although the removal of color is the prime objective in the use of activated carbon for sugar, other attendant benefits are of value. Carbon removes nitrogenous substances and lyophilic colloids from sugar juices and liquors.[1] Apart from direct analytical measurements, removal of colloids is demonstrated by favorable changes in many properties of the liquor. Carbon improves the filterability of juices and liquors, results in less foaming in the evaporators, and increases the speed of crystallization. Szende[10] found that the treatment of beet juices with carbon reduced the formation of color during evaporation and he ascribed this to the removal of iron and nitrogenous color generators.

Although not so effective as bone char for removal of ash, certain activated carbons do remove appreciable quantities of ash, particularly iron, calcium, alumina, and magnesia.[1] This action of carbon appears indirect. Some ash constituents in the untreated liquors are kept in solution by the peptizing action of lyophilic colloids. By adsorbing these protective colloids, carbon allows the mineral ash to precipitate and be mechanically trapped by the carbon.

Owen[11] observed that carbon treatment removes microorganisms from cane juice, particularly the gum-forming bacteria.

To be suitable for sugar operations, a suspension of the carbon in water should have a pH between 6.0 and 8.0, preferably as near 7.0 as is practical. Many data show that a neutral carbon will not cause inversion. Cases of inversion attributed to carbon have been traced to the use of carbons containing mineral acids. Activated carbon does not accelerate the inversion effect of acids present in the juice; in fact, carbon can reduce such inversion by adsorbing a portion of the acids. Even though an acid condition aids the decolorization, carbons with a low pH should never be used because of inversion.

3. SUGARS AND SYRUPS FROM THE CONVERSION OF STARCH

In America, the main source material of sugars and syrups from starch is corn starch. The starch is hydrolyzed to simpler chemical substances through the action of dilute acid—generally hydrochloric acid—under heat and pressure in a converter. The products of hydrolysis products are dextrin, maltose, and dextrose. As the hydrolysis progresses, the content of dextrose increases at the expense of the others. The conditions of the conversion are regulated according to the amount of dextrose required in the product to be produced.

Glucose

The most extensive application of activated carbon in this field is in the manufacture of the so-called glucose[1] —a mixture of dextrin, maltose, and dextrose in approximately equal parts. The treatment with activated carbon reduces the content of protein, hydroxymethyl furfural, iron, lime, and phosphate, and gives a stable, colorless syrup which does not darken with age.

The thin glucose liquor, as it comes from the converters, is treated with alkali to raise the pH to 4.5–5.5, and the fatty acids as well as the coagulated proteins are removed. The thin liquor is then treated with once-used carbon at 80°C. After filtration, the thin liquor is concentrated and treated with virgin carbon at 70°C.

Some large producers of glucose have substituted granular carbon for powdered carbon;

In Europe, three stages of countercurrent are recommended.[12,13] Powdered carbon is applied to the thick liquor, the once-used carbon to the thin liquor, and the twice-used carbon is added to the neutralized converter juice. To avoid rehandling the carbon filter cake, some manufacturers use a modification of layer filtration. In this procedure, after the thick liquor has been filtered, the thin liquor is pumped through the once-used carbon on the filter. Success in the operation depends on placing a carbon layer of uniform thickness on the filter.

In Europe, the industry is reported to prefer gas-activated carbons because they have a greater ability than chemically activated carbons to remove hydroxymethylfurfural (HMF). The HMF is known to be responsible for the redevelopment of color in decolorized syrups.

4. FATS AND OILS

Triglycerides are the principal ingredients of fats and oils from vegetable and animal sources. The crude oils contain minor quantities of free fatty acids and other substances.[14]

The conventional procedure for removing the impurities involves three steps—neutralization,* decolorization, and deodorization. The refining may include a preliminary washing of the oil with warm water, which is reported to remove proteins and gummy substances. Harris[15] found that benefits result from treating crude vegetable oils with small amounts of activated carbon. At this point, the carbon is not applied for removal of color, but to adsorb inhibitors which interfere with the neutralization of the fatty acids in the subsequent treatment with alkali.

The most important method of refining is by an aqueous solution of sodium hydroxide. The oil and the alkaline solution are mixed at 20°–24°C for 10–45 minutes, depending on the characteristics of the oil. In the operation, the free fatty acids form oil-insoluble soaps which settle to the bottom of the treatment tank together with other precipitated impurities. Considerable color is removed by the alkali treatment, but further processing is necessary to provide the light-colored oil desired by the consumer. Chemical methods are employed to some extent, but the use of adsorbents is more general. Natural bleaching clay (fullers' earth) is used for many oils, and clays activated by acid have become of considerable importance. An other important adsorbent for fats and oils is activated carbon. This is used alone for certain oils, e.g., coconut oil, for which it is more effective than clay. For oils such as cottonseed and soybean, activated carbon is used in conjunction with clay because the mixture gives better results than either adsorbent when used alone.

Decolorization is usually conducted as a batch process. The oil and adsorbent are mixed in the treatment tank or kettle, taking care to avoid oxidizing conditions, which impair the quality of the oil. The oil and adsorbent should not be mixed with turbulence. Sometimes open tanks are used, but a closed type is preferable because it permits operation under a vacuum or in the presence of an inert gas.

The oil-adsorbent mixture is heated to a temperature appropriate to the oil being processed, usually to 70° to 120°C. The mixing is continued for 10–30 minutes, then the batch is pumped through a filter to remove the adsorbents. Some of the oil retained by the adsorbent can be removed by blowing the press cake with

*The term refining is frequently applied to the alkali neutralization of fatty acids. Some investigators suggest that "refining" covers the complete conversion of an oil from the crude to the finished state.

steam or by extraction with a solvent. Complete recovery is seldom practical.

Although primarily used for removal of color, adsorbents serve other useful purposes. Harris[15] and Laptev[16] report that carbon removes soap and other substances which have a detrimental effect on the catalyst when the oil is subsequently hydrogenated. Harris and Welch[17] found certain active carbons are effective for removing substances giving a positive Kreis test.* Hagberg[18] found that better removal of peroxide bodies is accomplished with a carbon of low pH, but this was not necessarily accompanied by greater stability. Anderson[19] reported that carbon is effective for removing the bloom caused by traces of mineral oil which sometimes contaminate a vegetable oil. The removal of antioxidants, is a disadvantage, but this difficulty has been overcome by the development of antioxidants that can be added subsequently to the oil. In this connection, Hagberg found certain types of activated carbon have less adverse effect than others on the stability of cottonseed oil. Activated carbons are effective for removing odors and, when used in a mixture with clay, they prevent development of an earthy flavor which often results when clay is used alone.

In selecting a carbon, it is to be kept in mind that certain methods of activation are more effective in providing the specific adsorptive powers desired for an oil. The pH of the carbon should be between 5 and 8; carbons that are either too alkaline or too acid can cause destructive changes in the oil. The retention of oil is also a factor.

Oils that are to be used for shortening or margarine are hydrogenated to form firmer products which have increased resistance to reversion in flavor. During this process, a so-called "hydrogenation flavor" may develop. Hennessy,[20] working with fish oils, found that the hydrogenation flavor does not develop when the operation is conducted in the presence of activated carbon.

Although the treatment of all vegetable and animal oils follows the general pattern that has been outlined, modifications in the procedure are often necessary because of individual characteristics of the particular oil.

---

*Test for rancidity used in the edibile-oil industry.

Lard

Small quantities of activated carbon will remove the color and characteristic odor often present in lard rendered by conventional methods. Harris[21] has found that an excellent quality of lard is produced when the carbon is present during the rendering operation by incorporating the active carbon with the raw fats.

In the drip-rendering process, developed by Harris and Welch,[17] lard is rendered, decolorized, deodorized, and neutralized in a single operation. The raw fats are placed in the upper compartment of a specially designed cooker. The lard, as it is rendered, drips through a perforated floor into the lower compartment which contains activated carbon together with a solution of sodium carbonate or bicarbonate. The finished product is free from color and foreign odor, and has a very high smoke point. The high smoke point is due to the complete removal of free fatty acids. The normal reaction between fatty acids and soda is slightly reversible, but when carbon is also present, the soap is adsorbed as it forms and cannot participate in the reverse reaction.

5. ALCOHOLIC BEVERAGES

The use of carbon in alcoholic beverages is subject to Federal regulation. Certain applications are recognized as a routine operation, whereas other applications are considered to be rectification and require an additional tax. The Bureau of Internal Revenue should be consulted in all cases.

Whiskey

Freshly distilled whiskey has a disagreeable taste and odor. During aging, the *hog-track* taste and odor disappear and a characteristic bouquet develops. Certain ingredients in the raw whiskey contribute to the development of the bouquet. Aldehydes are oxidized to acids and these, together with the organic acids originally present, combine with the higher alcohols to form esters. This process is accelerated by the presence of activated carbon.

Not all of the ingredients in the raw whiskey contribute to the eventual development of desired properties. Various oils and proteins that cause the hog-track odor are especially objectionable. To some extent, the proteins are converted to simpler forms during the aging process, but the change occurs at a slow rate and does not always result in complete removal. Valaer and Frazier[22] observed a case in which almost 4 years elapsed before a whiskey lost the major portion of its slop taste and odor.

Activated carbon removes substances that cause the hog-track odor and yields a young whiskey that will show greater and more rapid improvement in storage. The characteristics of normal maturing become apparent in whiskey that has been treated with activated carbon. The treatment is applicable to all types of raw whiskey and also to rums and brandies.

The use of charcoal for purifying whiskey is an old practice. Formerly, the whiskey was percolated through a bed of charcoal—a method still employed in some operations. A more recent procedure is to mix the whiskey with a small dosage of powdered activated carbon for about 1 hour. Usually 0.1 to 0.2% of carbon is sufficient except for whiskies from sour mashes, which may require 0.3 to 0.5%. The carbon is added to the high wine (the distillate from the fermented mash) and after filtration, the whiskey is barreled and handled in the usual manner.

Very little is known of the actual chemical changes produced by treatment with carbon. Dudley[23] reports adsorption of furfural. Cafere[24] found that aldehydes and esters are partially removed, but the fusel oil content is only slightly reduced. Catalytic oxidation by certain types of carbon complicates the interpretation of the observed behavior; for instance, the amount of aldehydes formed by oxidation may exceed the quantity adsorbed. Glasenapp[25] observed that charcoal caused a portion of the fusel oil to oxidize to ketones and aldehydes—the latter being subsequently oxidized to acids. Bogojawlenskij[26] noted a decrease in fusel oil and an increase in aldehydes. Zaharia, Angelescu, and Motoc[27] studied the effect of carbon under conditions that preclude oxidation, and found that substances of higher molecular weight are more readily adsorbed. Atwood[28] found that treatment with activated carbon eliminates a tendency of whiskey to become cloudy during storage at temperatures below 5°C.

Studies by Tolbert and Amerine[29] on brandies indicate removal of a greater percentage of acids, furfural, and tannin than of esters, aldehydes, and higher alcohols. Williams and Fallin[30] reported adsorption of nitrogenous substances and reducing compounds. The latter are measured by the *permanganate time test* which is based on the time required to turn the color of a standard permanganate solution from pink to yellow.

The most significant change produced by carbon is in taste characteristics. Most observers agree that activated carbon produces a marked improvement in the flavor of all types of whiskey. It is also of similar benefit for brandies, rum, and vodka.

Greater adsorption usually occurs in lower-proof alcohols. All types of carbon do not have similar qualitative adsorptive powers; thus, tannin may be more strongly adsorbed than furfural by one carbon, whereas the relation is reversed with another carbon.

Although carbon is generally employed for the treatment of raw liquors, it can be used on matured whiskies and brandies. When applied to a matured liquor, the carbon should be employed in a small amount so that the natural flavor will not be impaired.

Goss[31] outlines a method of transferring substances from one liquor to another. Carbon is used to adsorb a desired flavor from one whiskey, and this carbon then is transferred to another whiskey.

A number of workers have developed methods for using carbon to accelerate the aging of whiskey.[32] In general, these processes involve prolonged contact of the carbon with aerated raw whiskey at somewhat elevated temperatures.

Neutral Spirits

This is the purest product of the distiller. To produce neutral spirits, the raw liquor is first passed through a continuous still from which a high wine is obtained. This high wine is redistilled through an aldehyde column, and the main portion of the distillate is reduced to about 101–proof and treated with activated carbon. After a contact period of 1 hour, the carbon is separated by filtration. A final distillation is made, taking precautions to remove traces of fusel oil and other impurities.

Wine

In Europe, *char* has long been used in the manufacture of high-grade wines. Formerly, bone char in paste form was used, but this has been superseded by activated carbon.[1]

The use of carbon to change one type of wine into another is not permitted by law; for instance, a red wine cannot be completely decolorized and then marketed as a white wine. Carbon can be used, however, to maintain quality standards, e.g., to treat a white wine that has become colored through contamination. Blumenthal[33] reported that the treatment of young wines with activated carbon causes them to age rapidly. He also found that cloudy, sick wines are improved by a 48-hour contact with small quantities of activated carbon. Astringent, musty, and unnatural flavors can be removed with activated carbon, but minimum quantities should be added lest the wine loses its natural flavor. Muller[34] found that, although charcoals are useful for improving off-flavored wine, they can impair the quality of a good wine.

The quantity of carbon to be added varies from 2–50 pounds per 1,000 gallons. Generally, the carbon is made into a thin slurry which is added to the wine in the treatment tank. The mixing should be gentle to avoid aeration. After several hours of contact, a clarifier, such as gelatin or bentonite, can be added and when the solids have settled, the liquid is filtered.

Beer

When beer is refrigerated, chill-sensitive proteins precipitate, causing a haze. Activated carbon is utilized as a supplementary purifier to remove undesirable proteins not removed by enzymes. An advocated dosage is 5–6 pounds of carbon per 100 barrels of beer.[35]

Treatment with activated carbon, applied to the water used in brewing, has a favorable effect on the quality of beer produced.

6.   PURIFICATION OF POTABLE WATER SUPPLIES[1,36,37,38]

Palatability is a basis on which residents of a community judge the purity of the water supply. A foreign taste and odor immediately suggests contamination and that the water may not be safe.

To provide palatable water is not always an easy task. In many communities, the raw water supply comes from rivers contaminated with industrial and municipal wastes—all of which must be removed to furnish a palatable water. Some forms of contamination are removed by the processing units in municipal water-works plants, *e.g.*; coagulation, sedimentation, filtration. The removal of the residual impurities has been the subject of numerous studies. The action of various chemicals has been well investigated and has included: chlorine, chlorine dioxide, chloramine, ozone, permanganate, lime. Although some degree of success has attended the use of these agents, the utility is restricted to certain taste and odor bearing substances, and even in such cases, careful control of all conditions is essential.

But the research discovered a dependable tool, namely: *activated carbon.*[1,36,37,38] Being insoluble, no harm can result from an overdose, and what is most important, the contamination is actually removed from the water. No special skills are needed to use it.

Powdered Activated Carbon

Because of the small dosages required (as small as 0.5 parts per million) the carbon must be finely pulverized to provide the scattering power necessary to contact all taste and odor bearing molecules. In this connection, effective distribution is aided by adding the carbon where there is sufficient turbulence to distribute the carbon particles throughout the body of the flowing water. The point of application should also permit sufficient time of contact to accomplish adequate adsorption.

The application of powdered carbon to water supplies involves departures from methods commonly employed for other products. Thus, instead of the contact-batch method, the carbon is fed continuously to the flowing water. This requires proportioning equipment to furnish uniform feed. For this, units for either dry or slurry feed are available.

In most water purification plants, powdered carbon can be added at any of several points. The most effective point depends on a number of factors: the intensity of odor and the ease of adsorption, the influence of other chemicals being used, and the effects of plant units such as sedimentation basins, and filtration. Because of the many factors involved and because their relative influence is not the same for all water supplies, the selection of the point of application requires an individual study.

For this, samples of water from different stages are treated with carbon in the laboratory under conditions that parallel those in the plant. The treated samples are then compared by the threshold odor test which measures the extent to which an odor bearing water is diluted to give a solution in which no odor is detected.

Aspects to consider in the application to various processing points are fully described in available publications.[1,36,38] We will however make brief mention of application to the *raw water* because of features of possible utility in other applications. When added to the raw water, carbon particles carry over to the coagulation chamber and aid floc formation. The presence of carbon stabilizes the floc and minimizes any release of odors when the floc settles in the sedimentation basin.

Powdered Carbon Dosage

The dosage required to produce a palatable water depends: on the type and concentration of the odor bearing substance; and on the adsorptive power of the carbon being used. In a survey of 137 municipalities, E. A. Sigworth[1,36] found the average carbon dose ranged from a high of 429 pounds per million gallons to a low of only 2 pounds per million gallons. The overall average of all plants in the survey was 23 pounds per million gallons.

Granular Activated Carbon

In contrast to decolorizing types of granular carbon which were not available until after World War II, supplies of granular carbon suitable for water purification have been available since about 1916. Therefore, it was natural for granular carbons to be included in the early research of ways and means to furnish palatable supplies of potable water.[1,36,37] Research by Bayliss and by Sierp led to construction of granular carbon filter units at Hamm, Germany in 1929, and at Bay City, Michigan in 1930. Both operations were entirely successful but attention was diverted to powdered carbon by the work of Spalding.[36,38] He demonstrated that with very minor cost for new equipment, powdered carbon will furnish palatable water at low cost. Moreover, the application is flexible and is readily adjusted to handle wide ranges of taste and odor concentrations. During the forty odd years since, most of this market has been supplied by powdered carbon.

However, recent years have witnessed a resurgence of interest in granular activated carbon in this field arising from successful installations. Most outstanding has been the experience at Nitro, West Virginia.[4,6] There, the raw water supply is contaminated with organic substances, and the threshold odor values range from 300 to 2,000. At times it was not possible with existing equipment to apply sufficient powdered carbon to deliver an acceptable finished water. The installation of granular carbon adsorption-filtration now furnishes a sparkling clear palatable water at all times.

The reactivation of the spent carbon brings a large reduction in the cost of using granular carbon, and moreover eliminates a disposal problem.

The success of the Nitro operation has led many to re-examine the relative merits of granular versus powdered carbon. A meaningful appraisal of their future relative potential is beyond the scope of this text. However, the probable position of extreme situations is rather evident. At one extreme are plants having carbon requirements large enough to warrant installation and operation of *on-site* reactivation. Such plants should certainly give consideration to granular carbon. At the other extreme are plants with small daily carbon needs; plants that have successfully used powdered carbon for many years—up to forty for some. They are not apt to begin a mass migration to granular carbon.

For operations between the extremes, correct decisions will be based on circumstances surrounding the individual situation. A new factor will enter if current research develops a practicable process to reactivate powdered carbon.

Potential Market

In an analysis of the markets for activated carbon, Parry[39] makes the following observation:

"It can be said with certainty that water treatment is a
continually expanding market for activated carbon. No
other chemicals appear to have replaced activated carbon
in this traditional market and there do not appear to be

any products on the horizon that might make significant inroads in the foreseeable future."*

Industrial Water Purification

Units for granular activated carbon are employed in certain manufacturing operations to remove chlorine and other taste-producing substances from water supplies.[1]

In the soft drink industry it is common practice to super-chlorinate water used in processing. It is then percolated through a bed of granular activated carbon to remove residual chlorine by catalytic action and also to adsorb any remaining taste.

Generally, the carbon is supplied in packaged form, *i.e.*, installed in filters furnished by the manufacturer. The service life, which depends on the quality of the water being supplied, is normally one or more years. It is important that the feed-water be free of suspended material. When turbidity is present the water should be prefiltered through suitable media.

A definite schedule should be followed for backwashing and disinfecting the carbon filter. Instructions are furnished by the manufacturer of the equipment.

7. ORGANIC COLLOIDS

Thus far the applications of activated carbon involving colloids have been in situations in which the colloid is a contamination to be removed in order to aid a processing operation and improve the quality of the product.

In some applications, however, such as the deodorization of a synthetic lacquer, instead of being merely a contaminant, the colloid is the main constituent, *i.e.*, the product itself, from which odor or color is to be removed. Some such applications necessitate departures from normal methods of treatment. To do this successfully requires an understanding of certain aspects of colloids.

The term *colloid* does not define a molecular species as do the names *acids, bases, salts,* and *esters.* Instead the name colloid describes a state in which practically any substance can abide if

*Reprinted by permission of Department of Mineral Economics, Pennsylvania State University.

suitable conditions are provided. In brief, COLLOID describes a characteristic manner in which one substance can be dispersed in another. Our interest is primarily in what are known as organic colloids. As this is a vast subject for which comprehensive texts are available, here we will give only a brief outline of aspects known to be factors in the industrial use of activated carbon.

From many points of view, a colloidal system can be regarded as a special type of solution. Indeed, the transition from a true to a colloidal solution is through a twilight zone in which it is difficult to say where a true solution ends and a colloidal state begins. A primary basis of differentiation rests on the size of the dispersed particles and for this there are several somewhat flexible yardsticks. According to one criterion, both colloids and true solutions will pass through filter paper, but only true solutions will pass through a semipermeable membrane. Another classification would view organic colloids as large molecules of at least a thousand atoms. When such substances as gelatin, glycogen, albumin, rubber are placed in a suitable solvent, they become solubilized in consequence of specific affinities. Thus the polar groups in gelatin provide a solubilizing force that is able to keep the entire large gelatin molecule dispersed in water.

When dispersed in colloidal form, the large organic molecules have shapes characteristic of the structure of the parent molecules. Some are globular or spherical, others resemble flat saucers; these two are less troublesome. A third type, however—the linear or thread type—can cause great difficulty. In some solvents the threads become interwined and interwoven to form a network comparable to a felted fabric. Color molecules can penetrate the loose structure and be adsorbed by it, and some fine carbon particles will be trapped and caged. The colloidal network immobilizes much of the liquid, resulting in a semi-rigid structure.

The felting action of linear colloids is minimized when they are fewer in number, *e.g.,* in low concentrations; hence such colloids should be treated in as dilute a solution as may be practicable. Another alternative that is sometimes feasible is the use of a different solvent. In some solvents the linear threads do not felt but instead roll into coils or balls in which forms they are more amenable to treatment with carbon. Thus, a synthetic lacquer that could not be treated effectively in aqueous solution was satisfactorily decolorized in a solvent consisting of equal parts

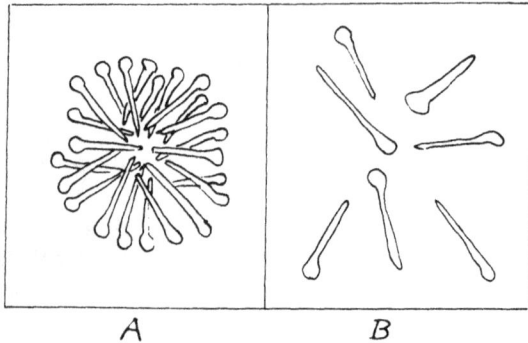

A   B

Fig. 7:1   Formation of micelles

of methanol and water.

A finely powdered carbon is usually preferable for the complete removal of traces of colloids from a crystalline product such as sugar; but the finely powdered carbons are less suitable when the main ingredient is a colloid. Then, to avoid peptization, a coarser carbon is desirable; and in some cases it is necessary to use a hard granular carbon.

Another form of organic colloid is provided by certain relatively small molecules, such as soap and detergents, that can clump together to form large clusters known as *micelles* (*A* in Figure 7:1). The ability to form micelles depends on specific relations existing between the solvent and the solute. Many detergents form micelles in aqueous solution, but when placed in certain other solvents, the detergent molecules will remain separated as individual molecules (*B* in Figure 7:1).

Because of this feature, some soaps and detergents that cannot be satisfactorily treated with carbon in aqueous solution can be effectively decolorized and deodorized in a solvent in which they do not assume a colloidal form.

Synthetic Resins

In the process of manufacturing alkyd and other polyester-type resins, minor side reactions often result in the formation of undesired color. The colored bodies can be removed by carbon, provided the carbon is added at the beginning of the processing and is present when the colored bodies are being formed. On the other hand, carbon will not decolorize if it is added after the resin has formed.

One explanation is that during the initial stages of processing, the ingredients have relatively simple molecular structures which do not interfere with the adsorption of colored bodies by the carbon. As the reaction procedes, however, the colored bodies become adsorbed on the newly formed colloidal structure—a behavior analogous to the dyeing of fabrics. After the colored bodies have been thus adsorbed by the resinous structure, they are no longer free and accessible for adsorption by the carbon. The situation is analogous to an attempt to use carbon to decolorize a dyed cloth immersed in water.

In this application, fine carbon particles are sometimes trapped in the resinous structure (*peptized*) and cannot be separated in the subsequent filtration. It is reported that this peptization can be avoided by interrupting the cycle and filtering off the carbon before it becomes caged. In another method, the processing liquid is recirculated through a short column of hard granular carbon.

8.   MISCELLANEOUS FOOD PRODUCTS

The use of activated carbon constitutes an integral part of the processing of many food products.[1] It is adaptable to foodstuffs that are in a liquid state or can be dissolved in a suitable solvent. Activated carbon should not be applied to foods such as milk, butter, mayonnaise, in which a colloidal state must be preserved. When such a product —*e.g.*, mayonnaise—is treated with carbon, the protective colloids are adsorbed and the emulsion is broken.

When a neutral flavor is desired the food product can be purified by adding sufficient carbon to remove odor, and other contaminants. The treatment of a product in which a natural flavor must be preserved presents greater difficulties. It is not always feasible to adsorb impurities without simultaneously removing desired constituents, although it is accomplished in a number of cases.

Gelatin

Activated carbon is employed to remove color and odor present in crude gelatin. Impurities that cause cloudiness are also removed. The treatment consists of adding the carbon to a 6 to 10% gelatin solution, maintaining the mixture at a temperature of 60° to 65°C for 30 minutes, and then filtering. Some operations provide a two-stage treatment in which the filtrate from the first step is concentrated to 20% gelatin content and again treated with carbon.

Pectin

This product is usually obtained by leaching the pomace of various fruits. The dark, cloudy appearance and characteristic odor are removed by treating a 1% solution with activated carbon for 1 hour at 50°C.

To prevent peptization of the carbon that often occurs, Bender, Douglas, and Cuthbert[40] developed a percolation unit. One part of cellulose pulp is intimately mixed with 1 to 3 parts of powdered activated carbon and a 2 to 4 inch layer of the mixture is placed in the filter unit. A 1% pectin solution at 50°C is then percolated through the unit. The filtrate usually has no smoky appearance which would indicate the presence of fine carbon particles.

Fruit and Berry Juices

Aehnelt[2] reports that decolorization should be conducted by allowing the carbon to remain in contact with the juice for at least 24 hours. To some extent, the required time depends on the type of juice, less time being needed for apple juice than for cherry juice. The decolorization of cherry juice must not be carried *too far* since the amount of carbon required to *do this* would impair the flavor. Similar treatment is employed to remove unpleasant tastes due to undesired false fermentation or other contamination. Off-flavors are removed successfully from pineapple juice without impairing the characteristic pineapple flavor.

Maple Syrup

Manufacturers of maple syrup and maple sugar have difficulty in maintaining a uniformly colored product throughout the season. A gradual darkening of color as the season progresses is generally due to the presence of caramel forms in the pan in which the syrup is concentrated. Activated carbon is applied to each batch to obtain a uniform color in the finished product.

Honey

The quality of honey is subject to variations beyond the control of the producer. By treating batches of strongly flavored dark honey with activated carbon and blending them with the main stock, it is possible to maintain a more uniform quality. Laws on labelling may require mention of the treatment with activated carbon, particularly if the characteristic honey flavor is altered by the treatment.

Scrap Candy

Sugar syrups, free from color and foreign flavor, can be produced from scrap candy. The treatment consists of dissolving the candy in water to make a 50 to 60% sugar solution which is treated with carbon at approximately 80°C for 30 minutes and then filtered.

9. ORGANIC CHEMICALS

Crystalline Organic Compounds

Faye and Hagberg[1] report that the use of carbon to remove adsorbable impurities before crystallization results in crystals that have a better form and a higher melting point than crystals formed from untreated liquors. The removal of adsorbable impurities also results in more efficient performance during the process of crystallization.

Studies of molasses provide an illustration. Coalstad[1] reported that in spite of improvements in vacuum pans, crystallizers, and centrifuges, the composition of molasses has remained the same for many years. An explanation of this behavior is that when the impurities exceed a critical concentration, they are preferentially adsorbed on the nuclei of sucrose crystals. Because of this, sucrose molecules in solution cannot reach the nuclei and growth of crystals does not occur. When such a molasses solution is further concentrated, the only effect is to cause more impurities to be adsorbed on the nuclei and still further retard growth of crystals. To demonstrate the validity of the theory, Coalstad diluted a sample of molasses from which no further sucrose could be crystallized, and treated the solution with a large quantity of activated carbon. When the filtrate was concentrated, the sucrose crystallized spontaneously.

Glycerin

Distillation improves the quality of glycerin, but even a double-distilled product often contains color and odor that can be removed only by activated carbon. The ease with which glycerin can be decolorized increases with distillation. A crude glycerin is relatively difficult to decolorize to a water-white product, but the task becomes easier after the glycerin has been distilled and is much easier on a twice-distilled glycerin. A low-soluble ash is required in carbons to be used for pharmaceutical glycerin.

For most purposes the treatment of glycerin with activated carbon is conducted at 60° to 80°C.

Lactates and Lactic Acid

Lactic acid is produced from lactose and other sugars by fermentation. As the bacterial action ceases at a low pH, the acid is neutralized as it forms by adding calcium carbonate or hydroxide. Therefore, calcium lactate is the end product of the reaction. When fermentation is complete, albumin and other proteins in the solution are precipitated by heating to 90°C and separated by decantation and filtration.

Various procedures have been developed for the purification of calcium lactate, and for conversion to other lactates or to lactic acid. In one procedure, activated carbon is added to the filtered liquor to remove much of the color. After filtration, the solution is made slightly acid and treated with another portion of carbon.

The filtrate is concentrated to 15° Baumé and then cooled to crystallize the calcium lactate. The crystals are dissolved in a small amount of water at 65°C and finally purified with carbon.

Weisberg[41] and his coworkers developed a somewhat different process. After precipitating the proteins in the fermented liquor. The filtrate is treated with activated carbon at 100°C for 15 minutes and lime is then added until the liquor shows a sharp *break*. The grayish precipitate settles rapidly and the supernatant liquid is filtered. The filtrate of *p*H 10 or over is neutralized with lactic acid and concentrated in a vacuum pan.

Lactic acid is formed by reacting calcium lactate with an acid, such as sulfuric, which simultaneously forms an insoluble calcium salt, the precipitate is removed by filtration. The filtered lactic acid—particularly when prepared from crude calcium lactate—contains impurities which cause color, odor, and haze. The acid is heated to 100°C and a predetermined amount of sodium or calcium ferrocyanide is added to precipitate iron and copper. Activated carbon is added simultaneously to assist in flocculating the precipitate and also to remove impurities that cause color and odor.

Tartaric Acid and Tartrates

These chemicals are prepared from crude potassium acid tartrates known as *argols,* which are precipitated during the fermentation of wines. The crude salt is slightly soluble and can be purified by recrystallization to produce the cream of tartar of commerce. The use of activated carbon gives a higher yield of crystals and usually eliminates the need of more than one re-crystallization.

Tartaric acid is prepared by converting the argols into an insoluble calcium tartrate which is treated with sulfuric acid to form tartaric acid and the relatively insoluble calcium sulfate. Tartaric acid is readily soluble and, therefore, does not lend itself to purification by repeated recrystallization. Purification is accomplished by treating the solution with activated carbon to remove color, and with sufficient calcium ferrocyanide to precipitate any iron salts.

Glutamic Acid, Monosodium Glutamate, Betaine

Waste water of Steffen's process from beet sugar factories contains 0.2 to 0.4% betaine and less than 0.2% glutamic acid. Nees and Bennett[42] developed a method to recover these substances. The first stage consists of controlled passage of the Steffen's waste liquor through a hydrogen-ion exchanger, whereby the glutamic acid and betaine are partially concentrated and simultaneously separated from most of the mineral constituents in the waste liquor. The partially concentrated liquor is treated with activated carbon to remove color and colloids, and the filtrate from this is concentrated by evaporation. The *p*H of the concentrate is adjusted to 3.2 and on cooling the glutamic acid crystallizes. The mother liquor is treated with sufficient hydrochloric acid to form betaine hydrochloride, which crystallizes after further concentration.

Glutamic acid is also prepared by the acid hydrolysis of gluten.[43] The crude, hydrolyzed product is filtered while hot to remove the insoluble humus, after which the filtrate is cooled to crystallize the glutamic acid hydrochloride. The crystals are dissolved in water. The solution (*p*H 1.5) is heated to 80° to 90°C and decolorized with activated carbon. When the filtrate is cool, sodium hydroxide is added to raise the *p*H to 3.2 and the solution is then allowed to stand until crystallization is complete.

Monosodium glutamate is formed by reacting purified glutamic acid with sodium hydroxide in an aqueous solution. Residual color is removed by a final treatment with activated carbon.

In a procedure described by Dorn,[44] crude *d*-glutamic acid is treated with hydrogen peroxide at 85°C for 45 minutes and then cooled to 25°C. The crystals that form are separated and dissolved in sodium hydroxide solution to form monosodium glutamate. The solution, heated to 55°C is purified with carbon.

10. INORGANIC CHEMICALS AND METALS

The purification of inorganic chemicals with activated carbon is not common practice. Usually it is more practical to procure source materials that do not require such purification. For the most part, applications in this field are limited to the treatment of materials containing impurities that are difficult to separate by other methods. Carbon performs a dual task of recovering a product and at the same time eliminating a disposal problem.

Alkali Hydroxides and Carbonates

Activated carbon will remove much of the pink color from commerical solutions of potassium hydroxide, sodium hydroxide, and sodium carbonate. Carbon is also useful for removing impurities that cause foaming. At times, difficulty is experienced in completely separating the carbon from caustic solutions because of a peptizing action.

Aehnelt[2] reported that a small amount of activated carbon effectively removes the dark color that develops in soda lye after use in mercerizing cellulose.

Alum

Activated carbon is used in small amounts for decolorizing alum. Alum liquors often have a characteristic brownish cast which can be removed by suitable grades of decolorizing carbon.

Iodine

Many petroleum brines contain iodides in concentrations of 50 to 250 parts per million. Iodides are not adsorbable, and the first step involves oxidation to iodine with nitrous acid, which is formed by adding sodium nitrite to the acidified brine solution. Activated carbon is added before the sodium nitrite in order to adsorb impurities that inhibit the oxidizing action of nitrous acid. Chlorine has been used as an oxidizing agent, but it has a disadvantage in that it tends to form some nonadsorbable iodate.

The liberated iodine is adsorbed by activated carbon, which is then separated from the brine and treated with sodium sulfite and sodium carbonate to re-form the iodide. The nonadsorbable iodide can be leached from the carbon in a concentrated form. Bierbaum[45] found that better extraction can be secured through a modification of the final step. After the adsorbed iodine is converted to iodide, the carbon cake is dried and heated above 250°C to remove organic impurities. The iodides can thus be recovered in a purer form and less water is required for the extraction.

Moreover, the heat treatment aids in regenerating the carbon for re-use.

Ohman[46] extracted the adsorbed iodine by an electrolytic process. The carbon containing the iodine is formed into a cake and made the cathode of an electrolytic cell. The adsorbed iodine is reduced to hydriodic acid which dissolves in the electrolyte and is oxidized to iodine at the anode. The iodine precipitates and settles to the bottom of the cell.

Chamberlain and Hooker[47] employ a vapor-phase system to recover the iodine. Air is passd through the oxidized brine to vaporize the iodine which is subsequently adsorbed by passing the air through granular activated carbon. The iodine is extracted by steaming or by treatment with a hot alkaline solution. The usual equation for this reaction is:

$$3I_2 + 6NaOH = 5NaI + NaIO_3 + 3H_2O$$

According to Chamberlain and Hooker, the carbon acts as a reducing agent and prevents the formation of the iodate.

Gold

In the conventional cyanide process, the gold-bearing ore is pulverized and leached with a solution of sodium cyanide to dissolve the gold. The solution, after separation by decantation and filtration, is treated with zinc to precipitate the gold. In an alternative method, the gold can be precipitated (adsorbed) by passing the clear solution through a bed of charcoal or activated carbon which is subsequently smelted to recover the gold.[1]

The process employing carbon was modified by Chapman,[48] who found benefits by adding the charcoal simultaneously with the cyanide solution to the ore while it was being pulverized. When charcoal is present during the dissolving process, the gold is adsorbed as it dissolves; therefore, the cyanide solution is almost free of dissolved gold at all times. Because of this, the gold not only dissolves more readily, but it is also more completely extracted from the ore. In normal cyanide operations, a thin pulp of ore is used to avoid large concentrations of dissolved gold which retard the dissolving process. Chapman points out that the presence of charcoal during the dissolving stage enables much thicker pulps to be used: in fact, a thick pulp is desirable because the charcoal is thus concentrated to a small volume.

The gold-laden charcoal is separated from the pulp by filtration or by flotation. The oil for flotation should not be added until the adsorption of gold is complete because the presence of oil on the surface of the carbon prevents subsequent adsorption of gold. Once the gold has been adsorbed, however, it is held irreversibly and is not displaced by oil added later. About 90% of the carbon is recovered by flotation, resulting in a recovery of over 85% of the soluble gold.

Although the carbon process can be used for all ores from which gold is to be extracted with cyanide, it offers no advantage in the case of ores in which the cyanide solution can be readily separated from the pulp by decantation or filtration. The carbon process is reported to be of value for ores that slime and those that are not amenable to other methods. It is also useful for tailings from other operations. The process gives a low recovery of silver and, therefore, it is not suitable for ores of high silver content.

It is reported that various practical difficulties are overcome in more recent processes.[1,49]

Other Metallic Ions

Most types of metallic ions are not appreciably adsorbed by activated carbon, unless special conditions are provided. Co-adsorbates are often useful. Thus, mercaptobenzothiazole will enable carbon to remove lead and zinc from an aqueous solution. Usually, the carbon is impregnated with the co-adsorbate before use, but, in some cases, the co-adsorbate can be added directly to the solution. Appropriate control of the pH is often necessary to accomplish selective adsorption and to provide a subsequent desorption.

It may be necessary to study a number of substances in order to find a co-adsorbate this is effective for a specific metallic ion. The approach can be simplified by observing the fundamental principle, that a suitable co-adsorbate forms a link between the carbon and the metallic ion. This narrows the study to substances that are adsorbed by carbon and that can also unite with the metallic ion to form a complex or compound. The efficiency is greater when this compound is not very soluble.

Co-adsorbates that are only slightly soluble in water can some-times be incorporated by recirculating a saturated aqueous solution through the bed of carbon. Another method is to apply the coadsorbate in a suitable organic solvent that can be subsequently evaporated.

Flotation

Flotation can be employed to separate the components of a pulverized ore. The success of the operation hinges on whether the flotation agent is more strongly adsorbed on certain components than on others; thus, oleic acid will adhere more readily to a metallic sulfide than to silica or to an oxide. Because of the adsorbed film of oil, the particles of metallic sulfide are not wetted by water and they collect as a froth at the surface when the suspension is aerated. Silica and other components of the gangue, which do not have an adsorbed film of oil, are wetted by water and remain in the aqueous phase. Care is required in adding the flotation agent because any excess may be adsorbed on particles of gangue and cause them to float. If such a condition develops, it can be corrected by adding sufficient activated carbon to strip the oil from the particles of gangue. Any excess of carbon should be avoided as it will remove the oil film from the metallic sulfides.

When several minerals having unequal adsorptive power for the flotation agent are present in an ore, they may be separated by a similar procedure. The separation is accomplished by adding a quantity of activated carbon that is just sufficient to strip the flotation agent from the less adsorptive mineral, and thus permit the other mineral to be separately floated.[1] Braunstein[50] reports the use of this method to separate blendes and galenas.

Nickel Electroplating Solution

Activated carbon will remove various impurities that accumulate in nickel electroplating baths. The purification can be accomplished by periodical treatment of the entire solution in a separate tank with 2 to 10 pounds of activated carbon per 100 gallons, or by continuously circulating a portion of the solution through a layer of activated carbon (0.2 to 1 inch thickness) deposited on a filter. Heltbig[51] suggests a filtration rate of 25 to 50 gallons per hour for each square foot of effective filter area. The carbon particles can be prevented from entering the plating tank by dressing the filter with heavy filter paper. When cloth is used, it is necessary to precoat with a filter aid. Activated carbon removes some of the brighteners and wetting agents and it is necessary to replace the amount adsorbed.

To be suitable for electroplating solutions, an activated carbon should have good filtration characteristics, be highly activated, and free of sulfides. Ease of wetting and the minimum tendency to create dust are important. Successful performance requires adherence to various operating details, and information on this can be obtained from the supplier.

11. DRY-CLEANING SOLVENTS[52,53,54]

In wet-wash laundries, the wash water is discarded after a single use and replaced with a new charge. In contrast, dry-cleaning solvents, such as perchloroethylene and petroleum solvents, are re-used indefinitely, and unless means of purification are provided the soil extracted from the garments will be deposited on later batches of clothes.

All impurities that accumulate, except odor, can beremoved by distillation, and distillation has long been a major method for reclaiming dry-cleaning solvents for re-use. Distillation, however, has become more costly since the introduction of the *charged* system in which special detergents are added to provide more effective cleansing. These detergents are lost in each distillation and must be replenished. To avoid the high cost of the replenishment,other means of purification have been developed.One method in general use combines several separations, and to understand the functions we can best begin with a description of the soil. This can be classified into three general groups:

  a) insoluble soil
  b) solvent-soluble soil
  c) water-soluble soil

a) The *insoluble soil,* consisting of lint, grit, and other solid substances is readily removed by filter frames or tubes covered with fine-mesh wire. To prevent the insoluble soil from plugging the screen, a *precoat* of diatomaceous earth is placed on the screen at the start of each day's operation, and this precoat serves as the actual medium for screening out the insouble soil. As the passage of dirty solvent will gradually saturate the precoat with soil, generally a small additional portion of diatomaceous earth is placed in the discharge from each washer for continuously maintaining a permeable filter.

b) The *solvent-soluble soil* consists of various oily substances, odorous compounds, dyes extracted from fabrics, and other contamination. This soil causes a dirty brown color that becomes deeper as the impurities accumulate, and any dirty solvent retained by the clothes leaves a gray cast that is very apparent on white fabrics.

Activated carbon is effective for removing many types of solvent-soluble soil. Two methods of application are in use:

In one method, the solvent is kept in use without purification until the contamination builds up to a certain point; then the entire quantity of solvent is brought to the desired purity by treatment with the necessary amount of activated carbon.

Much better is the preventive treatment, which avoids an accumulation of soil and maintains the solvent continuously in a good usable condition. The preventive method is known as *after-precoat* because a layer of activated carbon is applied to the filter *after* the diatomaceous earth precoat has been formed.

The over-all benefit of carbon is shown in the improved appearance of the solvent although some variation exists in the extent to which the separate items of soil are removed. The most troublesome sources of contamination, color, and odor, are effectively removed without appreciable loss of the detergent—an important feature. Fatty acids are only partially removed, and the reduction in content of mineral oil is negligible, but these impurities are seldom troublesome in the concentrations normally encountered. If and when their concentration becomes appreciable, they are removed by distillation.

c) The *water-soluble soil* is a consequence of the ability of the detergent to solubilize water into the organic solvent and then extract water-soluble substances, *e.g.*, caramel, from the garments. As in the case of mineral oil, water-soluble substances seldom cause serious difficulty, but if and when they do, they are separated from the solvent by distillation.

12.   ANALYTICAL APPLICATIONS

Activated carbon finds occasional use in analytical procedures. In some cases, the ingredient to be measured is selectively adsorbed from the mixture and subsequently eluted in a purer and more concentrated state. This method was used by Koch[56] to estimate the amount of gold in sea water. Giri and Naganna[57] used this procedure to measure the nicotinic acid content of foodstuffs, the nicotinic acid being adsorbed by carbon and subsequently eluted with a solution of sodium hydroxide in alcohol. Paget and Tilly[58] suggested the use of carbon to determine the concentration of barbiturates, the adsorbed compound being eluted with ether.

In another general type of application, carbon is employed to remove color and other interfering substances so that the solution can be conveniently analyzed by conventional methods. Chapin[59] used blood char to purify arsenic solutions before analysis. Graham,[60] in measuring the rotenone content of cubé powders, found that extraction in the presence of activated carbon gave an improved yield of rotenone crystals.

The use of carbon to clarify and decolorize urine before analysis can produce other changes that require consideration; for instance, Kolthoff[61] points out that glucose can be adsorbed. However, there is little or no loss of glucose when carbon is used in the minimum amount necessary to decolorize a urine. Woodyatt[62] suggested that the adsorptive power of each individual carbon for glucose should be determined before use. Bock[63] found that glucose is much less adsorbed from urine than from a pure aqueous solution of equal concentration. He reported that 5% of carbon removed appreciable amounts of phosphates and nitrogenous substances. Uric acid was completely adsorbed; creatinine and urea were adsorbed to a large extent. Only small amounts of ammonia, chlorides, and glucose were removed. Bohn[64] called attention to the change in pH that attends the use of carbon.

Care is necessary when using carbon to clarify specimens for toxicological analysis, particularly those that may contain alkaloids or other strongly adsorbable poisons.

Rooseboom[65] calls attention to precautions that should be observed when employing carbon as a clarification agent in the analysis of sugars. Rossee[66] states that when sugars are clarified with active carbon before analysis, it is necessary to apply a correction factor for the sucrose and invert sugar adsorbed by the carbon.

Chromatographic Analysis

As ordinarily used, adsorbents separate a mixture into two fractions, one containing the more adsorbable ingredient, and the other containing the less adsorbable ingredients. Chromatographic analysis takes advantage of the fact that the relative adsorbability of a number of substances on any one adsorbent is susceptible to fine gradations.

The method was devised by Tswett[1] who observed that when a petroleum-ether extract of plant pigments is percolated through a packed column of an adsorbent, *e.g.,* chalk, the pigments are retained in discrete bands or zones. Tswett found that he could increase the separation of the zones by adding only a small quantity of the pigment solution to the column and then allowing a portion of pure solvent to wash a pigment downward through the column —a procedure known as developing the chromatogram.

The following interpretation of the phenomenon (Figure 7:2) is taken from a discussion by Cassidy.[67]

In this latter procedure the adsorbent is placed in a tube and the fluid that serves as the second phase is passed upward or downward through the column of adsorbent, i.e., countercurrently to it. Here the more strongly adsorbed molecules (*A*), being held most tightly in the interface, congregate near where the fluid enters the adsorbent. The less tightly held ones (*B*), the available interface being first filled up chiefly with *A* molecules, move further on along the column where they may be adsorbed. Then if one passes pure solvent through the column, one can wash *B* molecules along the adsorbent farther and farther from the **A**, which are washed along the adsorbent with greater difficulty. This may lead, eventually, to complete separation of the molecules. The column of adsorbent may then be extruded from the tube and cut between the zones of adsorbed *A* and *B* molecules. These can then be desorbed from the separated sections of column and recovered.*

The development of the chromatogram is stopped when the zones are sufficiently separated. The adsorbent column is then pushed out of the tube and the individual zones are cut out of the

*Reprinted by permission from the *Journal of Chemical Education.*

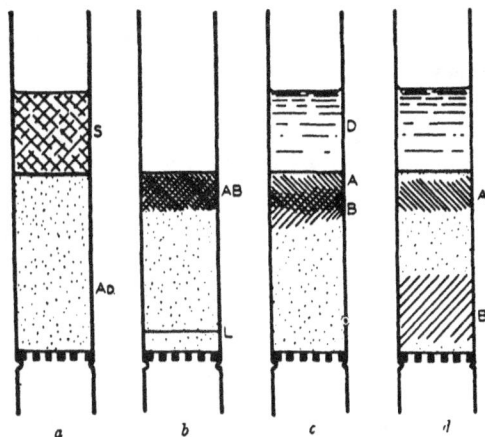

Fig. 7:2 Chromatography of a mixture of solutes A and B. a) Solution S containing A and B is placed above adsorbent AD; b) solution has been pressed into adsorbent, mixed zone AB is formed, and liquid level is at L; c) Developer liquid D is being passed into column and development of zones is beginning; d) Continued passage of developer liquid has produced a developed chromatogram, with separate zones shown at A and B. From H. G. Cassidy, *J. Chem. Educ.* 23: 427 (1946); reprinted by permission of the copyright owner.

column with a knife. Each zone is separately eluted with an appropriate solvent.

The term *chromatographic* has been so closely identified with the technique that it is retained even for processes involving the separation of colorless substances, also when using an adsorbent such as carbon that prevents color observations. When color distinctions are absent, it is necessary to employ other devices to locate the position of the individual zones. The liquid or floating chromatogram is often used when carbon is the adsorbent. In this method, the column, while still in place, is washed with an appropriate solvent and the percolate is examined from time to time. A *cut* is made each time the properties change.

Many varied fields of research find chromatographic adsorption to be a very useful tool, and a number of refinements in technique have been adopted to fit specific tasks. For information on this, the reader is referred to the original publications and to reviews on the subject.[1]

Carbon Adsorption Methods for Analysis of Organic Contamination in Water Supplies

In recent decades the quality of many water supplies has been adversely affected by factors associated with the tremendous growth in the production of industrial chemicals.[68,69,70,71,72] Some contamination accompanies normal manufacturing operations, but larger quantities often escape into ground and surface supplies as a consequence of use, *e.g.,* as insecticides or as detergents. Although present in but small concentrations—parts per million, or even per billion—they are often sufficient to damage the water, by taste and odor, by killing fish, and by foaming.

Many of the contaminants are highly stabilized organic compounds that are not destroyed by biological or chemical treatment of wastes, and they often survive the self-purification that normally occurs in moving streams. One synthetic chemical in waste, orthonitrochlorobenzene, was detectable in the Mississippi River for a distance of over 900 miles from the point of origin.[70]

The organic contamination is not appreciably removed by the steps of coagulation, chlorination, and filtration employed in water purification plants. Activated carbon is used in sufficient quantities to make the water palatable and eliminate foaming; but relatively few substances cause foaming, and only the more odorous ingredients will give a perceptible taste at the dilutions usually present (Table 7:1). Therefore we must recognize that the hazards from contamination may not be evident to sight, taste, or smell. In wastes of domestic and industrial origin, Wedgewood[72] detected the presence of polynuclear hydrocarbons possessing carcinogenic characteristics. Although the concentrations are usually minute, we cannot ignore the possibility of cumulative effects.

TABLE 7:1

CONCENTRATION OF SOME ORGANIC CHEMICALS CAUSING
TASTE AND ODOR IN WATER SUPPLIES

Substance	Concentration detectable, ppb*
Formaldehyde	50,000
Picolines	500–1,000
Phenolics	250–4,000
Xylenes	300–1,000
Refinery hydrocarbons	25–50
Petrochemical waste	15–100
Phenylether	13
Chlorinated phenolics	1–100

* Concentrations determined by taking the median of 4–12 observations. From Francis M. Middleton, Aaron A. Rosen, and Rice H. Burttschell, *J. Amer. Water Works Assoc.*, 50, No. 1 : 22 (1958); reprinted by permission of the copyright owners.

Each situation should be separately studied. In this the first step is to establish some means of concentrating the pollution into a form that can be analyzed, and then identify the ingredients and study them for possible hazards to health.

The personnel of the Robert A. Taft Sanitary Engineering Center** find that adsorption-desorption with activated carbon is an effective way of concentrating the pollution.[68,69,70,71,72] In equipment designed by Middleton and coworkers, the water to be studied is passed through a bed of granular activated carbon placed in a filter tube 3 inches diameter by 18 inches in length. The rate of flow is 1/4 gpm. After 3,000 to 5,000 gallons of water have passed through the filter, the carbon is removed and extracted with chloroform. The chloroform is then separated by distillation leaving a residue known as the *carbon-chloroform extract*.

The carbon is then extracted with alcohol in a similar fashion. Other treatments will yield still other substances, but the extraction process is usually ended after the elutions with chloroform and alcohol.

As natural causes—aquatic plant and animal life—contribute organic substances, it follows that all waters can contain some extractables. In clean waters the carbon-chloroform extractables

** U.S. Department of Health, Education, and Welfare. The original publications should be studied by those planning to conduct the procedure.

will often be less than 50, or even 25, parts per billion. In contrast, waters polluted with industrial waste may contain extractables in the hundreds or even thousands of parts per billion. The carbon-adsorption technique can be used to monitor the quality of drinking water supplies; thus, water with less than 50 parts per billion of extractables can be presumed to be clean. The *U.S. Public Health Service 1961 Drinking Water Standards* actually call for a recommended limit of 200 ppb of carbon-chloroform extractables.*

Because carbon holds many adsorbed substances very tenaciously, not all the pollution is desorbed and recovered for examination. The technique does provide tangible weighable quantities that can furnish an insight into the nature of the pollution which is not disclosed in any other existing procedure. It is not always easy to identify the separate ingredients, but this has been accomplished with success to assist in locating sources of pollution and in establishing cause of damage, *e.g.*, taste and odor. Sufficient quantities of the extracts can be obtained for studies of physiological properties.

REFERENCES

1. Deitz, V. R., *Bibliography of Solid Adsorbents,* United States Cane Sugar Refiners and Bone Char Manufacturers and National Bureau of Standards, Washington, D.C. Abstracts of the scientific literature during the years 1900–1942; supplementary volume for years 1943 to 1953 published in 1956.
2. Aehnelt, W. R., *Entfarbungs-und Klarmittel,* T. Steinkopff, Dresden und Leipzig, 1943.
   Operating with Activated Carbon, Counter-current Handbook, and *Adsorption from Solution,* Darco Corporation, New York. Helbig, W. A., p. 814 in *Colloid Chemistry,* Volume 6 (J. Alexander, Editor), Reinhold Publishing Corp., New York, 1946.

*This has been accepted and promulgated in the Federal Register, 26, No. 143: 6737, July 27, 1961

Kausch, O., Die aktive Kohle, W. Knapp. Halle, 1928, Supplement 1932.

Mantell, C. L., *Industrial Carbon*, D. Van Nostrand Co., New York, 1946.

3. Deitz, V. R., *Survey of Bone Char Revivification and Filtration*, National Bureau of Standards, Washington, D.C., 1947.

4. Literature supplied by manufacturers of activated carbon.

5. Personal communications: Jonathan C. Cooper, Franklyn M. Williams, *et al.*, Pittsburgh Chemical Company.

6. Cooper, Jonathan C., Williams, Franklyn M. *et al.*, Pittsburgh Chemical Company.

7. Linsbauer, A., *Z. Zuckerind. cechoslovak, Rep.*, 51:483 (1927).

8. Reichert, J. S., and Elliott, R. B., U.S. Patent 2,082,656.

9. Vincent, G. P., U.S. Patent 2,381,090; 2,430,262.

10. Szende, N., and Vadas, R., *Listy Cukrovar.*, 56:356 (1938).

11. Owen, W. L., *Inter. Sugar J.*, 26:207,255 (1924).

12. Visscher, M. J., *Decolorizing carbon and its application in the glucose industry*, United Norit Sales Corporation, Amsterdam, Holland.

13. Personal communication: W. C. Bokhoven and C. van der Meijden of N. V. Norit-Vereeniging Verkoop Centrale, Amsterdam, Holland.

14. Jamieson, G. S., and Baughman, W. F., *J. Oil and Fat Ind.*, 3:347 (1926).

15. Harris, J. P., and Glick, B. N., *Oil and Fat Ind.*, 5:263 (1928). Harris, J. P., *Active Carbon in the Decolorizing, Deodorizing, and Purifying of Oils, Fats, and Related Products*, Industrial Chemical Sales Division, West Virginia Pulp and Paper Co., New York, 1944.

16. Laptev, A., and Erzyutova, E., *Masloboino Zhirovoe Delo.*, 10: No. 11, 18 (1934).

17. Harris, J. P., and Welch, W. A., *Oil & Soap*, 14:3 (1937); U.S. Patent 2,105,478; Canadian Patent 370,570.

18. Hassler, J. W., and Hagberg, R. A., *Oil & Soap*, 15:115 (1938).

19. Anderson, J. H., *Cotton Oil Press*, 6: No. 10:33 (1923)

20. Hennessy, D. J., U.S. Patent 2,321,913.

21. Harris, J. P., U.S. Patent 2,035,126.

22. Valaer, P., and Frazier, W. H., *Ind. Eng. Chem.*, 28:92 (1936).

23. Dudley, W. L., *J. Am. Chem. Soc.,* 30:1784 (1908).
24. Cafere, J., *Ann. fals.,* 21:604 (1928).
25. Glasenapp, M., *Z. angw. Chem.,* 617,665 (1898).
26. Bogojawlenskij, A. von, and Humnicki, V., *Z. angew. Chem.,* 21:1639 (1908).
27. Zaharia, A., Angelescu, E., and Motoc, D., *Chimie & industrie,* Special No. 989 (1934).
28. Atwood, H. G., U.S. Patent 2,367,557.
29. Tolbert, N. E., and Amerine, M. A., *Ind. Eng. Chem.,* 35: 1078 (1943).
30. Williams, G. C., and Fallin, E. A., *Ind. Eng. Chem.,* 35:251 (1943).
31. Goss, W. C., U.S. Patent 2,271,797.
32. Barker, M. E., U.S. Patent 2,055,060.
    Bowden, F. S., and Bowden, R. B., U.S. Patent 2,097,545.
    Brown, K. R., U.S. Patent 2,114,331.
    Caywood, A. G., U.S. Patent 1,990,266.
    Hochwalt, C. A., and Carmody, W. H., U.S. Patent 2,027, 099.
33. Blumenthal, S., and Blumenthal, M. D., *Food Ind.,* 6:254,274 (1934).
34. Muller, W., *Mitt. Lebensm. Hyg.,* 16:77 (1925).
35. Ballos, C. J., *Brewers Digest,* August 1955.
36. *Taste and Odor Control in Water Purification,* Industrial Chemical Sales Division, West Virginia Pulp and Paper Co., New York, 1947; contains 1063 classified references.
37. Hassler, W. W., *J. Am. Water Works. Assoc.,* 33:2124 (1941).
38. Hassler, W. W., *Taste and Odor Control J.,* 8 Nos. 7, 9, 11, 13; contains 891 classified references.
39. Parry, Richard H., *Thesis* Pennsylvania State University; December 1961.
40. Bender, W. A., Douglas, R., and Cuthbert, L. H., U.S. patent 1,787,467.
41. Weisberg, S. M., Chappell, F., Stringer, W. E., Stevens, S., and Stebler, H. A., U.S. Patent 2,071,368.
42. Nees, A. R., and Bennett, A. N., U.S. Patent 2,375,165.
43. Waters, V. S., U.S. Patent 2,380,890.
44. Dorn, H. W., U.S. Patent 2,419,256.
45. Bierbaum, E. E., U.S. Patent 2,035,523.
46. Ohman, M. F., U.S. Patent 2,144,119.

47. Chamberlain, L. C., and Hooker, G. W., U.S. Patent 2,028,099.
48. Chapman, T. G., *Am. Inst. Mining & Met. Eng. Technical Publication No.* 1070, 15 p., 1939; U.S. Patent 2,147,009.
49. Crabtree, Jr., E. H., Winters, V. W., and Chapman, T. G., *Transactions AIME—Mining Engineering,* 187:217 (1950).
50. Braunstein, R., *Chimie & industrie,* 29:331 (1933).
51. Helbig, W. A., *Physical Removal of Impurities from Plating Solutions,* Darco Corporation, New York, 1941; *Proc. Am. Electroplaters' Soc.,* 68 (1941).
52. Literature is available from suppliers.
53. Conlisk J. R. U.S. Pat. 3,352,788.
54. Young J. R. U.S. Patent 3,203,754.
55. Schmitt, C., *Chem. Absts,* 73, 133787r.
56. Koch, H., *Kolloid-Z.,* 22:1 (1918).
57. Giri, K. V., and Naganna, B., *Indian J. Med. Research,* 29: 125 (1941).
58. Paget, M., and Tilly, F., *Trav. membres soc. chim. biol.,* 23: 1381 (1941).
59. Chapin, R. M., *Ind. Eng. Chem.,* 6:1002 (1914).
60. Graham. J. J. T., *J. Assoc. Off. Agric. Chem.,* 22:408 (1939).
61. Kolthoff, I. M., *Biochem. Z.,* 168:122 (1926).
62. Woodyatt, R. T., and Helmholz, H. F., *Arch. Internal. Med.,* 7:598 (1911).
63. Bock, J. C., *J. Am. Chem. Soc.,* 42:1564 (1920).
64. Bohn, H., *Arch. Exptl. Path. Pharmakol.,* 140:118 (1929).
65. Rooseboom, A., and Kreke, M. van de, *Proefsta. Java-Suikerind. Mededeel.* 1930, 48 pp.
66. Rossee, *Chem.-Ztg.,* 58:931 (1934).
67. Cassidy, H. G., *J. Chem. Educ.,* 23:427 (1946).
68. Middleton, F. M., *J. Amer. Water Works Assoc.,* 54:223 (1962).
69. Ruchhoft, C. C., Middleton, F. M., Braus, H., and Rosen, A., *Ind. Eng. Chem.,* 46:284 (1954).
70. Middleton, F. M., and Lichtenberg, J. J., *Ind. Eng. Chem.,* 52:99A (1960).
71. Middleton, F. M., Personal communication.
72. Middleton, F. M., and Rosen, A. A., U.S. *Department of Health, Education, and Welfare; Public Health Report,* Vol. 71, No. 11 Nov. 1956.

8

Purification of Domestic and Industrial Waste Waters

1. INTRODUCTION

Problems arising from pollution are part of the harvest of an affluent society. In a truly pastoral society there are no organic wastes. The *end products* of each form of life become nutrients for other forms of life.

Life exists through the utilization of energy which is conveyed from one form of life to other forms by a vehicle known as *matter*. The sun is the continuous source of the energy. Green plants use sunlight to convert carbon dioxide and water into oxygen and complex organic compounds such as carbohydrates in which the energy is stored. When animals eat plants, part of the food builds structural tissue for growth, and part furnishes functional energy.

The stored energy is released to become available for living activities when substances such as carbohydrates are oxidized during respiration. Some portions are completely oxidized to carbon dioxide and water, but other portions of plant food are only partially oxidized, and are discharged by the animal in solid and liquid forms to the soil. There, aerobic microorganisms take over and complete the oxidation to carbon dioxide and water, *nutrients for plant life*. In a purely pastoral society, life continues in this manner.

The development of villages and towns on the banks of rivers, and the advent of plumbing and sewers resulted in the discharge of much solid and liquid organic wastes into waterways.

There, the diluted wastes were oxidized to carbon dioxide and water by dissolved oxygen—a reaction generally described as *self-purification.* But as urban populations increased, the amount of dissolved oxygen was no longer adequate to oxidize the increased waste, and the existence of stream pollution became very evident.

2. SECONDARY TREATMENT

An obvious remedy was to employ ways and means of nature, *i.e.,* oxidation by the aid of aerobic microorganisms. To accomplish this, trickling filters and activated sludge basins were developed to utlize the oxidizing power of microorganisms on a massive scale, an operation known as *secondary treatment,* also *conventional secondary.**

For a number of years, secondary treatment provided much improvement; one might almost say—adequate treatment. But in recent years, dilution and secondary treatment no longer suffice to handle the waste from greatly increased urban populations and expanded industrial operations. To all this some newer forms of pollution have been added. These include:

    a) wastes from new types of commodities that cannot be metabolized by aerobic microorganisms

    b) toxic forms of waste that impair biological activity

To add to all this, a new malignancy has emerged, *eutrophication.*

3. EUTROPHICATION

The term comes from the Greek, eutrophos, meaning well nourished. As used today it refers to the overnourishment of waterways with inorganic plant nutrients—ammonia—nitrates—phosphates. When aquatic plant life is overnourished, the growth becomes too prolific and too abundant. Then upper layers of plant life keep sunlight from reaching lower layers which die as a result. A lack of sufficient dissolved oxygen in the water prevents aerobic microorganisms from taking over and oxidizing the dead plants. Instead, a putrifying bog-like mass forms and sinks to the floor

*Primary treatment consists of a prior removal of visible solids.

of the waterway—a condition reminiscent of the formation of fossil fuels.

Some plant nutrients enter waterways from drainage of fertilized farms but large quantities enter with the wastes of urban populations. To handle the twin problems of refractory organic wastes and inorganic plant nutrients, research developed what are known as Advanced Waste Treatments (AWT).

AWT comprises systems which of necessity must be flexible to take care of the diverse forms of contamination that can come from any and all forms of human activity-domestic and industrial.

4. TERTIARY TREATMENT

AWT became known as *tertiary treatment* when applied to the effluents from conventional secondary treatment. Tertiary treatment includes many patterns to fit specific conditions, but most include adsorption by activated carbon usually preceded by chemical clarification. Both powdered and granular forms of carbon have been and are being studied. Applications using powdered forms will be considered a little later.

Clarification

Coagulants studied include alum, ferric chloride, lime and organic polyelectrolytes. All of them form flocs that collect suspended solids and organic colloids. The inorganic coagulants also insolubilize and collect phosphates.

Lime can be added in concentrations adequate to raise the pH levels that facilitate ammonia removal by air stripping.

Floc formation requires about 15 minutes, then the water goes to a clarifier where the agglomerated flocs settle, usually within an hour.

When lime is used as the coagulant, it can be recovered for reuse by incinerating the sludge and calcining the ash. Incineration provides an advantage in localities where sludge disposal presents problems.

The supernatant wastewater may go directly to the carbon adsorbers, but a prior filtration through mixed or dual media removes particulate matter.

Carbon Filter Adsorbers

The design, construction, and operation of filter adsorbers are covered in Chapter 6, but to avoid the necessity of continual back reference, salient features will be repeated here.

The clarified wastewater is pumped at pressures sufficient to deliver a flow of 5 to 10 gal./min./sq.ft.* The optimum rate depends on local circumstances. Too slow a rate necessitates larger size equipment, and too fast a rate may not provide enough residence time for adequate adsorption. The residence time ranges from 15 minutes in some operations to 50 minutes in others. The removal of organics is usually greater than anticipated from laboratory tests, often 50 to 100% larger. Much of the increase may be attributed to oxidation of organics by visible biological growths that develop on surfaces of the carbon granules. After the intrinsic adsorptive capacity of the carbon is exhausted, the biological activity often continues.

It is of further interest that the biological growths on the carbon granules are able to oxidize organic substances that have resisted prior biological action in the secondary treatment. Such action could suggest that activated carbon can promote biological activity in carbon adsorbers.

Fixed Versus Expanded Beds

The flow in the adsorbers can be upward or downward through beds in which the carbon granules remain stationary in fixed positions, *fixed beds*. Fixed beds remove particulate material when present in the clarified waste water. Accumulations of particulate material are removed from the bed by backwashing.

Some investigators advise the removal of particulate matter by prior filtration through mixed or dual media. This is especially desirable when upflow is employed. This is because particulate material would separate where it enters at the bottom of the bed, a position from which removal by backwashing is difficult.

With upflow, the rate of flow can be increased sufficiently to slightly separate the carbon granules and form an *expanded bed*. Fixed and expanded beds given equivalent removal of dissolved organic constituents, but differences are experienced in other oper-

*Gravity flow is being studied, and is in operation.

ational aspects. Reported advantages for expanded beds include:
- a) they require less pumping pressure
- b) aeration is provided more readily
- c) less downtime is necessary for backwashing
- d) one investigator reports less attrition of carbon particles because gentle agitation will maintain an expanded state, whereas vigorous motion is needed to backwash a fixed bed.

Series Arrangement

Fixed and also expanded beds are usually placed in a series arrangement consisting of 2, 3, or more adsorbers. Two units are considered to provide optimum economic operation. The operation of a series can be described by reference to Figure 6:2. When the impurity concentration in the effluent from the guard position (adsorber C) approaches an unacceptable level, the lead adsorber A is taken off stream to regenerate the carbon. Adsorber B then takes the lead position and receives the incoming wastewater. An adsorber containing virgin or regenerated carbon is placed in the guard position.

Parallel Arrangement

This arrangement (Figure 8:1) can be used when partial removal of contaminination is required. Startup of the adsorbers is staggered so that the effluent of each adsorber is at a different degree of purification, and all effluents are blended. When the concentration of impurities in the blended effluents approaches an unacceptable level, the adsorber longest in service is taken off stream to be replaced by an adsorber containing regenerated carbon.

Pulse or Moving Bed

The upflow pattern led to the development of the moving bed, or as is generally termed, the *pulse bed*. In this, the wastewater flows upward countercurrent through a decending fixed bed of carbon. When the adsorptive power of the carbon at the bottom section of the bed is exhausted, that portion of the carbon is removed for regeneration. The adsorber is replenished by the addition of an equivalent quantity of virgin or regenerated carbon to the top of the bed. This method of operation has been very successful.

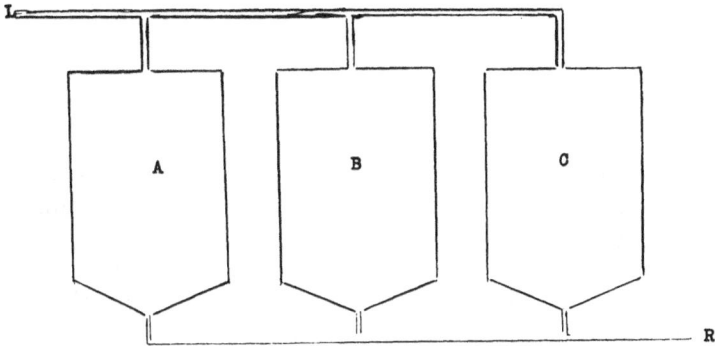

Fig. 8:1 Parallel Arrangement for Percolation Column: *A,B,C* contain regular carbon. In down flow, liquid enters system at *L* and leaves at *R*. In up flow, liquid enters system at *R*, and leaves at *L*.

Limitations of Tertiary Treatment

The efficacy of tertiary treatment for domestic and light industry wastewaters has been demonstrated in many bench and pilot plant studies, and amply confirmed by plant scale operations. However, tertiary treatment is not without limitations. To be fully effective, the secondary biological stage should deliver an effluent of uniform quality. The operation of the secondary stage can be adversely affected by: (1) changes in temperature or *p*H; (2) surges in the flow of the incoming wastewater (3) sudden inflow of toxic substances. Although troublesome, those difficulties have not prevented successful applications of tertiary treatment. However, there are wastewaters which poison biological activity to an extent that makes conventional secondary treatment inoperative, and consequently rules out tertiary. A search for ways and means to purify such wastewaters led to the direct applications of AWT to the raw wastewater. Initial successes led to extensive research which has been reported under various titles: Physicochemical Treatment Wastewater, Secondary Treatment with Granular Activated Carbon, Granular Carbon Treatment of Raw Sewage,

Clarification-Adsorption in the Treatment of Wastewater, Chemi-cal-Physical Treatment Process. Abbreviated references such as AWTRL and P-C are in use.

5.   PHYSICOCHEMICAL WASTEWATER TREATMENT

In general, the operation duplicates the chemical clarification and carbon adsorption stages as employed in tertiary treatment but with this difference: the application is made to the raw wastewater usually after primary separation of visible solids.

Existing data and present know-how are based on bench and pilot plant research, some of which has extended over many months.* The cumulative data are impressive. P-C gives greater removal of organic contamination than accomplished by conventional sec-ondary treatment, and the results are often on a par with tertiary systems.

The technological success has led to an examination of economic and other aspects. Comparison with conventional secondary treat-ment discloses that P-C requires much less land and capital in-vestment, causes less odor, and forms less sludge. The capital investment is much less than for tertiary except where a conven-tional secondary stage is already installed.

Aerobic and Anaerobic Facets

Some portions of organic contamination not adsorbable by carbon are often removed by biological growths that form on carbon granules in the adsorbers. The removal is accomplished by the conversion of organic substances to water and carbon dioxide by aerobic bacteria. The conversion can continue as long as sufficient dissolved oxygen is present in the wastewater.

If and when the supply of dissolved oxygen is depleted, anaerobic bacteria may take over. For that to occur, two types of con-stituents must be present: (1) oxygenated compounds such as nitrates, sulfates, carbohydrates; (2) readily decomposable or-ganic compounds. Anaerobic bacteria cause the oxygen in the

---

*Projected plant scale operations will soon be on stream.

oxygenated compounds to react with the organics. Some gases such as nitrogen, hydrogen sulfide, and methane may be formed; but most of the reaction products remain in the wastewater.

Aerobic conditions are obviously desirable in the carbon adsorbers because the conversion of organics to a gas (carbon dioxide) which leaves the system spontaneously, entails no further problem of removal.

Aerobic conditions are promoted by the presence of ample supplies of dissolved oxygen, and for this, more effective methods of aeration are needed Among the approaches being considered are: more extensive use of expanded beds use of open rather than closed type adsorbers and more positive aeration of the wastewater entering each adsorber.

6. INORGANIC PLANT NUTRIENTS

Phosphorous Compounds

Usually these are present in domestic sewage as phosphates, and as such are not adsorbed by carbon to any appreciable extent. They are effectively insolubilized and removed by the inorganic coagulants—alum, ferric chloride, lime—used in the chemical clarification stage of AWT. Removals are generally greater than 90%.

Nitrogen Compounds

Dependable methods have not yet been developed for the consistent removal of nitrogen compounds under diverse conditions. The nitrogen in raw domestic sewage is present mainly as ammonia and organic nitrogen. Part of the organic nitrogen is converted to ammonia during secondary treatment; and part is removed by chemical clarification and part by carbon. Ammonia is not removed by either chemical clarification or carbon.

In warm climates, air stripping is successfully used to remove over 90% of the ammonia from secondary effluents having a high pH. However air stripping is not effective at low temperatures.

Ammonia has been successfully removed by ion exchange, but the cost is appreciable. Nitrification-denitrification has also been tried and at present time it is the choice method. The use of two or three sludge systems has increased the reliability.

It is important to develop effective methods for removal of nitrogen compounds, but meanwhile too much damage may not ensue. Minimum concentrations of all plant nutrient are required to support aquatic plant life. A deficiency in any single nutrient will curtail plant growth. Therefore the effective known methods for phosphate removal should do much to check eutrophication.

7.   REGENERATION

Effective regeneration is essential to the economic use of carbon in AWT. Successful regeneration of granular carbon, at a fraction of the cost of virgin carbon, is now a routine operation. Recoveries range from 90 to 95 %. The regenerated carbon furnishes a finished water equal in quality to that obtained with virgin carbon. The adsorption capacity may be slightly less; in other words, regenerated is not kept on stream quite as long as virgin carbon Methods of regeneration are covered in Chapter 12.

8.   POWDERED CARBON

The rapid growth that began in 1929 in the use of powdered activated carbon for purifying municipal water supplies led to research on applications to sewage and other wastewaters. More than 30 studies were published. However, the enthusiasm that prevailed for purifying the incoming water did not extend to a willingness to pay for the purification of the water leaving for downstream. Concern about the environment was then dormant, and sales of powdered carbon for wastewater treatment never materialized.

With the advent of AWT, attention became centered on granular carbon for which a fully developed treatment system had been earlier established for sugar refining and other applications. The system was readily transplanted to AWT.

The neglect of powdered carbon was in part due to the absence of an established operational method of regeneration. In earlier days, powdered activated carbon had been regenerated in a number of applications, but the practice was later abandoned. However the advent of AWT has led to new methods of regenerating powdered carbon and these are outlined in Chapter 12.

Research is developing new ways and means of applying powdered carbon to AWT. The processes are still in early stages, but by the time this text is distributed, we believe much progress will have been accomplished. Therefore, we will now outline approaches being studied, and will list sources where up-to-date information will be available.*

In one approach, powdered activated carbon is employed to enhance the BOD removal by conventional wastewater treatment plants. The application is made in conjunction with ferric chloride and an organic polyelectrolyte. The addition is made to the influent to the primary or secondary clarifiers. The powdered carbon adsorbs dissolved organic compounds; and the ferric chloride and polyelectrolyte function collectively as coagulants to remove colloidal organics, residual suspended solids, suspended carbon, and phosphates. The sludge that forms is then removed along with the primary or secondary sludge from the bottom of the clarifier. BOD removals of 75% and greater are reported for primary treatment plants; and removals of 95% are obtainable from conventional secondary plants.

In another approach, the powdered carbon is added to the aerator of the conventional secondary basin. There, in addition to adsorbing organics and toxic chemicals that might poison biological activity, the carbon appears to enable the microorganisms to oxidize substances that are refractory to conventional secondary biological activity.

Other benefits include: greater hydraulic loadings, a more compact sludge, and less turbid effluents; the required capital investment is small.

Still another approach is being developed when carbon regeneration is planned. This consists of two stage countercurrent** application in which the powdered carbon is separated by being collected in a polyelectrolyte floc. Inorganic coagulants cannot be used when regeneration is to be employed because with each regeneration the inorganic content would increase and thereby dilute the carbon.

*See chapter 20, section 8.
**See Chapters 4 and 6. Countercurrent application is desireable in AWT because of the steep isotherm slopes.

9. MISCELLANEOUS FACETS

Existing technological know-how is adequate to make much improvement in many troubled waterways. However, as those who have studied AWT point out, there are no general rules that apply equally to all situations, and an individual bench study should be made of each projected application. Such studies are especially necessary prior to application to an industrial wastewater for which supplementary specific treatment may be needed.

The purification of wastewaters will cost money for the construction and operation of the processing facilities, all of which must ultimately be paid for by higher prices and/or taxes. Hence incentives exist to wait for the extensive research now underway to develop improved lower cost methods. But the condition of many waterways cannot await an indefinite future.

Moreover, a major change in the AWT pattern is unlikely. That belief is based on the history of the purification of municipal water supplies in which the contamination arises from sources similar to wastewaters. Any differences are primarily quantitative; in wastewaters the contamination is more concentrated and more abundant. The fact that no other agent has ever appeared to displace activated carbon since it entered the water field in 1929 is an encouraging harbinger of future use. Moreover, during those forty odd years, carbon has displayed a flexibility to adjust to changing ways of meeting new situations.

However, technological efficiency is only one factor in the cost of AWT. Much depends on the quality standards established for the finished water. Obviously, the standards should protect the health and general welfare, but the application of that guide line may not always be readily accomplished. The standards must cover not only the common everyday sources of pollution, but also diverse specific forms. Then too, quality standards as measured by scientific studies must be weighed against costs. If the costs become so high as to terminate some industrial operations, the resulting unemployment could generate adverse political pressures that would retard essential progress.

Answers may arrive gradually, first in regions where shortages of useable water begin to appear. There, reclamation of wastewaters for recycling will become necessary, and each unit of operation would establish its own quality standards. Recycling would require self-policing, and that could be productive of more effective processing. Thus, some relatively pure condensates and cooling waters could be rechanneled for other duties; and concentrated or hazardous wastes could be isolated and treated separately instead of becoming part of a general mix. Industrial units unable to purify their process water for reuse might apply it for other purposes such as irrigation. Otherwise, they would have to terminate operations, because the damaged water could not be dumped into waterways to become harmful to others. In such situations, operational difficulties would be responsible for the resulting unemployment. AWT could not be blamed.

Recycling will of necessity extend to domestic wastewaters. Those who shrink from the suggestion do so without realizing that it has already happened. Not deliberately, but indirectly as one of the facts of life. Towns and cities discharge their waste into nearby waterways because there is no other place. Those same waterways supply water to towns and cities down stream. Consequently, it is inevitable that residents of many communities must use and drink water—part of which has been through one or more wastewaters.

But such water can be perfectly pure. Water molecules do not themselves become impure by contact with contamination. All that is necessary is to separate them from the contamination. That can be done by removing the water molecules from the contamination by distillation as occurs when rain is formed. Or the separation can be accomplished by removing the contamination from the water molecules as occurs with AWT. We might draw a parallel to dining in a restauant. No one would consider using dishes, knives, forks, spoons as left by the previous diner, but no one hesitates to use them after the food residues have been washed away. And in a similar manner, AWT removes contamination from water molecules.

Because of AWT some of us will probably:

1) need to overcome some deep rooted prejudices
2) pay higher prices for some articles
3) do without some luxuries of an affluent society.

But if we look deeper, we will find it is part of the price to be paid for survival.

REFERENCES

Anon
   Clarification/Carbon Waste treatment
   Calgon Corporation Bulletin 20–9
Anon
   *"Powdered Activated Carbon for Enhanced BOD Removal from
   Conventional Waste Water Treatment Plants"*
   Nuchar Division, WESTVACO Inc.
Anon
   Activated Carbons for Waste Water Treatment
   Atlas Chemical Industries (1969)
Anon
   "Evaluation Granular Carbon for Waste Water Treatment"
   Atlas Chemical Industries (August 1969)
Anon
   Covington Sewage Treatment Plant with Granular Activated
   Carbon Westvaco Carbon Department (July 14, 1970)
D. F. Bishop, L. S. Marshall, T. P. O'Farrell, R. B. Dean,
B. O'Connor, R. A. Dobbs, S. H. Griggs, and R. V. Villiers
   "Studies on Activated Carbon Treatment"
   Journal WPCF Vol. 39 No. 2, p 188, (Feb. 1967)
J. M. Cohen
   "Dissolved Refactory Organics"
   Physical & Chemical Treatment Research Program (June 24,
   1970)
J. C. Cooper and D. G. Hager
   Water Reclamation with Activated Carbon
   Chemical Engineering Progress, Vol. 62, No. 10, p 85,
   (Oct. 1966)
R. L. Culp
   *"Wastewater Reclamation at South Tahoe Public Utilities
   District"*
   Journal AWWA, (Jan. 1968, p 84)
D. G. Hager and P. B. Reilly
   *"Clarification-Adsorption in the Treatment of Municipal and
   Industrial Wastewater"*
   Journal WPCF, Vol. 42, No. 5, p 794 (May 1970)
R. S. Joyce and D. G. Hager, Calgon Corporation

"Carbon Treatment Raw Sewage"
U.S. Patent 3,455,820 July 15, 1969
J. L. Kovach
Design Adsorber Using Activated Carbon
North American Carbon Company, Columbus Ohio, (1970)
P. A. L. Northcott
Cost, Benefit, and Reality
Effluent and Water Treatment Journal (August 1970, p. 453)
J. D. Parkhurst
Wastewater Reuse
J. Sanitary Engineering Division, Proceedings of American
Soc. of Civil Engineers, June 1970, p 653
J. D. Parkhurst, F. D. Dryden, G. N. McDermott, J. English
"Pomona 0.3 MGD Activated Carbon Pilot Plant"
U.S. Department Interior and County Sanitation Districts
Los Angeles County (Jan. 1967)
W. D. Pfoutz
Aqua Nuchar For Odor Control Waste Treatment
Westvaco Carbon Division (August 30, 1965)
M. W. Zukerman and A. H. Molof
*"High Quality Reuse Water By Chemical-Physical Wastewater
Treatment"*
Journal WPCF, Vol. 42, No. 3, p 437 (Mar. 1970)
W. J. Weber Jr.
Discussion
Journal WPCF, Vol. 42, No. 3, p 456 (March 1970)
J. L. Rizzo and R. E. Schade
"Secondary treatment with granular activated carbon"
Water and Sewage Works, August 1969, p. 307
A. F. Slechta and G. L. Culp
*Water Reclamation Studies at the South Tahoe Public Utility
District*
Journal WPCF, Vol. 39, No. 5, p 787 (May 1967)
D. G. Stephan and L. W. Weinberger
"Wastewater Reuse—Has it Arrived?
Journal WPFC, Vol. 40, No. 4, p 529 (April 1968)
L. R. J. Van Vuuren, M. R. Henzen, and G. J. Stander
*"Reclamation Purified Sewage Effluent for Augmentation Do-
mestic Supplies City of Windhoek"*
Presented at the 5th International Water Pollution Research

Conference (July-August 1970)

W. J. Weber Jr., C. B. Hopkins, and R. Bloom Jr.
Physicochemical Treatment of Waste Water
Journal WPCF, Vol. 42, No. I, p 83 (Jan. 1970)

Battelle Memorial Institute
Development of Fluidized-Bed Technique for Regeneration of Powdered Activated Carbon
U.S. Department of the Interior; Federal Quality Administration Water Pollution Control Research Series. ORD 17020 FBD 03/70

D. F. Bishop, T. P. O'Farrell and J. B. Stamberg
"Physico-Chemical Treatment Municipal Wastewater
U.S. Department of the Interior, Federal Water Quality Administration, Advanced Waste Treatment Research Laboratory Robert A. Taft Water Research Center, Cincinnati, Ohio (October 1970)

C. F. Garland and R. L. Beebe
AWT Using Powdered Activated Carbon in Recirculating Slurry Contactor Clarifiers
U.S. Department of the Interior; Water Pollution Control Series. ORD 17020 FKB 07/70

C. B. Hopkins, W. J. Weber Jr., R. Bloom Jr.
Granular Carbon Treatment Raw Sewage
U.S. Department of the Interior, Federal Water Quality Administration Water Pollution Control Research Series ORD 1705DAL 05/70

C. B. Hopkins, W. J. Weber Jr., R. Bloom, Jr.
"Comparison Expanded-Bed and Packed Bed Adsorption Systems"
U.S. Department of the Interior, Federal Water Pollution Control Administration, Robert A. Taft Research Center Report No. TWRC 2

A. F. Juhola and F. Tepper
Regeneration of Spent Granular Activated Carbon
U.S. Department of the Interior, Federal Water Pollution Control Administration, Robert A. Taft Water Research Center, Report N. TWRC 7

W. M. Kellogg Company
"Appraisal Granular Carbon Contacting Phases I, II, III"
U.S. Department of the Interior, Federal Water Pollution

Control Administration, Robert A. Taft Water Research Center, Report No. TWRC L"

J. F. Kreissal, S. E. Clark, J. M. Cohen and A. J. Alter
"Advanced Waste Treatment and Alaska's North Slope"
U.S. Department of the Interior, Federal Water Quality Administration, Advanced Waste Treatment Research Laboratory, Cincinnati, Ohio, August 1970

West Virginia Pulp and Paper Company
"Study of Powdered Carbons for Wastewater Treatment and Methods for their Application"
U.S. Department of the Interior, Federal Water Pollution Control Administration, Contract No. 14–12–75

Part III

OTHER APPLICATIONS

9

Gas and Vapor Phase Application

1. INTRODUCTION

The majority of gas and vapor phase applications involve the atmospheric environment. Even when—as in solvent recovery—the primary objective is to reclaim substances of value, the application simultaneously prevents contamination of the atmosphere.

Purification of the atmosphere is a primary objective as when gas masks are used to protect individuals against injury from existing pollution.

Recently much research has been devoted to prevent the escape to the atmosphere of harmful gases such as oxides of sulfur and nitrogen.

Therefore, this chapter will include the varied forms of gas and vapor phase systems.

Gas Masks

It was the use of gas masks to protect troops against gas warfare in World War I that first drew world wide attention to the purifying power of activated carbon.[1,2,3,4,5]

The use soon extended to protection against hazardous gases in industry. Today gas masks and other respiratory devices are commercially available for protection against almost all toxic vapors and gases, that may be encountered.

Other ingredients are often incorporated with the carbon in masks to take care of special situations. Thus the carbon is often impregnated with an appropriate catalytic promoter to change certain toxic vapors into harmless forms. Specific chemical reagents are incorporated in the mask when it is necessary to remove toxic vapors that are not adsorbable by carbon. Masks equipped with special type filters are available for protection against toxic dust and mists when and where smog conditions exist.

Situations arise in industrial operations that are not encountered in the outdoor conditions of gas warfare. Thus, localized concentrations of toxic gases may be excessive and, in other cases, there may be a deficiency of oxygen. Gas masks can be depended upon only when the concentration of toxic gases is less than 1 to 2%, depending on the kind of gas. Consequently, gas masks should not be used when high concentrations of toxic gas are encountered, as in closed tanks containing a volatile liquid, or in rooms where a large quantity of gas has suddenly been released. Moreover, gas masks should not be used when the oxygen content falls below a safe concentration, as they do not correct dangerous conditions caused by oxygen deficiency. The presence of 16% oxygen or more can be indicated by the burning of a flame safety lamp.

For operations in which men must enter confined spaces containing high concentrations of gas, or where there may be a deficiency of oxygen, helmets connected to a source of oxygen are available which assure the wearer a continuous supply of fresh air. Self-contained oxygen-breathing apparatus is of value when men must enter burning buildings.

Gas-protective equipment is employed under dangerous conditions and often in emergency situations. Therefore, adequate advance training should be given men who may be called upon to use such equipment.

Precautionary measures must be enforced to provide systematic renewal of fresh carbon in the canisters. Otherwise a gas mask may contain partially exhausted carbon at a time and in a situation that needs full reserve capacity.

2. INDOOR LIVING QUARTERS[1,2,8,9,10,11,12]

The open window has long been the symbol of wholesome indoor living quarters—and until recently it was the general belief that enclosed spaces require a continuous supply of outside air to maintain the oxygen at a proper level. Research has disclosed, however, that the depletion of oxygen is negligible in most living quarters under conditions of normal occupancy, and that any such loss is offset by leakage of air to and from the outside.

The staleness so often observed in unventilated enclosed spaces is caused by vapors that arise from the metabolism of the occupants, cooking, smoking, cleansing agents, and insecticides.

Most vapors have an odor, and the sense of smell is one of nature's ways of informing us as to the wholesomeness of our environment. Objectionable odors initiate a psychological chain reaction that urges us to take steps to get rid of an unpleasant odor. Sometimes this is attempted by masking with a perfume or chemical agent; but this only adds to the pollution and—what can be worse—it hides the true situation which in some cases may be toxic. A much better method is the traditional open window which, by sweeping out the odor, not only removes the odor but also eliminates the physiological harm that can result from toxic vapors.

However, the open window provides no relief when and where industrial pollution or smog is present. The difficulties presented by such situations can be surmounted by the use of activated carbon adsorbers for removing any and all contamination as soon as it appears. Complete freshness is maintained and the purified indoor air can be recycled for full ventilation. The potential is manifest by the successful use in submarines.[11] The recycling of recovered air provides economic benefits, for it makes it practicable to regulate the temperature with smaller units of heating and cooling equipment, and at a lower operating cost.

There is nothing essentially difficult in the design and installation of a carbon purification system. Means must be provided to collect the air from all contaminated spaces and pass this at a proper velocity through adsorbers containing a sufficient quantity of a suitable activated carbon. The purified recovered air is then returned to the living quarters.

Ready-built, self-contained cabinet units can give effective service in many situations, such as homes, offices, hospital rooms, laboratories. Complex building structures with a central system require special design and arrangement, but will usually be found to consist of a suitable assembly of prefabricated component units. Those without prior experience in ventilation will ensure a more efficient installation by the guidance of a qualified specialist.

Evaluation of Requirements

The size and arrangement of the components of an air-purification system will depend on the size and space arrangement of the occupied quarters, and on the rate at which the contamination forms and its adsorbability.

When, for one reason or another, it is impracticable to conduct an actual adsorption study, the experienced trained specialist can often make effective use of other guides. Thus, if the type or source of odor contamination is known, one can often learn the adsorption characteristics from published data, as in Table 9:1 Such data cover an average range of pollution, consequently in using such data one must take into consideration how the intensity of the odor in the unit under study compares with the average found for a similar type of occupancy.

TABLE   9:1

ODOR INDEX FOR TYPE OF SPACE*
(Condensed)

C—Aircraft	B—Department stores	B—Offices
D—Air raid shelters	C—Drug stores	C—Photo dark rooms
D—Animal rooms	D—Funeral homes	D—Pollution control
A—Apartment buildings	A—Homes	C—Public toilets
C—Apple storage	C—Hospitals	C—Recreation rooms
B—Auditoriums	B—Hotels	B—Restaurants
C—Bars	C—Kitchens	C—Schools
C—Beauty shops	C—Locker rooms	B—Super markets
A—Churches	D—Meat packing plants	C—Telephone exchanges
C—Conference rooms	C—Morgues	B—Theaters

* Data compiled by H. L. Barnebey; reprinted from ASHAE Transactions 1958, by permission of the American Society of Heating, Refrigeration and Air Conditioning Engineers.

The relative odor levels in the various types of locations where activated charcoal can be used for air purification are listed in the table as A, B, C, or D. Locations A contain the lowest amount of odor and D the highest; B and C represent levels in between. Many of the classifications are rather general so it was necessary to pick a typical average condition. The odor index for a specific situation could vary somewhat from that given in the table if special circumstances apply. In typical cases, one pound of 50-min activated coconut shell charcoal will purify the following cu ft of space for one year: A—2,000; B—800; C—300; and D—100.

Objective observations of the odor assist in the survey of a situation. Most vapors are characterized by an odor, and the sense of smell is without a peer for revealing the presence of any and all vapors even though it may not furnish individual identification. Inasmuch as minute quantities of a vapor (often less than 1 ppm) give a perceptible odor, it follows that the complete absence of odor can be considered as establishing the absence of any and all foreign vapors. Hence, in an air-recovery system, the complete absence of odor provides acceptable proof of satisfactory purification. This, however, is based on the assumption that carbon monoxide is not present. It is also reported that some war gases have no identifying odor.

The sense of smell is somewhat less serviceable for quantitative measurements when an odor is present. Impressions of odor are expressed as adjectives such as perceptible, definite, strong, and

TABLE 9:2

ODOR THRESHOLD CONCENTRATIONS*

Substance	Parts per million	Substance	Parts per million
Carbon tetrachloride	71.8	Cresol	0.056
		Ozone	0.05
Ammonia	53.0	Chloracetophenone	0.016
Phosgene	5.6	Benzyl sulfide	0.006
Sulfur dioxide	4.0	Pyridine	0.0024
Chlorine	3.5	Iodoform	0.0017
Acrolein	1.8	Diphenyl ether	0.0012
Amyl acetate	1.0	Valeric acid	0.00062
Carbon bisulfide	0.77	Isoamyl mercaptan	0.00043
Phenol	0.28	Ethyl mercaptan	0.00026
Hydrogen sulfide	0.18	Vanillin	0.00008
Methyl salicylate	0.066	Butyric acid	0.00006
Crotonaldehyde	0.062	Artificial musk	0.000004

* From Sleik, Henry and Turk, Amos, *Air Conservation Engineering*, published by Connor Engineering Corporation, Danbury, Conn. 1953. Reprinted with permission of the copyright owners.

the gravimetric concentration corresponding to each adjective depends on several factors:

a) The chemical nature of the vapor—some have a more intense odor than others;

b) The sensitivity of the individual observer, whether keen or dull;

c) The time of exposure to vapor—whether long enough to numb the sensitivity.

Table 9:2 gives the approximate amounts of various vapors that must be present in air in order for their odors to be detected. Even under the most favorable experimental conditions, it will generally be found that a tenfold change in gravimetric concentration must occur for an observer to become aware of a describable change in the intensity of odor. It is apparent that caution is needed in the use of direct odor impressions for the design of an air-recovery system.

By refinements in technique, the sense of smell can provide data that have greater meaning, and hence more utility. A procedure known as the *dilution method* gives significant information as to the amount of contamination to be removed. This is measured by diluting a portion of contaminated air with pure air and observing the proportions required to provide an odor-free mixture. At least several separate tests should be made at suitable intervals inasmuch as the rate at which odor is released may fluctuate, and hence the concentration of odor will vary from time to time.

Adsorbers

On reaching an estimate of the probable size of the installation required (in terms of CFM and pounds of activated carbon), the next step is the selection of a suitable type of adsorber cell. Although carbon can be purchased in bulk for use in adsorbers of one's own design, the more common practice is to purchase one of the ready-built types. Prefabricated adsorber cells are available in sizes ranging from 25 CFM to 1000 CFM and are of many designs, embodying a variety of features to fit specific situations. Detailed descriptions of the many available types of adsorber cells can be obtained from the suppliers of equipment. Here we shall limit the discussion to basic aspects as represented by the by-pass adsorber and the panel adsorber.

a) *By-pass adsorbers* are used in systems in which it is necessary to minimize the resistance to the flow of the circulating air. In by-pass adsorbers' the air passes around porous elements containing activated carbon. From 20–50% of the pollution is removed on each pass and this can suffice to prevent any accumulation if and when the contamination is being formed at a slow rate.

b) *Panel-type Adsorbers:* In situations in which the pollution forms at a rapid or even moderate rate, the air should pass through a bed of carbon as is done in panel-type adsorbers. Such type of adsorber is essential when toxic vapors are present in amounts that require complete removal. Panel adsorbers consist of flat perforated cells containing a thin bed of carbon. In an average situation the moving air-stream is passed through a half-inch bed of carbon at a rate of 40 CFM/square foot of face area. From 90–99% of the contamination is removed on each pass. Thicker beds of one or two inches of carbon are used in occupied spaces having a very heavy load of pollution.

The adsorbers should be protected with dust filters to prevent them from becoming filled with particulate material which would add resistance and restrict the flow of air. It is also desirable to make provision for the continuous removal of aerosols because these are not appreciably adsorbed by carbon. Smoke(a typical aerosol) is a colloidal dispersion of tars in air and, unless it is removed as soon as formed, the colloidal tars will deposit on all surfaces—walls, floors, ceilings—from which they will slowly evaporate to cause a persistent odorous environment. A typical example is a smoke-filled room on the morning after. All such after-effects can be avoided by placing a high-efficiency particulate filter or an electrostatic precipitator ahead of the carbon adsorbers to remove all smoke as it is produced.

Service Life

When placed in service, the apparent weight of the carbon will gradually increase with the take-up of adsorbed vapors. With the vapors normally encountered, a good grade of carbon will adsorb from 20–50% of its own weight before full service life is exceeded, at which time the carbon is replaced with either virgin or reactivated carbon. Some types of adsorbers are constructed so as to be returned to the supplier for regeneration and re-use; others—disposable types—are discarded when exhausted. So-called in-place regeneration is seldom practiced except when the service life is very brief, *e.g.,* in solvent-recovery.

The length of time elapsing between renewals of carbon will vary according to the quantity and quality of carbon used, and on characteristics of the pollution. When adequate adsorber capacity is installed in buildings having light pollution, several years may elapse before the carbon needs to be renewed. In contrast, the use of undersize adsorber capacity for heavily contaminated spaces may require renewal in six months or less.

From what has been previously mentioned, it is evident that considerable experimental error can exist in the data on which capacities are calculated and/or estimated. Hence it is important to know the nature of the possible hazards that can result if a projected installation should be undersize. In general, a system that is only slightly undersize will give satisfactory purification for an appreciable period of time; but the service life will be proportionately shorter.

If a larger unit is installed, the increased amount of carbon will not be wasted but will furnish longer service, and there will be longer intervals between carbon renewals. And of much importance is the fact that overcapacity will enable the system to handle more effectively conditions in which periodic surges of intense pollution occur.

Thus far we have considered purification as a means of recovering air for re-use. Situations exist, however, where the air collected at specific areas may be unacceptable for re-use, *e.g.,* where a high concentration of carbon dioxide is released from a processing liquor. Such air must be dissociated from the recirculated air, and instead is vented directly to the atmosphere. Good neighbor relations may require that such vented gases be free of obnoxious and toxic vapors, and such situations can be taken care of effectively by placing a separate carbon adsorber at each vent, or group of vents.

3. RECOVERY OF SOLVENT AND OTHER VAPORS

Organic solvents are an essential tool in many diverse industrial processes. They are used as extractors in dry-cleaning and degreasing operations, and for obtaining oil from seeds. They provide a processing medium in the manufacture of plastics, rayons, explosives, etc. They act as an agent for applying a product to its intended use in fluid form, *e.g.,* printing, painting, and impregnation of fabrics. Many solvents suitable for such purposes are very volatile and their vapors can contaminate the atmosphere of work rooms. The extent to which such contamination can occur depends on the volatility of the solvent and on processing conditions surrounding the operation.

The escape of solvent vapors into work rooms can create health, fire, and explosion hazards. Apart from this, the amount of solvent thus vaporized can represent a considerable portion of the operating cost, and recovery becomes desirable for economic reasons.[1,14,15,16,17]

In some operations, the vapors can be condensed by cooling or by absorption with suitable liquids, but the most general practice is adsorption on active carbon. An activated carbon system when properly designed and engineered is safe and easy to operate; in fact, the operation can be made almost completely automatic. The basic features of the process are simple and easy to understand; yet, in order to ensure high efficiency and low operating costs, proper correlation of the component units and careful attention to a number of details are essentail. Much will be gained by placing the design in the hands of an engineer qualified by experience for such work.

The discussion to be provided here will be limited to an outline of the general principles involved and certain precautions that must be observed. One important precaution involves fire and explosion hazards. Many solvent vapors, when mixed with air in certain proportions, are explosive, the degree of explosiveness depending upon the proportion of vapor to air. No explosion will occur when the vapor concentration is below a critical amount which varies from 0.5–10% according to the chemical nature of the vapor. As the vapor concentration increases above this lower critical point, the explosiveness increases to a maximum and then diminishes at still higher vapor concentrations. When the vapor concentration exceeds a second critical point, another nonexplosive zone is reached. Therefore, it is essential to provide conditions to maintain the vapor concentration in one of these nonexplosive zones.

There are practical difficulties in working with vapor concentrations in the upper nonexplosive zone, because there is always a hazard that the concentration may be diluted to a dangerous point through leakage of air. In the relatively few cases where it is necessary to use high vapor concentrations, it becomes desirable to employ a noninflammable solvent, or else, provide an inert atmosphere. The latter involves installation of completely enclosed equipment to prevent access of air.

Practical considerations usually require that the vapor concentration is kept in the lower safe zone. Such concentrations generally are too low to permit practical recovery of the solvent by condensation or by scrubbing with a high-boiling-point liquid. As a result, the general trend has been to install the active-carbon process which maintains a high efficiency even when the vapor concentration is low.

Although the concentration of vapor must be kept well below the danger zone, it should not be reduced to a needlessly low point as this involves handling unnecessarily large volumes of air. The vapor concentration is maintained at the optimum point by proper design of the hoods at which the vapor-laden air is collected. Devices are available which automatically maintain the vapor concentration at the proper point.

The vapor-laden air, collected at the hoods, is moved by a blower through a duct to the recovery units which can be at some distance. When the vapor-laden air is hot as it comes from the process, cooling coils or other means of cooling are provided to lower the temperature sufficiently to secure effective adsorption.

Corrosive ingredients if present should be removed. Scrubbing is commonly used but cannot be depended upon to ensure complete removal of corrosive substances.[18] For example, in one application where traces of hydrochloric acid are present, a water scrubber removes most of the acid but some corrosion of the adsorber can occur, and corrosion-resistant materials must be used. It should also be noted that many solvents that are not normally considered corrosive can form corrosive reaction products when adsorbed on activated carbon because of its catalytic activity.[18] For example, in the recovery of acetone, acetic acid is formed; and in the recovery of chlorinated substances, hydrochloric acid is formed.

Lint and dust are removed by a filter. All electrical equipment is explosion-proof, and points at which static electricity might develop are grounded. Continuous gas analyzers are available to sound an alarm or shut down the operation if the concentration of vapor should rise above a predetermined safe point.

The clean, cool, vapor-laden air, on reaching the recovery equipment room, is passed through a bed of granular activated carbon contained in the adsorber, where the vapor is adsorbed and the denuded air discharged to the atmosphere.

The heat developed by the adsorption of the vapor will cause a rise in temperature in a dry cabon bed. Difficulties resulting from this are avoided by having the carbon in a moist condition when the adsorption cycle starts. Then, as the vapors of solvent are adsorbed, they displace some moisture, which on being evaporated absorbs the heat of adsorption. This prevents a rise in temperature.

At the beginning of the cycle, most of the vapor is adsorbed by the carbon near the vapor inlet, but as this portion becomes saturated, the zone of adsorption moves father along. Finally, at the so-called break-point, measurable quantities of vapor will be found in the air that goes out. Should the flow of air be continued, the entire bed will become saturated with respect to the concentration of vapor in the entering gases. When this stage is reached, no further adsorption occurs and the exit gases have the same composition as those entering. If a mixture of vapors of different solvents is passed over carbon, all ingredients may not be equally adsorbable, and censequently all will not break through simultaneously Moreover, at the saturation point, the less adsorbable ingredients may be displaced by those which are more strongly adsorbable.

In order to maintain maximum efficiency in recovery, the adsorbers are usually operated on a time cycle. This is adjusted to provide a reasonable factor of safety so that no solvent will be lost at the end of the cycle. Break-point control is used on recycle-cooling systems wherein the air leaving the first adsorber is subsequently passed through a second adsorber; thus any vapor of a solvent that comes through the first bed is picked up by the second. This results in a substantial increase in the capacity of the bed and reduces the requirements for steam in the later stage of desorption.

The adsorbed vapor is extracted from the carbon by admitting steam at the point indicated in Figure 9:1 It passes through the bed in a direction opposite to that of the flow of vapor-laden air during the adsorption stage of the cycle. Low-pressure steam is generally advantageous. With high boiling solvents, however,

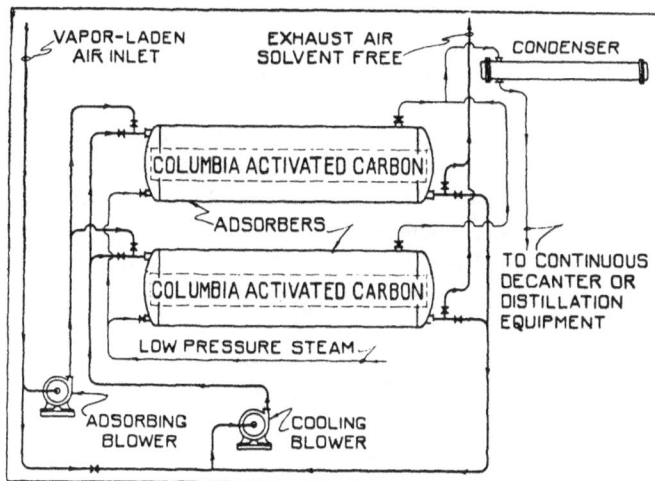

Fig. 9:1   Flow chart of a solvent-recovery plant, operated with cooling period, Courtesy of Carbide and Carbon Chemicals Corporation, New York

steam at higher pressure may be needed to get the temperature high enough to desorb the solvent.[18]

In passing through the carbon bed, the steam evaporates the adsorbed solvent and conveys it to the exit where the steam is condensed and the solvent is recovered in a liquid state. A solvent that is not appreciably soluble in water can be separated from the water layer by continuous decantation. Water-soluble solvents are separated by fractional distillation.

The removal of the adsorbed solvents simultaneously regenerates the adsorptive power of the carbon. After passing steam through 15 to 60 minutes, the carbon in the adsorber may be cooled by passing air through it before again placing it in service. Moisture is simultaneously partially removed, but complete removal is not desirable when using a carbon that can effectively adsorb vapors of the solvent in the presence of moisture.

In the horizontal type of adsorber shown in Figure 9:1 the carbon is usually placed in a single bed 12 to 48 inches thick. Special fittings are required to support the bed properly, and to provide proper distribution of the vapor-laden air as well as of the steam used in the subsequent desorption of the solvent. The arrangement must be such as to prevent excessive amounts of steam from condensing and wetting the bed of carbon. Stainless steels or other suitable materials are used when corrosive conditions exist. Although generally horizontal units are installed when relatively large volumes of vapor-laden air are to be handled, vertical types can also be used in small plants or where floor space is limited.

Generally, two adsorbers are used, as shown in the arrangement in Figure 9:1 the adsorption proceeds in one adsorber while the other is being regenerated. In large plants, several units are arranged in parallel to provide the required capacity. In processes that are operated intermittently, a single adsorber of adequate capacity may suffice. The adsorber is placed in service during the processing operation and is regenerated during shut-down periods.

As early as 1927, arrangements existed for placing activated carbon adsorbers in series.[19] Such a setup requires at least three adsorbers, one of which is being regenerated while the other two are in service. The adsorbers follow a definite order of rotation. The vapor-laden air enters one adsorber, and the exhaust from this goes to a second adsorber. When the first adsorber is saturated, it is taken out of service to be regenerated, and the entering vapor is diverted to the second adsorber which thus becomes first in the new series. The adsorber which had been regenerated during the previous run is placed in service to become the second unit in the new series.

The efficiency of an activated carbon system is high, generally 98% of the vapor collected at the hoods is recovered. Typical over-all recoveries range from 85 to 95 per cent. Operating costs are low. Some of the solvents that are now being recovered by this process are the following:

Acetone	Methyl ethyl ketone
Benzene	Petroleum naphtha
Ethanol	Solvent naphtha
Ethyl acetate	Tetrachloroethylene
Ethyl ether	Trichloroethylene
Isopropanol	Toluene
Methanol	Xylene
Methyl acetate	

Recovery of Benzene from Artificial Gas

Light oil, which consists of benzene together with smaller quantities of toluene, xylene, and other organic substances, is obtained from manufactured gas.[1,14,15,20,21,22]

Formerly, when gas was extensively used for illumination, the presence of benzene in gas was desirable. Today, gas is rated more on its calorific value and there are economic advantages in extracting the benzene for sale as motor fuel, or for use in the manufacture of explosives. The general principles of the operation are similar to those involved in solvent recovery. In adsorbing benzene, the temperature should be kept as low as possible, but not below the temperature of the gas at the inlet. An appreciable rise in temperature is to be avoided as the adsorption is then less; moreover, higher temperatures accelerate the formation of gums. In some operations, the adsorbers are equipped with internal coils which perform a dual function of temperature control. Cooling water circulates through the coils during the adsorption stage, whereas steam is passed through them during the desorption stage.

The use of activated carbon to recover light oil involves difficulties that seldom exist in solvent-recovery systems. Manufactured gas contains a variety of adsorbable compounds, some of which cannot be completely desorbed by steam; consequently, there is a gradual decrease in adsorptive power with use. The effective life of the carbon depends upon a number of factors, including the extent to which tar, naphthalene, and hydrogen sulfide are removed from the gas before it reaches the carbon adsorbers. The portion of carbon nearest the entering stream of gas picks up residual tars and naphthalene. In order to prevent these substances from being distributed over the rest of the carbon bed during the desorption stage, the steam is admitted so as to flow in a direction opposite to that of the original gas stream. From 30 to 40 minutes are required for the steaming operation, after which the carbon is cooled and the adsorber is placed in service.

4. SEPARATION OF HYDROCARBON GAS MIXTURES[1,23,24,25]

Shortly after World War 1, activated carbon was employed to recover gasoline from natural gas. The plants were operated manually in a manner that involved high cost. Moreover, unsaturates, heavy hydrocarbons, sulfides, and other contaminants greatly reduced the effective life of the carbon. In recent years, many of the earlier difficulties have been surmounted and interest in this field is now active.[1,23,24]

Although many aspects of a solvent-vapor system apply to gasoline recovery, there are some fundamental differences. The recovery of a solvent is a relatively simple process in which an adsorbable vapor is removed from non-adsorbable air; whereas the recovery of gasoline involves a selective separation of ingredients that are more adsorbable from ingredients that are less adsorbable. Because of this, the arrangement and design of the equipment and the operational procedure must be modified to fit the needs of the individual operation. The operation can be adjusted to include the recovery of propane, butane, pentane.

The recovery plants include a desiccant to dry the gas and prevent the formation of hydrate in the transmission lines. The desiccant is used in the same adsorber with the carbon on the inlet side; or two adsorbers are used in series, one containing the desiccant and the other, activated carbon. The desiccant also adsorbs the higher-boiling-point hydrocarbons so that the total adsorption is a combined effect of the two. Operating cycles are fairly fast, the adsorption period varying from 20 to 45 minutes.

5. MISCELLANEOUS APPLICATIONS

This section will report briefly on a few applications selected to illustrate the scope and diversity of the utility of activated carbon in vapor-phase systems.

Some valuable applications of gas-and vapor-adsorbing carbons are:

> To protect scientific instruments, manuscripts, photographic film in storage.
>
> To prevent tarnishing of silver
>
> Control of volatile toxic vapors in laboratory storage spaces

Odor tents in hospitals

In vents from laboratory hoods to prevent contamination
of surroundings

Filters for cigarettes

Removal of sulfur from synthesis gases

Removal of arsenic from hydrogen chloride

Gas chromatography

Aid in operation of vacuum pumps

Placed in shipping and storage containers to protect
odor-sensitive substances; and to prevent odorous materials
from creating a nuisance

Building boards for protective shelters

Kitchen hoods

To control odors of fermentation

Storage of fruits and flowers

Smock and Southwick[26] have studied methods of preventing
"apple scald" which occurs during storage and lowers the market
value of the fruit. They found that the organic emanations arising
from stored apples appear to be a contributing factor in the development of scald, and that they are adsorbed by activated
carbon. Studies on a commercial scale show that the use of
activated carbon to decontaminate the air in apple storage rooms
controls apple scald, improves the keeping quality, and lengthens
life in storage. Moreover, the fruit is free from all foul odors
usually found in apples when first taken out of storage.

The process has been very successful and it has been adopted
for a number of warehouses.

Sealed electrical instruments containing plastic components
often yield organic vapors that deposit on contact points and are
then decomposed at the heat of the arc into carbon and carbonaceous deposits. Such developments arc prevented by incorporating activated carbon—in molded forms called *getters*—into the
instrument.[27]

Carbon dioxide for dry ice or carbonated beverages should be
free from foreign odors. It can be purified by passing the gas
through a bed of active carbon.

Protective Fabrics for Vesicant Vapors

Finely powdered activated carbon is incorporated in fabrics
for garments designed to protect the wearer against injuries

from vesicant and other poisonous vapors. Kingan, Phillips, and Wilsdon[28] describe various methods for retaining the activated carbon in the fabric by suitable binders such as casein, rubber, cellulose derivatives, and synthetic resins. In a typical procedure, the fabric is dipped in or sprayed with a mixture of benzene-acetone containing 1% cellulose acetate and 5% powdered activated carbon. The solvent is subsequently evaporated from the fabric. In an alternative procedure, the carbon is initially incorporated in the fabric and the binder is applied subsequently.

Storage and Transportation of Gases

Briggs[29] reported that the capacity of commercial cylinders when filled with dry nitrogen, is increased by placing dry activated carbon in the cylinders.

Barnebey[30] describes the use of activated carbon in a sheet metal container for storing gases, such as dimethyl ether and ethyl chloride, at atmospheric pressure.

Low Temperatures[1]

Heat is evolved when a gas is adsorbed and conversely, the removal of an adsorbed gas results in a cooling effect. The latter behavior has been utilized to produce low temperatures. In a typical procedure, a glass bulb containing degassed activated carbon is cooled by being inserted in a Dewar jar containing liquid air. The carbon is then permitted to adsorb hydrogen gas, the heat evolved by the adsorption being dissipated by the liquid air in the outer container. When thermal equilibrium is established, the bulb containing the carbon is thermally insulated and the adsorbed hydrogen is desorbed by a vacuum pump. The evaporation of the adsorbed hydrogen causes a further lowering of the temperature. Using this process, Andrews[31] and coworkers attained a temperature of 20° K. By employing refinements in the technique, Mendelsshon[32] reached temperatures as low as 1.6° K.

High Vacua

Dewar[1,33] utilized the adsorptive power of coconut charcoal to improve the vacuum in a vessel exhausted by a mercury pump. He later found that much greater efficiency is provided by cooling the carbon to low temperatures.

Blythswood and Allen[33] describe the use of activated carbon to produce high vacua in X-ray tubes without the use of a vacuum pump. The X-ray bulb is attached to a glass bulb containing degassed coconut charcoal. The apparatus is first heated to remove much of the gas in the system, after which it is cooled. The gas remaining in the system is adsorbed by placing the charcoal bulb in liquid air. They report the method to be particularly useful when it is desirable to avoid the presence of mercury in the vacuum tube, as in the Geissler tubes used for spectroscopic analysis.

Permanent Gases

Some mention should be made of adsorption methods as employed to separate mixtures of permanent gases, e.g., nitrogen from helium, and carbon dioxide from air.[1] Satisfactory separation of permanent gases can depend on providing a suitable temperature to fit the individual case.

Dewar[34] was the first to employ coconut charcoal to separate rare gases from air. By passing air at a slow rate through a carbon bed at the temperature of liquid air, the oxygen, nitrogen, and argon were adsorbed; whereas the less condensable neon and helium were not adsorbed and appeared in the exit gases. To recover krypton and xenon, Dewar passed a slow stream of air through char at $-183°$ C for 24 hours, after which the temperature of the char was raised to $-78°$ C to evaporate the adsorbed nitrogen, oxygen, and argon. The less volatile krypton and xenon remained on the carbon and were subsequently removed by heating the char to a higher temperature. Peters and Weil,[35] by careful control of the temperature and pressure, separated krypton from xenon.

Correction of Atmospheric Pollution

In recent years much attention has been given to elimination of pollution of the atmosphere. One problem centers in the release of SO_2 from stack gases. In Europe and Japan several processes are in use. Little acceptance has been found in the United States primarily because the products recovered are in general either dilute sulfuric acid or ammonium sulfate. There is only a relatively small market for ammonium sulfate, and most industries releasing large amounts of SO_2 have little or no need for dilute sulfuric acid. However, increasing demands to correct air pollution must lead to measures of control—and a promising approach is via activated carbon. Successsful elimination of SO_2 by carbon is dependent upon a number of factors: time, temperature, SO_2 concentration, ratio SO_2 to oxygen, moisture concentration. The relative influence of each factor varies according to conditions prevailing in each situation. Therefore, this is another instance where the reader is advised to search out the original sources of information listed in Chapter 20, section 9. A process patented by Torrence that produces sulfur from SO_2 may provide marketing advantages.[36,37]

Another very recent development in connection with air pollution relates to the use of carbon to control gaseous emmissions from automobiles. Here again the reader is referred to the original publications listed in Chapter 20, section 9.

REFERENCES

1. Deitz, V. R., *Bibliography of Solid Adsorbents,* U.S. Cane Sugar Refiners and Bone Char Manufacturers and National Bureau of Standards, Washington, D.C., 1944. Abstracts of the scientific literature published during the years 1909—1942. Supplementary volume published in 1956 with abstracts for years 1943—1953.
2. Chaney, N. K., Ray, A. B., and St. John, A., *Ind. Eng. Chem.,* 15:1244 (1923); *Trans. Am Inst. Chem. Eng.,* 15, i: 309 (1923).
3. Fieldner, A. C., Oberfell, G. G., Teague, M. C., and Lawrence, J. N., *Ind. Eng. Chem.,* 11:519 (1919).
4. Fieldner, A. C., Teague, M. C., and Yoe, J. H., *Ind. Eng. Chem.,* 11:622 (1919).
5. Lamb, A. B., Wilson, R. E., and Chaney, N. K., Ind. Eng.

Chem., 11:420 (1919).

6. Information on specific problems can be obtained from manufacturers of respiratory protective devices approved by the United States Bureau of Mines.

7. *ASHRAE Guide,* published annually by American Society of Heating, Refrigerating and Air Conditioning Engineers, United Engineering Center, New York, N. Y.

8. Barnebey, H. L., *ASHAE Transactions* of 1958.

9. Barnebey, H. L., *Air Pollution Control Assoc.,* May 22, 1956.

10. Sleik, Henry, and Turk, Amos, *Air Conservation Engineering,* published by Connor Engineering Corporation, Danbury, Conn., 1953.

11. Ratcliff, J. D., *Readers Digest,* July 1958; *Today's Living,* New York Herald Tribune, Inc., June 22, 1958.

12. Literature furnished by suppliers

13. Richardson, N. A., and Middleton, W. C., *ASHAE Jour.,* Nov. 1958, *Air Conditioning and Refrigeration News,* April 28, 1958.

14. Mantell, C. L., *Industrial Carbon,* D. Van Nostrand Co., Inc., New York (1946); *Adsorption,* McGraw-Hill Book Co., New York, 1945.

15. Kohl, A. L., and Riesenfeld, F. C., *Gas Purification,* McGraw-Hill Book Co., New York, 1960.

16. Barry, H. M., *Chem. Eng.* Feb., 8, 1960.

17. Treybal, R. E., Helbig, W. A., Browing, F. M., Merims, R., Lapidus, L., Mantell, C. L., *Chem Eng.,* October, 1952.

18. Pennington, H. E.: Personal communication.

19. Barnebey, H. L., Barnebey-Cheney: Personal communication.

20. Robson, G. W., *Gas World.,* 123:399 (1945).

21. Engelhardt, A., and Ruping, H., *Gas-u. Wasserfach,* 76:478 (1933).

22. Hollings, H., and Hay, S., *Chemistry & Industry,* 1934, p. 143.

23. Russell, G. F., *Petroleum Refiner,* 40 No. 4:103 (1961).

24. Dow, W. M., *Petroleum Refiner,* 36 No. 4:141 (1957).

25. Berg, C., *Trans. Am. Inst. Chem. Engrs.,* 42:665 (1946).

26. Smock, R. M., *Cornell Univ. Agr: Expt. Sta. Bull. No. 799* (1943).
    Smock, R. M., and Southwick, F. W., *Cornell Univ. Publication of Dept. of Pomology,* 1946.
    Gross, C. R., and Smock, R. M., *Refrig. Eng.,* 5:535 (1945).

Smock, R. M., and Southwick, F. W., *Agr. Expt. Sta. Bull., No. 843*:3 (1948).

27. *ASTM Bulletin* October, 1959.
28. Kingan, R., Phillips, J. W. C., and Wilsdon, B. H., British Patent 575,379.
29. Briggs, H., and Cooper, W., *Proc. Roy. Soc. Edinburgh,* 41:119 (1920—21).
30. Barnebey, O. L., U.S. Patent 1,608,155.
31. Andrews, D. H., Witt, R. K., and Crigler, E., *Refrig. Eng.,* 19:177 (1930).
32. Mendelssohn, K., *Z. Physik,* 73:482 (1931).
33. Blythswood, L., and Allen, H. S., *Phil. Mag.* (6), 10:497 (1905).
34. Dewar, J., *Chem. News,* January 10, 1908, p. 16; *Proc. Roy. Soc.* (London), 74:127 (1904).
35. Peters, K., and Weil, K., French Patent 708,930; *Z. angew. Chem.,* 43:608 (1930).
36. U.S. Patent 3,563,704.
37. P. Wiley and A. Hyndshaw of Westvaco personal communication

10

Diverse Applications

1. INTRODUCTION

This chapters includes types of application which because of their characteristics do not naturally fit into the mold of the types of application described in previous chapters. Together with other applications listed in Chapter 20, Section 10 they illustrate manifold forms of potential utility that have been studied with varying degrees of success.

2. MISCELLANEA

Anticorrosion Paints

Prill[1] reports that paints, containing activated carbon charged with ammonia protect iron surfaces from corrosion. Other studies indicate that activated carbons containing organic amines provide greater protection than lead oxide against certain vapors such as sulfur dioxide, hydrogen sulfide, and moisture.

Galvanizing Baths

Spowers[2] found that a layer of char prevents oxidation of a molten zinc bath used for galvanizing, activated carbon being more effective than ordinary charcoal. By impregnating the activated carbon with boric acid, the burning tendency is minimized and such treated carbon will not burn when spread over a molten zinc bath at 450° C.

Chewing Gum

Flint[3] adsorbs flavoring materials on powdered activated carbon which is then incorporated with the gum. The flavor is released gradually, and the gum does not lose all its flavor in the first few minutes of chewing as do gums prepared by conventional methods.

Perfumes

Activated carbon was used by Gordon[3] to remove undesired color, and by Welter[5] to eliminate unpleasant odors arising from unfavorable conditions of storage. Some essential ingredients were also removed and required subsequent treatment.

Liquid Carbon Dioxide

Although carbon dioxide is usually purified in the gas phase, McKee[6] reports that purification can be accomplished in the liquid form with activated carbon.

Fuel Cells

These cells convert chemical energy directly into electrical energy without the losses of energy that are involved when the transformation is accomplished by intervening stages of the heat engine and the dynamo. Studies reported of using activated carbon in forming electrodes for fuel cells, are listed in Chapter 20, section 10.

Nuclear Technology

Activated carbon is employed in varied aspects of nuclear technology. An important use is for the adsorption of fission products such as radioactive methyl iodide. Reports on this and other aspects of the use of carbon in nuclear technology are listed in Chapter 20 section 10.

3. RECLAIMED RUBBER

Considerable difficulty has been encountered in producing an entirely satisfactory white-sidewall rubber tire. The white or light-colored rubber compositions forming the sidewall are subject to staining from migrating substances present in the carcass of the tire. The staining materials include accelerators, softeners, oils, and other compounding substances, which are soluble in rubber and will migrate from one layer of rubber to another. Phillips[7] found that this staining can be prevented by interposing a buffer layer between the white sidewall and the carcass. The buffer layer consists of a rubber composition containing 10 to 30% by volume of activated carbon.

Castello and Dixon[8] find that staining of light-colored rubber products is often caused by the incorporation of reclaimed rubber. They correct this condition with activated carbon; the staining substances are fixed by the char remaining in the rubber and are unable to migrate. Rubber so reclaimed is similar in appearance and in physical properties to ordinary reclaimed rubber except that it is nonstaining.

The process has been improved in recent years. In present practice, the cellulose fiber and free sulfur are removed in the initial stages of the process. Then the reclaimed rubber is mixed with activated carbon in a Banbury mill. The blended product is nonstaining.

Keilen and Doughetry[9] reported that carbons with a high iodine number are usually effective for anti-staining.

Acidic carbons appear somewhat more effective than alkaline carbons for GR-S reclaim. However, the pH of a carbon had little effect on the activity of a carbon for natural rubber reclaim.

Rubber processors check the effectiveness of activated carbon by a simulated end-use test. A sample of white-sidewall stock is placed over reclaim rubber containing the activated carbon that is being tested, and it is then exposed to light from a lamp or to sunlight. The sample of white-sidewall rubber, is then examined for discoloration.

4. AGRICULTURAL STUDIES

Several investigators report that the addition of active carbon to a soil promotes plant growth. Active carbon is not a direct fertilizer and any benefits result from indirect effects. Verona and Ciriotti[10] report active carbon results in greater biological fixation of nitrogen and a more rapid conversion of organic nitrogen to ammonia and nitrates. Adsorption of toxins and other harmful substances is indicated by Perotti,[11] and Verona.[12] Other reported benefits include higher soil temperatures, increased water capacity, and improved aeration. It is also claimed that active carbon regulates fertilizer consumption by the growing plant. However, the beneficial effects on plant growth are not confirmed by all investigators.

Arle, Leonard, and Harris[13] report the use of active carbon to protect sensitive field crops against injury from a weed killer— 2, 4-D (2, 4-dichlorophenoxyacetic acid). In one of several experiments, a field was sprayed with 2, 4-D—1.3 pounds free acid equivalent per acre. One-half of the plot was planted with sweet-potato sprouts that had been previously moistened and then dusted with active carbon; the other half of the plot was planted with untreated sprouts. In the case of the untreated sprouts, only 2.5 % survived; whereas 95.0 % of the active-carbon-treated sprouts survived and produced plants that were normal in every respect. Weaver[14] found that the toxic effects of 2, 4-D on soybean, red kidney bean, white mustard, or marigold were eliminated or decreased by dusting or spraying active carbon in aqueous suspension on the plants before the 2, 4-D treatment.

5. IMPROVING THE RECOVERY OF METALS

The use of powdered activated carbon in metals winning operations has been limited to certain specific applications where it has contributed to the efficiency and economy of the operation. Recently two new uses have been put into commercial use.[15]

Wild Frothing

During bulk flotation, excessive frothing may occur, caused by organic materials that are present in the ore deposit, such conditions can result in:

(1) Poor separation of the metal from the gangue.

(2) Spill-overs which cause housekeeping problems and reduced metal recovery. This frothing can be controlled, by adding

powdered carbon very early in the bulk float state. In one operation the carbon dosage was varied from 0.1 to 3 lbs. per ton of ore. The volume and size of the froth could be controlled by regulating the amount of carbon. When sufficient carbon was added, the froth was completely eliminated.

Cleaner Separation of Metals

A benefit often associated with activated carbon is the ability to preferentially attract one substance to its surface and to separate it from the rest of the system. This benefit was demonstrated at two different unrelated operations—at which cleaner separation of metals resulted.

In a different application involving another metal, powdered carbon gave a "cleaner" separation of the two metals and better recovery of the primary metal. These applications for powdered carbon are now in commercial operation.

Electro-Refining—Electro-Winning

The use of granular activated carbon is particularly beneficial for the recovery of metals by electro-refining or electro-winning. For these operations the electrolyte solution is passed through the granular carbon just before entering the cell (tank house). Grease, solvents and other organics, which could adversely effect the "plating out" of the metal, are adsorbed.

REFERENCES

1. Prill, P., Korrosion u. Metallschutz 17, 345 (1945).
2. Spowers, W. H., U.S. Patent 92, 2,092,595.
3. Flint, C. H., U.S. Patent 2,138,804.
4. Gordon, M., Chime & Industrie 53, 314 (1945).
5. Welter, E., Seifensieder-Ztg. 58, 275 (1931).

6. McKee, R. H., U.S. Patent 2,039,330.
7. Phillips, W. E., U.S. Patent 2,240,885.
8. Castello, A. D., and Dixon, H. L., U.S. Patent 2,278,826.
9. Keilen, J. J., and Dougherty, W. K., India Rubber World, 129:199 (1953).
10. Verona, O., and Ciriotti, P., Boll. ist. super. agrar. Pisa, 11:401 (1935).
11. Perotti, R., and Ferretti, C., Boll. ist. super. agrar. Pisa, 6:147 (1930).
12. Verona, O., Ann. Facolta agrar. univ. Pisa (N. S.), 1:383 (1938).
13. Arle, H. F., Leonard, O. A., and Harris, V. C., Science, 107 No. 2775:247 (1948).
14. Weaver, R. A., Botanical Gazette, 109 No 3:276 (1948).
15. Hyndshaw, A., WESTVACO, NUCHAR Division, Covington Va. Communication.

PART IV

PREPARATION OF ACTIVATED CARBON AND PHYSICO-CHEMICAL PROPERTIES

11

Manufacture of Activated Carbon

1. GENERAL

In discussing the manufacture of activated carbon it is important to keep in mind that the term *activated carbon* comprises a family of substances. None of the members of the family is characterized by a definite structural formula nor can any be separately identified by chemical analysis. Our only basis for differentiating is by adsorptive and catalytic properties. As is well known, carbons

TABLE 11:1

SOURCE MATERIALS THAT HAVE BEEN STUDIED
FOR THE PRODUCTION OF ACTIVATED CARBON

Bagasse	Kelp and seaweed
Beet-sugar sludges	Lampblack
Blood	Leather waste
Bones	Lignin
Carbohydrates	Lignite
Cereals	Molasses
Coal	Nut shells
Coconut shells	Oil shale
Coffee beans	Peat
Corncobs and corn stalks	Petroleum acid sludge
Cottonseed hulls	Petroleum coke
Distillery waste	Potassium ferrocyanide residue
Fish	Pulp-mill waste
Flue dust	Rice hulls
Fruit pits	Rubber waste
Graphite	Sawdust
	Wood

from such different source materials (Table 11:1) and/or prepared by dissimilar activation processes, will have dissimilar assortments of adsorptive properties (Table 11:2). We do not know the full significance of the variations; tentatively we can assume that they typify different varieties of activated carbon—different members of the family.

TABLE 11:2

SPECIFIC ADSORPTIVE POWER OF DIFFERENT
ACTIVATED CARBONS IN AQUEOUS SOLUTION*

Carbon code[a]	Methylene blue[b]	Phenol[c]	Iodine[d]	Molasses[e]	Caramel[e]
A	0.84	1.35	0.65	0.8	1.3
B	0.74	1.24	0.59	0.6	0.9
C	0.70	1.30	0.57	1.1	1.6
D	0.37	0.60	0.36	0.7	1.2
E	0.30	0.60	0.32	0.6	1.3
F	0.44	1.20	0.55	0.2	0.4
G	0.83	0.83	0.39	1.0	1.0
H	0.73	0.57	0.37	1.1	1.0
J	0.50	1.60	0.59	nil	nil

a) The carbons in this code do not necessarily correspond to carbons appearing with similar code letters in other tables.
b) Millimoles adsorbed per gram carbon at an equilibrium concentration of 0.1 millimole per liter.
c) Millimoles adsorbed per gram carbon at an equilibrium concentration of 1.0 millimole per liter.
d) Grams adsorbed per gram carbon at an equilibrium concentration of 0.127 gram per liter.
e) Color units adsorbed per gram carbon at 90% decolorization.

* Data from *Ind. Eng. Chem.* 37 : 645 (1945); reprinted by permission of the American Chemical Society.

Just how many varieties exist is uncertain. The data on adsorption of different carbons could suggest the existence of very many varieties, but this is not necessarily so. The situation may be analogous to that of color, in which a few primary colors can be blended to furnish an infinite variety of hues and shades. Similarly, the many variations in adsorptive characteristics revealed in different carbons may be but different permutations of a relatively few basic adsorptive properties.

During the first half of this century, many factories were built in

Europe and America to manufacture commercial brands of activated carbon. A number of different source materials and activation processes were used; consequently the various brands furnished many forms of adsorptive power. Many of the operations did not long survive, and today only a few remain. The shrinkage in numbers was inevitable because the market was never of sufficient size to support more than a fraction of those that started.

It can be helpful to those that may contemplate entering the manufacturing field to have some knowledge of factors that caused the demise of the less successful past endeavors. For the most part we lack knowledge as to the exact cause of the fold up of any particular unit, but in some instances we can deduce the probable reason from circumstantial evidence. Thus, the reason is obvious with those carbons that lacked the adsorptive properties needed in industrial applications. Equally obvious is the cause of failure of those brands that were priced out of the markets they might have served. Some operations failed because the costs of source materials and processing were unsuitably high. And there were those that because of inadequate testing procedures were unable to furnish customers with consistent and uniform quality in all shipments. Several operations arrived too early, and at least one arrived too late. On of the less obvious causes of failure can be traced to a lack of *"market know-how"*. An activated carbon—no matter how good—cannot be successfully bought and sold as a commodity through brokers. Instead it must be viewed as a specialty that requires trained technical personnel to provide effective sales and service.

A more positive insight for newcomers may be gleaned by considering the characteristics of the manufacturing units that have survived. It is significant that each of the current manufacturers in America uses a different source material and employs specialized methods of processing. Many commercial brands are interchangeable in some applications, but some have distinctive properties that render them more suitable for certain application.

Many patents have expired and may be freely used, but the newcomer will do well to consider that to employ a patent as written can be quite different from applying it as now practiced. Moreover, the veteran operator has developed skills and know-how that are not readily or quickly acquired.

This chapter is not prepared for the staff of established manufacturing operations; that would indeed be carrying coals to Newcastle. But the production of activated carbon embraces more than the creation of adsorptive properties. Those properties must be utilized so as to provide benefits desired by the user. A knowledge of manufacturing processes may at times aid a potential user in the search for an appropriate carbon. However, a more direct and positive utility of activation know-how is associated with the increasing number of applications that require *on-site* regeneration. For workers in those operations, an understanding of activation know-how can be helpful.

Much published data are available on activation of carbon and methods of manufacture,[1,2,3,4,5] and various specific ways and means are described in hundreds of patents.[6,7] However most of the patented methods are but various permutations of a basic operational procedure. This consists of carbonizing the source material under appropriate conditions, after which the resulting char is often subjected to oxidation in a controlled environment.

The specific conditions under which the carbonization and activation are conducted can have much influence on the type of adsorptive capacity that is formed. This aspect is important for some applications, but not for others. Thus, specially prepared carbons are required for sugar refining, whereas carbons made by various processes are effective for adsorption of iodine or phenol. Some source materials are inherently suitable for certain applications; coconut shells shrink on carbonization to yield a char suitable for adsorbing gases and vapors, whereas bones are readily converted to a char useful in sugar refining. However, inherent characteristics can often be modulated by appropriate processing.

2. CARBONIZATION

Carbonization (pyrolysis) is usually conducted in the absence of air. Chaney[8] and others report that temperatures below 600° C are preferable for producing chars suitable for steam activation. However, this is not universal.[9,10] McBain produced an activated char from sugar carbonized at 900°C. In a study of activation of anthracite, carbonization at 850°C in an atmosphere of steam yielded a more porous structure that was readily activated.[12] The presence of oxygenated constituents in the source materials aid in forming a porous structure.

Carbons with a broad spectrum of adsorptive capacities, *e.g.*, decolorizing carbons, are more readily prepared from source materials that contain suitable inorganic constituents.

The presence of some specific inorganic constituents can make it possible to prepare a broad spectrum of activated carbon by carbonization without a supplementary stage of oxidation.

Carbonization with Metallic Chlorides

The effectivness of carbonization is enhanced when the source carbonaceous materials (*e.g.*, wood chips, peat) are impregnated with a solution of a metallic chloride.[1,6,7,14,15,17,18] Calcium and magnesium chlorides were among the first to be so used. Zinc chloride is employed on an industrial scale in Europe and Japan. In that process, one part of the source carbonaceous material is mixed with 1 to 4 parts of zinc chloride in solution. After drying, the mixture is carbonized at 600°C to 700°C—although temperatures from 400° to 900°C have been reported. The finished char, after washing with acid and water has suitable adsorptive capacity for many applications.

The zinc salts are recovered for reuse.[1,4,15,16]

A loss of adsorptive capacity has been observed when the washed zinc chloride carbons are dried. It is reported that much of this adverse effect is avoided if the activation is conducted in two stages the first stage at 250°C, and the second stage at 400°C.

Carbonization with Potassium Carbonate

Potassium carbonate has been used since a very early date for production of blood char. A mixture of 8 parts of dried blood with 1 part potassium carbonate is ignited at 800°C in the absence of air. The charred product is washed and dried.[37]

3. ACTIVATION WITH OXIDIZING AGENTS

Chars prepared under appropriate carbonization conditions can be used successfully without further processing in many applications. However, in the case of chars, prepared by carbonization processes that do not confer adequate adsorptive capacity, such chars are subsequently subjected to some form of controlled oxidation. Many processes involve a reaction with oxidizing gases— steam, air carbon dioxide at elevated temperatures.[1,4,6,7,19,20,21,22,23]

$$H_2O + C_x \rightarrow H + CO + C_{x-1} \qquad \text{(at } 800° - 900°C)$$
or
$$CO_2 + C_x \rightarrow 2CO + C_{x-1} \qquad \text{(at } 800° - 900°C)$$
or
$$O_2 + C_x \rightarrow 2CO + C_{x-2} \qquad \text{(at } 800° - 900°C)$$
or
$$O_2 + C_x \rightarrow CO_2 + C_{x-1} \qquad \text{(below } 600°C)$$

In activation by oxidation, the kind of adsorptive powers developed are determined by:

1) the chemical nature and concentration of the oxidizing gas
2) the temperature of the reaction
3) the extent to which the activation is conducted
4) the amount and kind of mineral ingredients in the char.

In any discussion of proper conditions we must always keep in mind that optimum conditions are specific for each situation and to a large degree depend on prior history of the char. Thus, the optimum environment for oxidation for a char prepared with zinc chloride will not be the same as that for a char formed by roasting wood with an alkali. In other words, preparatory stages leave an imprint that affects each succeeding stage of processing. This we describe by saying that *"carbon has a memory"*.

For the activation of many types of chars, steam is preferable to carbon dioxide and much better than air (Table 11:3). With steam, the temperature of activation must be high enough to provide a reasonably rapid rate of oxidation, but temperatures above 1000°C are to be avoided because they impair adsorptive powers. Some adsorptive powers, such as for iodine and phenol, develop equally well over a wide range of temperature, whereas other adsorptive powers, such as for molasses, may develop within a narrow range of temperature. The optimum range is often modulated by characteristics of the char being processed. Thus, to produce a decoloriz-

TABLE 11:3

ADSORPTION BY ACTIVATED PINE WOOD CHAR

Activation conditions			Substance adsorbed, g/g carbon		
Gas	Temperature °C	Chryso- idine R	Ponceau R	Aniline Blue	Iodine
Air	600	0.34	0.10	0.05	0.36
Air	740	0.16	0.08	0.05	0.40
Air	790	0.15	0.08	0.06	0.42
Air	860	0.14	0.08	0.06	0.42
Air	910	0.13	0.10	0.06	0.40
Steam	770	0.37	0.19	0.16	0.60
Steam	825	0.37	0.17	0.17	0.60
Steam	880	0.36	0.16	0.21	0.62
CO_2	880	0.32	0.12		

All processes of activation conducted to 30% yield based on weight of original char.
Adsorption evaluated at an equilibrium solution concentration of 0.10 g/liter.

ing carbon from anthracite roasted with alkali, it was necessary to lower the temperature of steam-activation to 750°C.[12]

Activation with carbon dioxide is conducted at temperatures 800°C to 900°C. Activation with air involves an exothermic reaction and measures must be taken to keep the temperature from rising above proper limits—usually not over 600°C. Activation with air is seldom very practicable except for certain chars such as those prepared by roasting lignin or sawdust with an alkali. For such a char, air at 600°C will provide decolorizing power equal to that from steam at 800° to 900°C (Table 11:4).

TABLE 11:4

ADSORPTION BY ACTIVATED LIGNIN CHAR

Activation conditions			Substance adsorbed		
Gas	Temperature C°	Chrysoidine R	Ponceau R	Methylene Blue	Molasses Color
Air	600	0.34	0.33	0.23	1.6
Air	840	0.21	0.10	0.14	0.5
Steam	840	0.36	0.24	0.29	1.4

Lignin char prepared by carbonizing equal parts of lignin and sodium hydroxide. Char washed before activation.
Molasses adsorption based on color units; adsorption of dyes expressed as g/g carbon at an equilibrium solution concentration of 0.10 g/liter.

Even greater decolorizing power can be developed on alkali-roasted chars if they are given a brief activation with steam at 800°C, followed by further activation with air at 500° to 600°C (Table 11:5) Curiously enough, this two-stage activation procedure, which is so effective for alkali-roasted chars, is of no practical value for other chars thus far studied.

When proper activation conditions are provided, the oxidizing action of the activating gases does not consist of an indiscriminate removal of successive layers of atoms from the surface. Instead, the oxidation selectively erodes the surface so as to increase the surface area, develop greater porosity, and leave the remaining atoms arranged in configurations that have specific affinities. All adsorptive powers do not develop simultaneously nor at a uniform rate. Some develop during the initial period and than cease to increase further. Certain adsorptive powers develop only during late stages of activation; whereas still others show a rather consistent rate of increase throughout the entire period of activation (Figures 11:1 and 11:2)

The steam-activation procedure can be conducted very satisfactorily in the laboratory, but many difficulties attend the industrial application of the process. Numerous patents cover methods of surmounting the large-scale difficulties, and of these, typical ones will be discussed.[4,24]

Early processes employed externally heated vertical tubes containing the char, the activating gas being admitted at the bottom

TABLE   11:5

ADSORPTION BY SPECIALLY PREPARED WOOD CHAR ACTIVATED IN TWO SEPARATE STAGES

| Activation conditions | | Substance adsorbed | |
First stage	Second stage	Aniline Blue	Molasses Color
Steam at 800° C	Steam at 800° C	0.12	1.50
Steam at 800° C	Air at 550° C	0.22	1.70
Air at 550° C	Air at 550° C	0.15	1.50

Char prepared by boiling wood chips in caustic soda solution, and then carbonizing chips at 600° C. Char was washed with acid before activation.

Molasses adsorption based on color unit scale; adsorption of Aniline Blue expressed as g/g carbon at an equilibrium solution concentration of 0.10 g/liter.

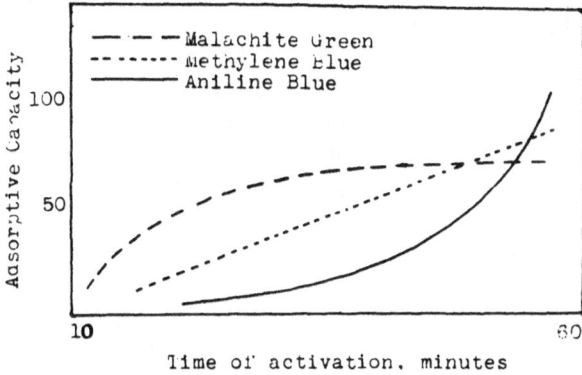

Fig. 11:1   Rate of development of specific adsorptive capacities during steam-
activaion of a chemically treated sawdust.

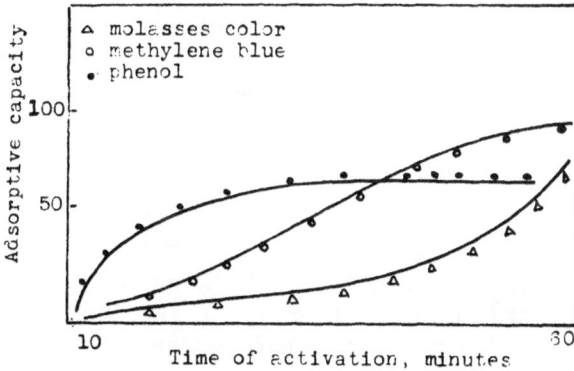

Fig. 11:2   Rate of development of specific adsorptive capacities during steam-
activation of chemically treated peat.

and passing upward. Batch activation was first attempted but was
abandoned when it was found that only the portion of the carbon
near the entering gases became activated. The reaction:

$$H_2O + C = H_2 + CO$$

reaches equilibrium rapidly, after which no further reaction occurs. To extend the zone of oxidation, it became desirable to admit additional quantities of activating gases at several points at successively higher levels in the tube. Subsequently, the process was improved by being made continuous, the carbon moving downward countercurrent to the upward-moving gas stream. The exit gases which contain carbon monoxide and hydrogen can be subsequently burnt to superheat the intake steam, or to aid in carbonizing the source materials.

The reaction between carbon and steam or carbon dioxide is endothermic; *i.e.,* it has cooling effects which may be compensated by heat applied to the external surface of the activation tubes. This method is only partially successful because carbon is a poor thermal conductor and the temperature falls rapidly in passing to the center of the tube. To some extent, this is offset by convection currents in the moving gases, and the situation is also ameliorated by designing the tube so as to be narrow in one dimension of its cross section, *i.e.,* of oval shape. Some air can be admitted to provide an exothermic reaction at the expense of recovering carbon. Wickenden and Okell[25] have utilized electrical heating. In the process, the carbon, acting as a resistor, provides a uniform temperature throughout the carbon bed.[26] The cost of electrical power for this purpose is appreciable.

In order to permit free upward passage of the activating gases through the bed, the particles of char initially were prepared in sizes as large as 1 inch by 2 inches. It soon was found that the activating reaction occurred mainly on the external portion of these large particles, leaving the kernel inactive. Therefore, it became necessary to employ a smaller particle size. The difficulty of passing gases through a bed of fine-mesh char made it necessary to use thinner beds, and later led to a study of other procedures. In one system, the carbon, instead of moving downward as a solid bed, falls in a relatively thin stream through a vertical tube and, while falling, is continuously deflected by baffles from side to side of the tube. This arrangement provides effective contact with the upward stream of activating gases.

In methods developed by Wickenden and Okell,[25] and by Sauer,[27] the activating gases are passed upward through the bed at sufficient velocity to keep the carbon particles in a suspended or fluidized state. This provides a uniform distribution of temperature and also continuously brings different carbon particles into contact with the entering gas stream. It was found that many carbon particles were carried away in the effluent gases, and advantage was taken of this feature. The individual carbon particles diminish in density as they become active and by suitable control of the velocity of the gas stream, the carbon particles can be continuously floated away as they reach a desired activity. The active particles are separated from the effluent gases by a series of collectors.

Proper size of the particles is an important factor when they are activated in a state of suspension. The particles should be uniform—small enough so as not to require a high velocity of gas to keep them suspended, yet not so small as to introduce difficulties in separating them from the exit gases. Very small particles of carbon can be collected by a mist of water in the final collector.

In some types of activation equipment, *e.g.,* Herreshoff furnaces and rotary kilns, the oxidizing gases pass over instead of through the carbon bed. In such operations, mechanical means are provided which bring fresh carbon particles continuously to the surface where they are exposed to the oxidizing gases. This is accomplished by stirrers in Herreshoff furnaces; whereas with rotary kilns, the rotation of the drum furnishes the needed agitation. Direct-fired kilns have been designed to permit the admission of oxidizing gases at several points to maintain better activation environment throughout the length of the rotary kiln.

The procedure of having the gases flow over instead of through the carbon bed is generally employed when air is the oxidizing agent. Activation with air is an exothermic reaction, and measures must be provided to prevent the temperature from rising above proper limits—usually not over 600°C. This may be accomplished by impregnating the carbon with a substance, such as phosphoric acid, which reduces the combustibility, or by diluting the air stream with an inert gas to repress the rate of oxidation.

Dolomite Process

A process that is utilized to provide a uniform distribution of oxidizing gases throughout the entire mass of carbon involves incorporating substances that continually release such gases at the activating temperature.[28] In a typical method, 1 part of powdered dolomite and 1 or more parts of pulverized lignite are mixed with sufficient milk of starch to form a paste. After drying, the mixture is carbonized and then heated at 600° to 900°C; at this temperature the dolomite evolves carbon dioxide uniformly throughout the mass.

In a modification of this process, a wet mixture of sawdust or peat and magnesium carbonate is subjected to the action of carbon dioxide under pressure. A soluble bicarbonate is produced which penetrates the mass uniformly. After activation, the magnesium oxide remaining in the char can be extracted with water and carbon dioxide under pressure.

Sulfate Process

This is similar in principle to the dolomite process—sodium, potassium, or other sulfates being incorporated with the raw material, either before or after carbonization.[29] At the activating temperature, a portion of the carbon is oxidized by the sulfate which is reduced to the sulfide. The sulfide formed has an erosive action on the carbon and continues the activation.

Phosphoric Acid Process

Phosphoric acid can be used to provide the oxidation environment.[1] This process, invented independently in America by Hudson[30] and in Europe by Urbain,[31] is represented by several modifications.[32]

In one method, finely pulverized peat or sawdust is saturated with 25° to 30° Baumé solution of phosphoric acid. A high ratio of phosphoric acid is employed when a decolorizing carbon is to be manufactured. The mixture is dried and then heated between 400° and 600°C.

Some activation is due to the dehydrating power of the phosphoric acid and, to this extent, the method is analogous to the zinc chloride process. Frequently, the charred product is subsequently ignited for 2 to 8 hours in retorts at 800° to 1,000°C. During this latter stage, the carbon is eroded by being partly oxidized by the phosphoric acid which is reduced to phosphorous and hydrides.

These vaporized products are subsequently oxidized to phosphoric acid to repeat the cycle with a fresh batch of raw material.

In other modifications of the process, some sulfuric acid is substituted for part of the phosphoric acid. A number of patents describe the use of sodium, potassium, and calcium phosphate as other substitutes for phosphoric acid.

Unlike zinc chloride, which is effective only when added before carbonization, phosphoric acid will assist the activation when added after carbonization. For this purpose, 2 to 25% of phosphoric acid is added to a charcoal which is then ignited or steam-activated.

Caustic, Thiocyanate, and Sulfide Processes

When char is impregnated with caustic soda or potash and heated above 500°C, an energetic reaction occurs[33] which results in erosion of the carbon accompanied by an increase in adsorptive power.[34] The process has a destructive effect on equipment, and caustic alkali by itself is seldom used commercially.

A similar but less energetic reaction produced by sulfides[35] or thiocyanates[36] has found industrial application. In one method, 100 parts of char are mixed with 15 parts of potassium sulfide and 30 parts of caustic potash in a slurry which is dried and ignited at 900°C with the exclusion of air. In another process, sawdust, impregnated with a 35% solution of potassium thiocyanate, is dried and heated to a temperature of 300° to 350°C in the absence of air for 1/2 hour, after which the temperature is elevated to 800°C. The cost of thiocyanate can be reduced by partial substitution with cheaper chemicals such as carbonates or sulfates. Further economies are effected by adding the thiocyanate after the carbonization stage, in which case, the impregnated char is heated to about 800°C.

4. STUDIES OF OTHER ACTIVATION AGENTS AND PROCESSES

Other chemicals studied as potential activation agents are listed in Table 11:6. Other chemical reactions reported to form activated carbon are listed in Table 11:7.

TABLE   11:6

CHEMICALS STUDIED AS POTENTIAL ACTIVATION AGENTS[49,50,51]

Ammonium Salts[39]	Hydrochloric acid[44]
Borates	Manganese dioxide[45]
Boric Acid[40]	Nickel Salts[46]
Calcium Oxide[41]	Nitric acid[47]
Cyanides[42]	Sulfur[48]
Ferric and Ferrous Compounds[43]	

TABLE   11:7

CHEMICAL REACTIONS REPORTED TO YIELD ACTIVATED CARBON

Action of oxygen on organic substances dissolved in hydrofluoric acid[58]
Passage of carbon monoxide over hot iron catalyst[54,55]
Reacting mercury with organic halides[56]
Extracting charcoal with selenium oxychloride[53]

Specialties

A *pseudo-activated carbon*[69,70] is prepared by reacting wood, peat, and similar materials with concentrated sulfuric or phosphoric acid and heating to 120°–300°C.

The residue, after washing, has adsorptive and ion exchange properties if used in a wet state. Drying causes a loss of adsorptive power.

A recent development is a new type carbon adsorbent having an open network of 5 angstroms in uniform size known as a molecular sieve carbon. Details of applications are available from the supplier.[100]

Processes are described in which the raw material before carbonization is incorporated with mineral substances such as kieselguhr[65], silica,[66] clay,[67] and bone.[68]

5. COMMERCIAL POWDERED AND GRANULAR ACTIVATED CARBONS

Washed Carbons

Carbons are washed with water for applications that require a low soluble ash. When removal of acid soluble ash is required, the carbon is reacted with a suitable mineral acid. When this dissolving action is complete, the carbon is thoroughly washed with water. The carbon is then brought to a neutral pH with dilute sodium carbonate, and then given a final wash with water. The neutralization step is necessary because a free acid cannot be completely eluted with water, but sodium salts are readily extracted.

Powdered Carbon[59,60,61,62]

The preparation of powdered carbon should be accomplished by the mildest possible pulverizing action. A powerful crushing action as by heavy weights of balls in a ball mill can damage the filterability. It can also impair the adsorptive power of decolorizing types of carbon.

Carbons should preferably be very dry when pulverized because the presence of moisture augments adverse effects on filterability.

For the preparation of some powdered washed carbons, it is advisable to pulverize prior to washing because any adverse effects of pulverizing are more pronounced with carbons that have been washed.

For some applications, the pH of alkaline carbons is adjusted without a washing stage. To do this, the pulverized carbon is mixed with the required amount of acid in a blender. Sufficient moisture should be present to enable the neutralization to take place without delay. Otherwise, the presence of free acid on the carbon surface can have unfavorable effects, especially if and when the carbon is applied to a non-aqueous liquid, *e.g.*, an edible oil.

Granular Carbons[1-4,63,64]

Granular carbons must have effective adsorptive capacity combined with great mechanical strength. Commercial brands have been available for gas phase applications since World War I. The production of granular carbons for liquid phase applications was long delayed because of the additional activation necessary to provide the required types of adsorptive capacity. The additional activation oxidized the walls of pores and thereby weakened the structure. As a result, the finished carbons lacked the mechanical strength to withstand the abrasion incident to continual recycling required of granular carbons. The difficulty was finally surmounted about the time of World War II. Granular carbons with effective adsorptive capacity combined with adequate mechanical strength have been available for liquid systems for a number of years.

Coconut shells are a very satisfactory raw material for production of granular carbon for gas phase adsorption. Few other substances in their natural state are suitable for the manufacture of effective granular carbons. However, processes have been developed whereby some materials can be converted into forms from which effective granular carbons can be manufactured. In a typical process, coal is pulverized and mixed with sufficient binder to form a plastic mass which is briquetted or extruded at pressures variously described ranging from 100 to 2,000 lbs. per square inch. The pellets or spaghetti-like strings are carbonized slowly to avoid rapid evolution of gas, after which the char is steam activated. Morgan and Fink[10] suggest the binder (*e.g.*, tar, pitch) should swell just enough during carbonization to provide proper porosity.

In some chemical activations, the raw material is plasticized by the chemical agent, and auxiliary binders are not needed. In the zinc chloride process, pulverized peat is added to a concentrated zinc chloride solution. When the mass becomes plastic it is passed through an extrusion press; and then dried and carbonized below 700°C. After an acid wash to remove the zinc, the dried granules are suitable for recovery of solvent vapors. For use in gas masks, it is given a supplementary steam activation.

A similar process has been developed using phosphoric acid.

6. THEORIES OF ACTIVATION

We are far from having a clear picture of just what occurs during activation aside from the fact that the surface area is increased and a very porous structure is developed. This lack of knowledge is not surprising in view of the various methods by which an activated carbon can be produced. One is often perplexed to find a step described as necessary in one process and omitted from another. In some cases, this is because the value of a particular stage may depend upon the previous history as well as on the treatment that follows. Then, too, the value of certain steps is the fact that they confer certain specific powers that may be needed in one application and not in another. It is to be remembered that each-method of activation leaves a characteristic imprint on the adsorptive powers.[71] Therefore, it is not uncommon to combine several different processes to provide a diversity of properties.

Nature of Amorphous Carbon

Cokes, chars, and activated carbons are frequently termed *amorphous carbon*. X-ray studies have shown that many so-called amorphous substances have crystalline characteristics, even though they may not show certain features, such as crystal angles and faces, usually associated with the crystalline state. An amorphous-looking powder may be composed of crystals of submicroscopic dimensions, so-called *crystallites*. This is true of chars and cokes.

Studies, by Riley,[72,73] Warren,[74] Berl,[75] Hofmann,[76] and others, have shed much light on the structure of these carbon crystallites. Although interpretation of the X-ray diffraction patterns is not free from ambiguities, there is general agreement that amorphous carbon consists of flat plates in which the carbon atoms are arranged in a hexagonal lattice (Figure 11:3), each atom, except those at the edge, being held by covalent linkages to three other carbon atoms. The crystallites are formed by two or more of these plates being stacked one above the other. Although these crystallites have some structural resemblance to a larger graphite crystal, differences other than size exist. In graphite, the plates are more closely together; moreover, the manner in which the plates are stacked above one another follows a definite crystalline pattern in graphite, a feature that is absent in the carbon crystallites. In addition to the evidence provided by X-ray examination, differences between chars and graphite have been shown by chemical methods.[77,78] Chars and cokes react with gaseous oxygen at lower temperatures than does graphite, however, graphite is oxidized much more readily than char by a mixture of chromic and phosphoric acids at a temperature of 100°C.[73]

The size of the crystallites is influenced by the temperature of carbonization and, to some extent, by the composition and structure of the raw material. In chars prepared from such sub-

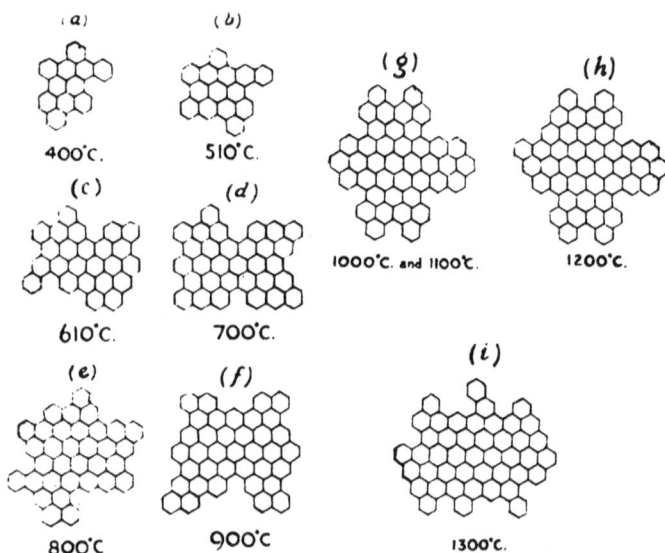

Fig. 11:3 Arrangement of carbon atoms in a single-layer plane of crystallites formed at different temperatures. From H. L. Riley, *Chemistry and Industry*, 58 No. 17:391 (1939); reprinted by permission of the copyright owners.

stances as cellulose, the c dimension (height of the crystal) shows little change until a temperature of 1,300°C is reached, whereas the a dimension (diameter of each layer) shows a continuous growth up to 700°C.[72] This latter fact is of interest because many properties such as electrical conductivity and the ease of subsequent activation show a change at this temperature.

The crystallites may be formed through several mechanisms.[10,72,79] During pyrolysis, the original organic substance may be split into fragments which regroup to form the thermo-stable aromatic structure existing in the hexagon. It is also possible that suitable nuclei initiate a transformation in which the hexagonal lattice grows gradually at the expense of the original substance. The transformation is seldom complete and residual hydrocarbon chains and rings remain.

As these residual hydrocarbons cannot be extracted with solvents or removed by degassing they are presumed to be attached by

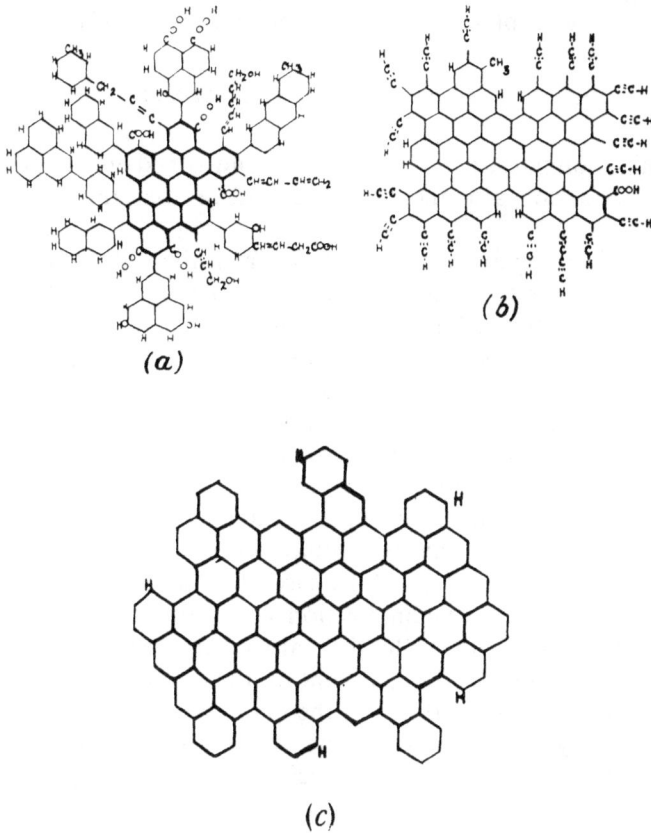

(a)

(b)

(c)

Fig. 11:4 Theoretical structures of crystallites. From H. L. Riley, *Chemistry and Industry*, 58 No. 17:391 (1939); reprinted by permission of the copyright owners.

chemical bonds to the border atoms of the crystallites. A hypothetical structure is shown in Figure 11:4. It is to be noted that the term *hydrocarbon*, usually applied to the cementing substances, may also cover compounds containing other elements in addition to hydrogen and carbon, such as oxygen. It is believed by some that the hydrocarbons may cement the crystallites into clusters to form secondary structures. Differences in the size, shape, and arrangement of the crystallites in the secondary structures could affect the adsorptive power and other properties of the char.[76]

Action of Oxidizing Agents

Oxidation burns away hydrocarbons adsorbed during carboniza-tion. Oxidation also increases the surface area and enlarges the pore volume:
  a) by burning away carbon atoms from the walls of open pores
  b) by perforating closed pores and thereby providing access
      to pores formed initially without an inlet.
All this explains the increase in total adsorptive capacity, but it leaves unexplained the development of diverse specific adsorp-tive powers by different oxidation conditions. Why, for example, are specific adsorptive powers furnished by steam at 850°C unlike those conferred by air at 450°C?
An understanding of the phenomena involved in the combustion of carbon is veiled by the complexities surrounding chemical reactions that include a solid phase. The equilibrium conditions that govern the final composition of products are known, but this sheds little light as to the path of the reaction. That such knowledge would shed much light on the development of adsorp-tive power is questionable, but the inability to obtain information clarifies why it is difficult to interpret the relation between con-ditions of activation and the specific types of adsorptive powers that are formed.

Action of Mineral Salts

The development of specific properties in carbon is affected by the presence of mineral salts and other noncarbon elements. In attempting to understand the influence of the noncarbon con-stituents on activation, one must distinguish between effects produced during the carbonization or pyrolysis and those pro-duced during the subsequent ignition or oxidation.[81]
One view of the action of the mineral ingredients during the carbonization is that they provide a skeleton on which the carbon is deposited,[75,82] the freshly formed carbon becoming bonded by

adsorption forces to the mineral elements. Evidence of the strength of such bonding is shown when carbonaceous substances burn on cooking utensils or on the valves of gasoline engines. Such carbon is difficult to remove by mechanical means. As expressed by Alexander and McBain:[83] "It seems as if the nascent bonds, appearing as the more or less profound chemical changes occur, tend to be satisfied by whatever is nearest." When the mineral ingredients are subsequently dissolved by acid or water, the exposed carbon surface becomes free to attract other substances.

The presence of certain inorganic substances increases the yield of char.[84] This suggests that they may alter the course of the reactions in pyrolysis* so that less of the objectionable tarry products are formed. This appears to be so when dehydrating salts are used.[85] Zinc chloride causes hydrogen and oxygen atoms in the source materials to be stripped away as water rather than as hydrocarbons or as oxygenated organic compounds.

Although the mechanisms that have been mentioned indicate how noncarbon ingredients could enhance adsorptive power in general, they fail to explain specific effects. Why, for example, should a carbon prepared with calcium chloride be more adsorptive for caramel, whereas a carbon prepared with zinc chloride is more effective for iodine?[71] A convenient interpretation is that each chemical exerts a specific influence on the molecular architecture of the surface that is formed.** In this, as always, we must keep in mind that because a given picture can explain a process, it does not necessarily follow that it must be so.

Many noncarbon ingredients exert specific influences during oxidation with steam or air.[1,86] Some of these effects may be due to

TABLE   11:8

FLOW PATTERN FOR ACTIVATION OF ANTHRACITE

  a) Steam activate 20 × 60 mesh anthracite at 850°C
  b) Mix char residue with sodium sulfate
  c) Roast at 850–900°C
  d) Wash residual char with water, then acid
  e) Steam activate at 750°C

From proceedings of Anthracite Conference, Oct. 18, 1956; reprinted by permission of Pennsylvania State University.

*Perhaps this explains the well known trick that a lump of sugar can be made to burn by previously dipping it in cigarette ash.
**Some inorganic ingredients, for example, alter the size of the pores formed during activation.[98]

an influence on the temperature at which the activation occurs. Thus, carbon will burn at lower temperatures when potassium carbonate is present, whereas orthophosphoric acid raises the ignition point. The difference in the properties imparted by potassium carbonate as compared to orthophosphoric acid may be in part traceable to this influence.

Another possbility is that noncarbon atoms are held or adsorbed on certain areas of the surface and thus influence positions from which carbon atoms are preferentially etched away by the oxidation process. That activation involves oxidation at the point of bond between the carbon and the inorganic ingredients is illustrated by processes in which a normally soluble inorganic substance becomes bonded to the carbon during carbonization and cannot be extracted with water until after activation.

An experience with anthracite coal is of interest in connection with the influence of inorganic ingredients.[12] Activation with steam produced a carbon with adsorptive power for substances such as iodine and phenol, but this method failed to develop decolorizing power (e.g., for molasses). By following the procedure shown in Table 11:8 however, it was possible to produce a carbon with great decolorizing power. Conditions during the roasting stage had to be closely controlled. If the roasting temperature were below 850°C, the char would not develop decolorizing power when subsequently activated with steam. On the other hand, a roasting temperature above 910°C caused the roasting chemicals to become bonded to the carbon in such a manner that they could not be eluted in the washing stages. This resulted in a char that ashed excessively during the subsequent activation.

Another possible mechanism is suggested by the work of Berl[75] who found evidence that some elements, such as potassium, can penetrate between the hexagon plates of the crystallites and spread them apart, enabling erosion to occur at surfaces otherwise unexposed.

So far, the influence of the mineral ingredients has been traced to the manner in which they alter the mechanism of an activating process. Apart from being a ladder to achieve a more effective spacing and structural arrangement of the carbon atoms, some noncarbon elements become part of the molecular architecture of the activated carbon. This is indicated by the tenacity with which many foreign elements are held by carbons.[5,88,89,90] There is considerable evidence that oxygen can be attached by covalent bonds to the carbon structure to form stable surface oxides. When sulfur is heated with carbon, only part of the sulfur can be separated subsequently by extraction with toluene, by oxidation with bromine, or by heating to 900°C in a stream of nitrogen; but heating in a stream of hydrogen will remove the sulfur as hydrogen sulfide. All this indicates a chemical union between the carbon and sulfur atoms, a union which Wibaut[91] suggested to be similar to the surface oxides. Ley and Wibaut[92] found evidence that nitrogen is combined with carbon in a form suggestive of nitrile groups. Miller[93] studied a commercial carbon from which repeated boiling with acid failed to extract appreciable quantities of the mineral ingredients, but when the carbon was ashed, the ash dissolved readily.

It is at least possible that many of these strongly held foreign elements provide adsorptive bonds. Support for this view is found in the fact that specific adsorptive and catalytic powers have been traced to the presence. of noncarbon atoms—oxygen, iron, and nitrogen. The influence of noncarbon atoms may extend to adjacent carbon atoms in a manner analogous to the way in which a polar group introduced into an organic compound affects the chemical properties of remote atoms.

At one time, the adsorptive powers of animal char were believed to reside in chemically held nitrogen;[5,94] later when very active carbons were produced from nitrogen-free source materials, the importance of nitrogen was discredited.[95] In this connection, however, some workers have observed that carbon from nitrogen-free sources may contain appreciable quantities of combined nitrogen[96]—indicating that carbon can fix nitrogen during the activation process.

Carbon, prior to activation, contains hydrogen in the form of hydrocarbon chains and rings attached to border atoms of the hexagon plates. Much of this hydrogen is removed during activation at temperatures below 950°C, but some hydrogen is still held after activation and is not released unless much higher temperatures are reached.[97] It is to be noted that the evolution of this latter portion of hydrogen at very high temperatures is paralleled by a simultaneous decrease in adsorptive power.

REFERENCES

1. Deitz, V. R., *Bibliography of Solid Adsorbents,* United States Cane Sugar Refineries and Bone Char Manufacturers and the National Bureau of Standards, Washington. D.C., 1944. Supplementary volume published in 1956.
2. Courouleau, P. H., and Benson. R. E., *Chem Eng.,* 55 No 3:112 (1948).
3. Krezil, F., *Kolloid Z.,* 59:109 (1932); 64:99 (1933); 68:381 (1934); 84:122 (1938); 87:327 (1939).
4. Kausch, O., "Die active Kohle". W. Knapp Halle 1928. Supplementary volume 1932; *Brennstoff. u. Warmewirts.,* 8:35 (1926); *Chem. App.,* 11:173 (1924),
5. Bancroft, W. D., *J. Phys. Chem.,* 24:127, 201, 242 (1920).
6. Patents not otherwise classified:
    Austrian Patents 134,282, 137,001, 158,209.
    Belgian Patents 358,993, 385,944, 393,382, 444,407, 445,981.
    British Patents 2037 (1855), 1287 (1857), 3164 (1857), 1355 (1858), 2578 (1858), 1522 (1859), 2104 (1860), 3033 (1860), 1409 (1865), 683 (1866), 2031 (1868), 45 (1869), 1103 (1869), 58 (1870), 524 (1871), 1406 (1871), 2569 (1872), 168 (1873), 967 (1879), 5286 (1879), 9569 (1886), 6687 (1888), 3978 (1891), 8112 (1894), 162,117, 116,253, 117,828, 165,788, 176,476, 189,148, 197,971, 198,328, 201,163, 202,639, 206,862, 208,555, 211,886, 213,935, 215,327, 216,761, 217,566, 218,242, 224,521, 225,160, 228,582, 228,812, 231,466, 233,840, 238,889, 239,744, 246,130, 246,954, 247,560, 255,816, 260,567, 262,278, 265,916, 266,673, 267,240, 269,477, 273,683, 273,761, 279,104, 283,267, 283,485, 283,573, 287,982, 288,148, 289,170, 291,725, 292,213, 301,330, 302,774, 306,490, 309,855, 310,908, 315,810, 316,870, 317,047, 322,160, 322,185, 329,630, 341,861, 357,812, 358,110, 358,940, 359,546, 365,685, 422,521, 425,611, 425,891, 446,889, 446,892, 465,148, 474,237, 484,197, 499,956, 509,859, 527,213, 576,044, 577,792, 591,060, 599,949.
    Canadian Patents 186,545, 242,850, 257,964, 264,535, 266,507, 269,466, 272,729, 275,980, 278,517, 285,459, 366,287.
    Dutch Patents 13,161, 21,200, 27,031.

French Patents 357,432, 423,070, 475,184, 483,104, 634,415,
635,781, 635,832, 642,360, 645,602, 646,481, 650,256,
655,092, 657,545, 658,638, 666,246, 666,472, 666,655,
666,744, 668,677, 669,863, 673,387, 673,426, 674,545,
679,370, 683,299, 690,684, 695,212, 698,752, 700,276,
700,511, 715,364, 716,939, 742,153, 764,289, 765,716,
778,922, 778,929, 782,165, 782,829, 789,776, 792,814,
800,298, 805,997, 809,922, 811,051, 821,375, 826,469.
German Patents 1,774, 6,446, 22,948, 24,341, 53,380, 55,922,
248,571, 267,443, 255,348, 305,895, 463,772, 469,277,
482,175, 482,412, 483,061, 485,769, 485,825, 486,076,
486,078, 487,026, 488,247, 488,526, 488,572, 489,278,
491,971, 499,791, 500,582, 500,981, 502,040, 504,636,
506,552, 507,524, 510,065, 512,484, 512,798, 516,881,
517,966, 518,514, 524,613, 524,614, 525,648, 528,239,
528,505, 533,936, 534,191, 543,674, 546,560, 547,516,
547,639, 553,235, 558,295, 558,748, 568,127, 570,590,
571,947, 571,969, 583,053, 583,206, 589,255, 619,137,
645,222, 654,906, 684,774, 704,292, 711,523, 712,503,
715,874, 742,499.

Japanese Patents 31,604, 34,755, 40,301, 42,343, 42,382,
42,510, 91,748, 94,257, 95,502, 100,771, 128,514, 163,030.

Norwegian Patent 28,579.

Swiss Patents 75,240, 75,971.

Swedish Patents 59,741, 89,568.

U.S. Patents 1,087,486, 1,137,852, 1,151,553, 1,195,720,
1,249,041, 1,262,770, 1,290,002, 1,447,452, 1,452,166,
1,474,120, 1,479,851, 1,505,496, 1,510,131, 1,518,072,
1,518,289, 1,525,770, 1,530,536, 1,535,797, 1,558,137,
1,565,129, 1,575,561, 1,592,599, 1,614,913, 1,623,598,
1,638,070, 1,639,356, 1,641,281, 1,663,000, 1,699,243,
1,712,930, 1,713,347, 1,743,975, 1,763,102, 1,774,341,
1,849,503, 1,856,302, 1,882,916, 1,927,244, 1,989,107,
2,037,257, 2,040,931, 2,148,827, 2,166,225, 2,245,579,
2,280,611, 2,354,713, 2,359,910, 2,352,932, 2,362,463,
2,365,729, 2,377,063, 2,378,246, 2,403,140, 2,405,206,
2,405,206, 2,441,125, 2,448,051, 2,464,902.

7. References in *Chemical Abstracts* to literature since 1950 on
activation:

Vol. 44: 1680e, 2208i, 2734d, 4658b, 5567d, 6606e, 8096b,

       9138i, (1950).

   45: 1274d, 4912i, 5913b, 5913d, 6368g (1951).

   46: 8351f, 7281h, 6816a, 3781d (1952).

   47: 2960c, 4586b, 7763b, 8998a, 10828c (1953).

   48: 350h, 3666f, 6101c, 10310f, 13202f (1954).

   49: 1310a, 4268e, 4268f, 6510g, 6581b, 9194i, 14298e, 16403d, 16404a (1955).

   50: 1294a, 3742b, 5251a, 5276i, 6773e, 8177g, 8180, 8980d, 9720b, 9722h, 13405f, 17392e (1956).

   51: 12474d, 12474f, 14237e (1957).

   52: 2384h, 5796e. 6768a, 6768c, 7809b, 9574b, 19095b (1958).

   53: 699a, 4791e, 5671h, 9651a, 22868c (1959).

8. Chaney, N. K., *Trans. Electrochem. Soc.*, 36:91 (1919). U.S. Patents 1,497,543; 1,497,544; 1,499,908.

Chaney, N. K., Ray, A. B., and St. John, A., *Trans. Am. Inst. Chem. Eng.*, 15 i:309 (1923).

9. Alekseev, V. N., *Colloid J.* (U.S.S.R.), 3:667 (1937).

Philip, J. C., and Jarman, J., *J. Phys. Chem.*, 28:346 (1924).

10. Morgan, J. J., and Fink, C. E., *Ind. Eng. Chem.*, 38:219 (1946).

11. McBain, J. W., and Sessions, R. F., *J. Phys. Chem.*, 40:603 (1936).

12. Hassler, J. W., Proceedings of the Anthracite Conference, Pennsylvania State University, Oct. 18, 1956.

Hassler, J. W., Westley, Oren A., and Fillicky, Joseph J., U.S. Patent 2,894,914.

13. Dacey, J. R., and Cadenhead, D. A., *Proceedings of the Fourth Conference on Carbon*, Pergamon Press, New York, 19 60, p. 9.

14. British Patents 14,224 (1900); 125,230. Dutch Patent 23,669, German Patent 275,973.

15. Austrian Patent 3,718. Belgian Patents 396,962; 443,709. British Patents 10,126 (1914), 291,043, 437,400. French Patents 471,295, 766,091, 776,328. U.S. Patents 1,951,538, 2,025,367.

16. Broderick, S. J., and Hertzog, E. S., *Report U.S. Bur. of Mines 3548* (1941).

Mukerjee, S., and Bhattacharya, S., *J. Am. Chem. Soc.*, 71:1725 (1949).

Turncock, L. C., and Lowdermilk, F. R., *Combined Intel-*

ligence Objectives Sub-Committee: Item 22, File 29–14, 78 p. (1945).
17. Mukherjee, S., and Bhattacharya, S., *J. Sci. and Ind. Research* (India), 4:235 (1945).
18. U.S. Patent 2,146,024.
19. U.S. Patents 1,517,543, 1,541,099, 1,573,509, 1,591,235, 1,614,707, 1,751,612, 1,753,507, 1,753,984, 1,778,747, 1,810,871, 1,923,918, 1,933,579.
20. Alekseevskii, E. V., and Alekseev, V. N., *Colloid J.* (U.S.S.R.) 2:333 (1936).
21. U.S. Patents 2,003,278, 2,026,355.
22. Kozakevich, P. P., and Izmailov, N. A., *Kolloid Z.*, 48:241 (1929); 57:294 (1931).
23. French Patents 442,476, 649,043. German Patent 488,779. U.S. Patents 1,617,533, 1,709,611, 1,778,343, 1,819,314.
24. British Patents 106,089, 166,202, 167,195, 173,624, 230,293. 242,659, 257,766, 463,226, 487,819, 489,360, 496,942, 583,113.
    French Patents 459,828, 746,543, 784,079, 784,115, 784,139.
    German Patents 136,792, 467,928, 486,077, 488,418, 488,944, 490,537, 491,403, 506,424, 517,316, 535,064, 665,509, 681,153, 699,551, 708,888.
    U.S. Patents 739,104, 1,362,064, 1,368,987, 1,413,146, 1,440,194, 1,440,195, 1,502,592, 1,505,517, 1,517,523, 1,563,295, 1,582,718, 1,759,138, 1,815,525, 1,826,209, 1,843,941, 1,848,946, 1,859,450, 1,913,340, 1,920,172, 1,927,459, 1,939,678, 2,205,011, 2,257,907, 2,258,818, 2,296,438.
25. U.S. Patents 1,634,477, 1,634,478, 1,634,480, 1,686,100, 1,854,387, 1,863,361.
26. U.S. Patents 1,593,879, 1,597,208, 1,601,222, 1,701,272, 1,731,473, 2,003,278, 2,270,245.
27. British Patents 228,954, 239,694, 243,801.
    U.S. Patent 1,641,053.
28. British Patents 1395 (1856), 8738 (1888), 10,622 (1915), 133,759, 122,465.
    U.S. Patents 1,250,228, 1,286,187, 1,287,592, 1,359,094, 2,180,735.
29. British Patent 213,195.
    French Patent 641,908.

German Patents 567,604, 737,332.
U.S. Patents 1,547,037, 1,383,755.

30. U.S. Patent 1,438,113.

31. British Patents 249,138, 254,262, 257,269
    U.S. Patents 1,610,399, 1,659,931, 1,689,647, 1,709,503,
    1,735,096, 1,755,156.

32. British Patents 835 (1868), 2661 (1877), 121,035.
    German Patents 488,669, 517,428, 710,800.
    U.S. Patents 1,621,195, 1,694,040, 1,819,165, 1,875,795,
    1,921,297.

33. Haber, F., and Bruner, L., Z. *Elektrochem.*, 10:697 (1904).

34. British Patents 2887 (1862), 179,108, 292,039.
    German Patents 44063, 267,346, 477,372, 478,945.
    U.S. Patents 1,588,868, 1,704,765.

35. French Patent 778,854.
    U.S. Patents 1,575,703, 1,586,106.

36. German Patents 489,633, 719,168.
    U.S. Patent 1,821,117.

37. U.S. Patents 1,505,496, 1,845,815, 2,377,063.

38. Boyk, S., and Hass, H. B., *Ind. Eng. Chem.*, 38:745 (1946).

39. German Patent 440,769.

40. U.S. Patents 1,438,113, 2,437,174.

41. British Patents 4421 (1874), 18,040 (1900), 122,698, 124,638,
    231,935, 246,954.
    U.S. Patents 1,402,007, 1,510,284, 1,520,801.

42. British Patent 277,714.
    German Patent 490,399.

43. Belgian Patent 444,407.
    U.S. Patent 1,551,074.

44. U.S. Patent 2,036,380.

45. U.S. Patents 1,286,187, 2,091,696.

46. British Patent 541,403.
    U.S. Patent 1,559,054.

47. Firth, J. B., Farmer, W., and Higson, J. *J. Chem. Soc.*,
    125:488 (1924).

48. Japanese Patents 41,969, 101,277.
    U.S. Patents 1,777,943, 2,234,769.

49. Hirschkind, W., and Lowdermilk, F. R., *Combined Intel-
    ligence Objectives Sub-Committee: Item* 22, *File* XXVII 83,
    14 (1945).

50. British Patents 197,971, 213,252, 274,538, 277,987.
    Japanese Patents 37,827, 40,529, 110,697
    U.S. Patents 1,521,541, 1,839,735.
51. French Patent 639,242.
52. German Patent 433,524.
    U.S. Patents 1,774,585, 2,312,707.
53. U.S. Patents 1,385,081, 1,423,231.
54. German Patents 570,028, 725,166.
    U.S. Patents 1,789,507, 2,162,763, 2,358,359.
55. Krczil, F., *Kolloid Z.*, 67:253 (1934).
56. French Patent 802,765.
57. Stratton, G. W., and Winkler, D. E., *Ind. Eng. Chem.*, 34: 603 (1942).
58. Simons, J. H., and McArthur, R. E., *Ind. Eng. Chem.*, 39:364 (1947).
59. British Patent 292,213.
    German Patent 485,824.
60. U.S. Patents 2,076,645, 2,076,646, 2,111,436.
61. British Patents 166,229, 225,891, 285,386.
    U.S. Patents 1,778,466, 1,856,571.
62. Other purification procedures: French Patent 672,731.
    German Patents 482,173, 486,109.
    U.S. Patents 1,502,896, 1,537,286.
63. British Patents 166,202, 167,740, 269,961, 303,669, 434,461, 445,342, 446,385, 576,044, 591,060.
    French Patents 666,647, 777,364, 785,427, 829,627.
    German Patents 482,348, 546,805, 568,876, 574,740, 635,279, 712,081, 737,803.
64. U.S. Patents 1,478,985, 1,530,392, 1,530,393, 1,736,051, 1,885,141, 1,899,810, 1,918,467, 1,966,553, 1,968,846, 1,968,847, 2,008,144, 2,055,755, 2,083,303, 2,304,351, 2,339,742, 2,362,463, 2,354,713, 2,402,304, 2,407,268.
65. British Patents 196,002, 228,512, 499,992.
    U.S. Patents 1,251,546, 1,314,204, 1,448,846, 1,498,708, 1,528,370.
66. British Patents 247,241, 440,411.
    French Patents 635,916, 822,338, 827,523.
    German Patents 466,358, 523,668.
    U.S. Patent 1,876,435.
67. British Patent 3541 (1877), U.S. Patent 1,899,339.

68. British Patent 530,809.
69. Keppeler, G., and Radbruch, G., *Int. Sugar. J.*, 43:317 (1941).
70. British Patents 453,627, 498,201.
    U.S. Patents 1,251,546, 2,067,985, 2,216,756, 2,216,757.
71. Hassler, J. W., and McMinn, W. E., *Ind. Eng. Chem.*, 37:645 (1945).
72. Blayden, H. E., Riley, H. L., and Taylor, A., *J. Am. Chem. Soc.*, 62:180 (1940).
    Riley, H. L., *Quarterly Review Chemical Society*, 1:59 (1947).
73. Riley, H. L., *Trans. Faraday Soc.*, 34:1011 (1938).
    Balfour, A. E., Blayden, H. E., Carter, A. H., and Riley, H. L., *J. Soc. Chem. Ind.*, 57:1T (1938).
74. Biscoe, J., and Warren, B. E., *J. Applied Physics*, 13:364 (1942).
75. Berl, E., *Trans. Faraday Soc.*, 34:1040 (1938).
76. Hofmann, U., *Wien. Chem. Ztg.*, 46:97 (1943).
77. Bhatnagar, S. S., Kapur, P. L., and Luthra, R. K., *Kolloid Z.*, 80:265 (1937).
78. Thiele, H., *Trans. Faraday Soc.* 34:1033 (1938).
79. Ishikawa, H., *Elec. Rev.* (Japan), 19:419, 493, 690, 726, 824 (1931).
80. Scott, G. S., *Ind. Eng. Chem.*, 33:1279 (1941).
    Strickland-Constable, R. F., *Fuel in Science and Practice*, 19 No. 4:89.
81. McKee, R. H., and Horton, P. M., *Chem. and Met. Eng.*, 32:13 (1925).
82. British Patent 133,759.
83. McBain, J. W., and Alexander, J., in *Colloid Chemistry*, Vol. 3 (edited by J. Alexander), Chemical Catalog Company, New York 1931, p. 14 (by permission of the copyright owners).
84. Blayden, H. E., Gibson, J., Riley, H. L., and Taylor, A., *Fuel in Science and Practice*, 19:24 (1940).
85. Herbst, H., *Kolloid-Beihefte*, 42:184 (1935).
86. Day J. E., and Robey, R. F., *Ind. Eng. Chem.*, 28:564 (1936).
    Lambert, J. D., *Trans. Faraday Soc.*, 34:1080 (1938).
87. Roychoudhury, S., Nandi, S. K., and Banerjee, J. K., *J. Indian Chem. Soc.*, 13:410 (1936).
88. Day, Jesse E., *Ind. Eng. Chem.*, 28:234 (1934).

89. Emmett, P. H., *Chem. Revs.* 43:69 (1948).
90. Oswald, M., *Chimie et industrie, Special No.* 251–66 (March 1931).
91. Wibaut, J. P., Z., *anorg. allgem. Chem.*, 211:398 (1933); *Proc. Acad. Sci. Amsterdam*, 24:92 (1921).
92. Ley, P. H., van der, and Wibaut, J. P., *Rec. trav. chim.*, 51:1143 (1932).
93. Miller, E. J., *Mich. Agr. Exp. Sta. Techn. Bull. 73* (1925).
94. Zerban, F. W., and Freeland, E. C., *Ind. Eng. Chem.*, 10:812 (1918).
95. Honig, P., *Kolloidchem. Beihefte*, 22:345 (1926).
96. Anderson, R. B., and Emmett, P. H., *J. Phys. Colloid Chem.* 51:1308 (1947).
    Mixter, W. G., *Amer. J. Sci.*, 45: (3), 369 (1893).
97. Lowry, H. H., *J. Phys. Chem.* 33:1332 (1929); 34:63 (1930).
98. Holmes, J. H., and Emmett, P. H., *J. Phys. Colloid Chem.* 51:1276 (1947).
99. Yamada, D., *Kogyo Kagaku Zasshi* 62, 161–3 (1959).
100. Eguchi, Y., Takeda Chemical Industries Ltd. Osaka, Japan communication.

12

Regeneration

1. INTRODUCTION

Processes employed to restore the adsorptive capacity of spent carbon are known by various terms: reactivation, regeneration, revivification. Of these regeneration is in frequent use.

2. DESORPTION OF VOLATILE ADSORBATES

Regeneration of carbon has been conducted in varied ways. Carbons used for vapor phase adsorption are regenerated by passing low pressure steam through the carbon bed to evaporate the adsorbed solvent and convey it to the exit where the steam is condensed and the solvent is recovered in a liquid state. In most cases the adsorptive capacity of the carbon is thereby restored.

The principle has been embodied in efforts to regenerate carbon used to remove volatile substances from liquid systems, *e.g.*, the recovery of phenol from coke oven waste liquors. It was found impractical to use steam to extract the phenol. One difficulty is that a large quantity of absorbed water must be evaporated before the carbon can reach a temperature at which phenol will be desorbed. Then too, the phenol concentration in the recovered aqueous solution is very dilute unless superheated steam is used, and this introduces other difficulties. The high temperature involved in the use of super-heated steam causes chemical changes in adsorbed phenol and in other substances adsorbed from the waste liquors. These changes lower the phenol recovery and also render the carbon unfit for reuse.

3.   DESORPTION IN LIQUID PHASE
APPLICATIONS

Desorption, the extraction of the adsorbate with a solvent, can restore much adsorptive capacity in some applications. Unfortunately desorption seldom restores the full capacity. Generally a variety of ingredients are adsorbed from an industrial solution, and because an effort is usually made to obtain selective extraction of the desired substance, it follows the other substances remain as residuals on the carbon. Apart from this, one must consider that desorption is often accomplished by replacement with a strongly adsorbable substance that remains on the carbon. However, it is to be mentioned that instances are known in which better recoveries are obtained by adsorption-desorption when a carbon is reused. A possible explanation is that the initial adsorption saturates areas of the surface on which irreversible adsorption occurs, and therefore the reuse involves only those areas from which complete reversible adsorption occurs.

4.   THERMAL REGENERATION

In general, regeneration from liquid phase applications is accomplished by thermal means, and for many years this method was employed by those using powdered carbon for sugar refining. Initially, it was often combined with desorption. Much of the adsorbed organic substances were initially removed by washing with dilute NaOH solution, after which the carbon was washed with water and finally neutralized with dilute HCl. An alternative method was fermentation of the excess adsorbed organic substances. Thermal treatment then followed in either case.

Later those preliminary steps were discontinued and the carbon was directly subjected to oxidation, either with air at 300°–600°C or with steam and/or CO_2 at 800°–900°C. The carbon recovery was satisfactory and the adsorptive capacity was restored.

However, the type of equipment employed caused further pulverizing of the powdered carbon, all of which increased the difficulty of filtering subsequent batches of syrup. Somewhat later, inorganic filter-aids were added and created further problems because the filter aids were impaired and damaged by the thermal regeneration operation. The net result was the gradual abandonment of the regeneration of powdered carbon.

The development of broad spectrum types of granular carbon useful for sugar refining and other industrial liquors made it desirable to develop methods of regeneration to offset the higher initial cost per pound of granular carbons. Two successful general methods have developed: the rotary kiln, and the multiple hearth furnace.

In the rotating kiln, the carbon moves countercurrently to a mixture of combustion gases and superheated steam. The recovery of carbon is reported to be over 90 to 95% and to have an adsorptive capacity approximately equal to new carbon.

The multiple hearth furnace as designed by Herreschoff is heated internally by gas burners—usually more than one to each hearth. Supplementary heat is supplied by admitting sufficient air to burn the carbon monoxide and hydrogen produced by the regeneration reaction.

A vertical shaft carries arms on each hearth equipped with rabble blades to agitate the carbon and continuously bring fresh granules to the surface. The rabble blades also move the carbon in a spiral path across each hearth to drop-holes to the hearth below.

The wet carbon enters at the top hearth. Evaporation of moisture and carbonization of much of the adsorbed organics is accomplished on the upper hearths. The activation reaction takes place on the lower hearths.

It is important to select a proper size regeneration unit. If too small, it will not be possible to regenerate the full carbon requirements, and if the unit is too large—it cannot operate continuously. Consequently carbon remaining on the hearths during shut-down periods will be completely oxidized and wasted.

Recoveries of 90–95% are reported, and the regenerated carbon has similar adsorptive capacity to that of new carbon.

Some report that the regeneration efficiency is greater when the spent carbon is not overloaded with adsorbed organic impurities.

It is possible that more effective regeneration could be accomplished if the drying and carbonization were conducted in units separate from those in which the activation reaction occurs.

5. RECENT DEVELOPMENTS IN REGENERATION POWDERED CARBON

In recent years much research has been devoted to studies of regeneration of powdered forms of activated carbon. Most studies are still in early stages of development and it would be premature at this time to attempt to evaluate them. The various patents and published reports are listed in Chapter 20, section 12 to enable the reader to contact the original sources for the latest detailed information.

We will, however, briefly outline two approaches.

In one method, the spent carbon in the form of an aqueous slurry, is subjected to the action of oxidizing gases at high temperatures and elevated pressures. Under such conditions, oxidation of adsorbed organic substances restores the adsorptive capacity of the carbon.

In another approach, the spent carbon is dewatered to approximately 50% moisture, and a high velocity low pressure air stream carries the moist carbon in a fluidized form into a patented regenerator[1]. Within the regenerator, the spent carbon is exposed to an oxidizing atmosphere at a controlled high temperature. The gas stream on leaving the regenerator is immediately quenched to a low temperature and no further reaction occurs, and the regenerated carbon is separated.

In contrast to earlier regeneration methods of powdered carbon the carbon particles are not pulverized. Regeneration recoveries range from 85-90%, and the adsorptive capacity is restored.

The method is now being operated on an industrial scale. A unit regenerating 10 tons per day has been in continuous operation for a year.

6. FINISHING TREATMENT OF REGENERATED CARBON

When the liquids being purified contain appreciable concentrations of inorganic substances, it may be necessary to remove any inorganics retained by the carbon. This can be done by washing with water and/or acid. When the inorganics cannot be so removed, the usefulness of regeneration is thereby limited.

REFERENCE

1. Personal communication P. Wiley, A. Hyndshaw of Westvaco.
    Corporation.
    See Chapter 20, section 12, for further references.

13

Nature of Activated Carbon

1. INTRODUCTION

In this chapter we shall contemplate theories that have been offered to interpret varied aspects of adsorption on carbon.[1,2] For this, it is well to comment briefly on the role of a theory as we envision it. Theories are to be viewed, not as blue-prints to portray the inner working of a phenomenon, but rather as mental inventions that enable us to make more effective use of the material world.

Most theories have their inception on meeting a new and unfamiliar phenomenon. We become curious and seek to explain the phenomenon to ourselves; that is, we theorize. When these theoretical explanations are examined we find that they usually explain new phenomenon with words that are symbols of some earlier and more familiar experience. Having done that, we feel we have gained a better understanding of the new phenomenon.

But what is the nature of that understanding? Making a phenomenon more familiar does not enable us to see the inner working because we do not know the inner working of any phenomenon whether entirely new, or as old and familiar as gravity. This does not mean that it is idle day dreaming to give a new experience a familiar aspect. Even though we do not know the inner working of familiar phenomena. we have learned how to live with them and often how to use them effectively. Hence a theory built on lessons learned from past experiences with familiar phenomena can furnish clues to search out ways and means to utilize new phenomena. Eventually a new phenomenon becomes familiar in its own right and will be used to explain some still newer experience. In the process of tying new phenomena with old, knowledge is integrated into a more compact form.

As is evident, the foregoing is but another way of saying that much of our theorizing is based on analogies in which we assume that when different phenomenon are similar in some ways, they may be similar in other ways.* In this we are on safe ground as long as we keep in mind that deductions from theories are but clues to be verified by working experience.

We must also recognize that the particular past experience brought to mind by a new phenomenon will depend on the history of the observer. Different individuals will recollect different prototypes of a new experience, and this can result in a number of theories being offered to explain an identical phenomenon. Although this could be confusing, actually such multiplicity of interpretation enlarges our field of view and adds avenues on which to approach a working know-how. And it saves us from becoming prisoners of a single fixed idea.

2. SURFACE TENSION

Many early studies explored the role of surface tension in adsorption from solution.[2] As is well known, a drop of liquid tends to become spherical, this being the shape having the smallest area of surface for a given volume; the behavior is as though the liquid were enclosed by an elastic skin. The analogy is quite descriptive but should not be taken literally because surface tension is fundamentally different from the tension exerted by an elastic container. Stretching an elastic membrane makes it more taut, but spreading a liquid over a greater area does not increase the surface tension.

Basically, surface tension is a symptom of the cohesive forces that hold molecules of a liquid together. Molecules in the interior of a drop of liquid tend to stay there because of being held by mutual attraction to the surrounding molecules; whereas molecules at the surface are only partially surrounded and consequently are subjected to an unbalanced inward pull. Because of the inward pull

*This concept is implicit in the use of language to describe any new phenomenon. This is obvious in the case of direct analogies, but even when abstract phrases are used they convey meaning only to the extent that they recall to the reader or listener some item of previous experience.

any surplus molecules are drawn to the interior, and this process continues until the surface has become as small as possible, subject to external forces such as gravity and the shape of the container.

The phenomenon of surface tension is manifest in other ways: If two small drops of unequal size are in a closed container they will merge into a single drop of which the total surface is less than that of the two original smaller drops. Upon direct contact the smaller drops will flow together; but even if they do not touch, the merger will take place by transfer through the vapor phase. To understand the mechanism of the merger via distillation, consider the clusters of molecules forming drops *A* and *B* in Figure 13:1.

A   B

Fig. 13:1   Clusters of molecules in large and small drops.

Within small clusters of molecules, the mutual molecular attractions are additive, hence the greater number of molecules in drop *A* will be held together more firmly than the fewer molecules in drop *B*. Consequently, in a closed system, molecules that evaporate from drop *B* will be captured by the stronger forces emanating from drop *A*; The transfer will continue until all of drop *B* has moved to drop *A*.*

A similar manifestation of cohesive forces associated with surface tension is observed in crystallization. When both large and small crystals are present in a saturated solution, surface molecules will be held less firmly by the smaller crystals and so they will escape to be seized by the stronger forces of the larger crystals. This familiar phenomenon is usually expressed by saying that small

*The vapor pressure of a convex surface increases with curvature; as smaller drops have greater curvature, they have correspondingly greater vapor pressure and a greater escaping tendency.

crystals are more soluble than larger crystals.

Still other behavior can result when more than one ingredient is present in the system. In pure liquids, the cohesive forces influence only the size of the surface; but when several components with unequal surface tensions are present, the cohesive forces that cause surface tension can produce changes in the composition of the surface.

Consider a dilute aqueous solution of ethanol: Because of surface tension, the molecules of water, with their greater cohesive force, are pulled into the interior of the liquid; and the ethanol molecules, having weaker cohesion, are left at the surface. Or as it is sometimes expressed: The strong mutual attraction between molecules of water can squeeze to the surface any molecules having less attractive force.

Many of the earlier investigators thought that the foregoing phenomenon accounted for the purifying power of carbon. They believed that impurities with a low surface tension* formed a ready-made film at the surface of a liquid, and that the function of the carbon was to furnish a large surface area on which such a film could gather. Such a mechanism is an important factor but it does not suffice to explain all situations, because many substances that do not lower the surface tension of water are removed from an aqueous solution by carbon. We now recognize that adsorption by carbon can involve mutual affinities between the surface of the carbon and the substance to be adsorbed.

3. CAPILLARY CONDENSATION

Studies by Zigsmondy[3] in 1911 led him to believe that the ability of silica gel to take up water vapor is associated with the presence of small capillaries in the gel. As is well known, a liquid that *wets* the walls of a capillary will rise in that capillary; and the meniscus at the upper surface will be *concave* and have a vapor pressure *lower* than that of the bulk liquid.[1,2] From this, Zigsmondy deduced that in small capillaries a vapor would condense to a liquid at pressures below those required for normal condensation. The greater ease of such condensation depends on the diameter of

*Substances that lower the surface tension of an aqueous solution are often designated as *capillary active*.

the capillary, and as this becomes smaller condensation occurs at a lower pressure.[1,2]

According to this theory of capillary condensation, then, when capillaries of unequal diameters are present in a solid structure, the narrowest capillaries fill at the lowest pressures. With an increase in vapor pressure there is a stepwise filling of the larger and still larger capillaries.

Today it is recognized that capillary condensation is a factor in adsorption only when vapors are present in relatively high concentrations. Inasmuch as most industrial applications deal with low vapor concentrations, the theory of capillary condensation is seldom directly associated with the industrial use of carbon. However, the theory draws attention to other aspects of capillaries: they furnish much surface on which conventional adsorption can take place; and being of different sizes and shapes they provide a screening action that is a factor in selective adsorption.

As with surface tension, capillary condensation is associated with forces that bind like molecules in a liquid—forces commonly known as *van der Waals*. As has been long recognized, these attractive forces can also often cause molecules of different species to unite; *e.g.,* salt dissolved in water, chloroform adsorbed on a carbon surface.

Much thought has been given to whether an adsorption depends on fortuitous collisions of wandering molecules with a solid surface, or is caused by forces that can capture molecules from a distance in a manner analogous to gravitational and magnetic forces.

4. LANGMUIR THEORY

Langmuir's concept is so clearly expressed that it can be readily comprehended; and it has the elasticity to cover homogeneous and heterogeneous surfaces, chemical and physical forces, mono- and multi-molecular layers.

The surface of many adsorbents can be viewed as a patchwork assortment of different kinds of adsorptive spaces, with each kind of space having characteristic affinities. No far-reaching forces are envisioned, but when a wandering molecule[7] of vapor collides with a suitable unoccupied surface space, the molecule will adhere.

It does not adhere indefinitely, however; sooner or later, depending on the thermal energy and the strength of the attractive forces, the adsorbed molecule evaporates.*

At first when the surface is bare, the number of molecules that condense exceeds the number of those that evaporate. As the surface becomes covered, other gas molecules have greater difficulty in finding unoccupied spaces, but, there is also an increase in the number of molecules escaping from the surface. When the rate of evaporation equals the rate of condensation, equilibrium is reached. The amount adsorbed at equilibrium is a function of the following factors, some of which are interrelated:

1. The time lag, *i.e.*, the average period between the moment the molecule condenses and its subsequent evaporation.

2. The total area of the solid surface.

3. The proportion of the total surface which has specific attractive power for the molecules in the gas phase also the accessibility of such surface.

4. The number of adsorbable molecules in the gas state, *i.e.*, the pressure.

5. The number of layers of adsorbed molecules.

Considered individually, each factor provides a clear approach to an understanding of adsorption. Collectively, they can result in observed phenomena of the utmost complexity. No single mathematical equation other than a purely thermodynamic one can cover all cases, and much attention has been given to different limiting equations for certain types of adsorption.

The Langmuir theory regards most of the adsorption as occurring in a layer one molecule deep. although it is recognized that the molecules adsorbed in this layer may have their fields of force altered and thus be able to attract a second layer of molecules, which in turn could hold a third layer. This induced attraction would be weaker and, therefore, molecules in the second layer would have a shorter staying time. Consequently, fewer molecules will be present in the second layer than in the first. For a similar reason, the number of molecules in a third or higher layers would be still less.

---

*The same adsorption pattern applies to solutions.

5. POLANYI ADSORPTION
POTENTIAL THEORY

This theory of the nature of adsorptive forces, suggested by de Saussure in 1814, was later developed quantitatively by Polanyi[4] and others.[5,6] It is based on a concept that cohesion forces can reach out and attract molecules from distances somewhat greater than molecular diameters. The *adsorption potential**, *i.e.*, the intensity of the attraction, diminishes rapidly with distance to become zero within small finite spaces. The theory envisions that the captured molecules form a miniature atmosphere in which the outward density diminishes in a fashion analogous to the atmosphere on a planet. The theory, originally based on a concept of multilayers, does not exclude mono-layers.

When an adsorption isotherm for a system is known at one temperature, the theory provides a means of predicting isotherms for other temperatures.

In recent years, a number of fruitful studies have been made of various facets of the theory. Dubinin[66] has drawn attention to the action of the adsorption potential in pores of molecular diameters. Molecules that enter such pores are held by attractive forces radiating inward from all sides. This reinforced attraction results in *volume filling*. In larger pores, adsorbed molecules lose contact with some sides of the pores, and then are held less firmly as adsorbed layers.

Other studies of the Polanyi theory are reported in Chapter 20, section 13

6. PHYSICAL AND CHEMICAL
ADSORPTION

Much study has been given to exploring the relative roles of physical and chemical forces in adsorption phenomena.[1,2] Physical adsorption is caused by forces similar to those that cause vapor molecules to condense to a liquid; the adsorption of nitrogen on carbon at low temperatures is an example. In physical adsorption, the attachment is weak; adsorbed nitrogen will leave the carbon

---

*Polanyi later also employed the term *adsorption affinity*.

surface if a vacuum is applied or if the temperature is slightly elevated.

Although physical forces (van der Waals) suffice to explain many adsorption phenomena, there are adsorptions in which the behavior indicates that chemical action is involved. Consider the initial portions of oxygen adsorbed on carbon at temperatures above 0°C: Oxygen so adsorbed can be removed from carbon only by elevating the temperature, and then it comes off not as oxygen, but drags away carbon atoms to form carbon monoxide and dioxide. This indicates that the oxygen is held to the carbon by bonds stronger than those existing between the carbon atoms. Adsorption of this type is designated as *chemical adsorption or chemisorption.*

Physical adsorption usually involves a smaller energy change than does chemisorption. The adsorption of nitrogen on carbon evolves about 5000 calories per mole, this heat being somewhat greater than the heat of liquefaction, whereas the initial adsorption of oxygen on some carbons at 0°C liberates over 100,000 calories per mole—an amount that is greater than the heat of formation of carbon dioxide.

Chemisorption is specific and depends on the chemical nature of both the adsorbent and adsorbate. Although some selectivity is apparent in physical adsorption, usually this can be traced to purely physical properties; thus the adsorption of greater amounts of nitrogen than of hydrogen can be traced to the greater ease with which nitrogen can be condensed or liquefied.

Physical adsorption diminishes at higher temperatures, whereas some elevation of temperature is often necessary to initiate chemisorption. For instance, oxygen is physically adsorbed by carbon at −190°C, whereas chemisorption does not occur until higher temperatures are reached. Taylor,[8] who drew attention to this phenomenon, suggested that molecules may be activated by heat or other energy to participate in chemisorption—a phenomenon he termed *activated adsorption.* Although the concept has been questioned, it finds a parallel in many chemical reactions which are initiated by an external source of energy.

The physical adsorption of nitrogen and the chemisorption of oxygen are clear-cut examples of each type of behavior, but a number of adsorptions are not so easy to classify. Many properties show no sharp demarcation point to distinguish physical from chemical adsorption, and the classification of borderline cases is difficult. In many chemisorptions, the heat evolved is much less than with oxygen on carbon; then too, some chemisorbed substances can be extracted in their original form. There is evidence that both physical and chemical adsorption may occur simultaneously; thus, although the initial adsorption of oxygen at 0°C is chemical, subsequent portions of adsorbed oxygen are held by weaker forces.

At times, a classification as chemical or physical may be quite arbitrary, depending on which properties are measured. Thus, an adsorption on the basis of the criteria usually applied might be defined as physical, but if it should be found later that the substance in the adsorbed state acquires enhanced chemical activity in certain reactions, we could define it as chemisorption.

This dependence of the interpretation of a phenomenon on the particular properties that are measured is true of much scientific knowledge. Take for example the reaction of zinc with hydrochloric acid: If our measurements were limited to a chemical analysis, we would consider the reaction as due to a greater chemical affinity of chlorine for zinc than for hydrogen. However, if the reaction takes place in a primary cell, the resulting electromotive force shows that electrical forces are involved.

No doubt, the varied phenomena can be traced to the same ultimate force which proceeding via different paths becomes manifest in different forms—forms which often merge by imperceptible degrees. Therefore, difficulties of interpretation often arise from our efforts to classify phenomena into rigid mental compartments that do not have exact counterparts in nature. This is said not to belittle efforts to classify, because classification may be necessary for practical purposes. When we build a power plant, it does not suffice to know that the energy comes originally from the sun. We shall not be able to proceed with our plans unless we know whether the source to be used will be in the form of today's sunlight or that stored ages ago as coal.

That many difficulties in interpreting nature originate in our mental processes, becomes evident when we endeavor to weave older theories with newer ones. Consider the concepts of chemical and physical forces: Chemical forces deal with specific properties of the atoms forming the molecule, and the manner in which the atoms are combined with one another. Unlike truly physical properties such as gravitational attraction, chemical properties are specific. We cannot definitely predict how *A* will react with *B* from a knowledge of how *A* or *B* will separately react with *C*. A chemical reaction implies some structural rearrangement of the atoms within the molecules, whereas in physical changes, the molecules remain intact and participate as whole units. Therefore, as Langmuir has suggested, the distinction between chemical and physical adsorption hinges on the definition of the molecule. Until about 1915, the chemical formula defined the structure of the molecule—a molecule of NaCl being regarded as consisting of a single atom of sodium combined with a single atom of chlorine. This view is still valid for gases, but studies by Bragg[9] show that it does not apply to crystalline solids. We no longer consider a crystal to be an aggregation of individual molecules held together by an indefinite mutual cohesion. Instead, each atom is combined by chemical bonds with its neighboring atoms and these in turn are combined with those beyond, the entire crystal being regarded as a single molecule.

Langmuir has suggested that the chemical forces that provide the cohesion to hold atoms together in a solid crystal can also extend from the surface to hold adsorbed molecules. This view, as developed, classifies chemical adsorption into two groups based on Werner's concept of primary and secondary valences.[7] According to Werner, primary valence applies to simple compounds in which two different elements are united, e.g., CO_2, SO_3, NO_2, NaCl. These simple compounds can be united by residual or secondary valence to form compounds of a higher order: thus CO_2 and Na_2O yield Na_2CO_3, which in turn can form compounds of a still higher order—for instance, a hydrate, $Na_2CO_3 \cdot 10\ H_2O$. There is no sharp limit to the size of the aggregates that can be formed, but specificity exists as is shown by the fact that barium chloride will take up water of crystallization whereas sodium chloride will not.

This approach would regard an adsorption, such as of oxygen on carbon at temperatures above $0°C$ as due to primary valence forces; and by a somewhat liberal interpretation, secondary valence applies to adsorptions that are not clearly attributed to primary valence. The adsorption of a dye could be regarded as a union analogous to that between barium chloride and its water of crystallization, or between potassium cyanide and ferrous cyanide in potassium ferrocyanide.

Perhaps a closer parallel is to be found in the growth of crystals. When a molecule is in the interior of a crystal, the attractive forces are satisfied on all sides; whereas molecules at the surface have a free side that is able to attract and hold wandering molecules of solute. When a crystal is in a saturated solution, the attractive forces of the surface molecules take hold of the solute molecules and use them as building blocks to enlarge the crystal. If an appropriate dye is placed in such a system—some of the dye will be attracted to become bonded to the surface and color the face of the crystal.

Adsorption by Crystals

Studies with crystals indicate that the specificity, so often encountered in adsorption, can depend on mutual relations between the lattice structure of the solid and the configuration of the adsorbed substance. methylene blue is adsorbed by crystals of lead nitrate. but not by potassium sulfate. Nellensteyn[1] found that diamond powder adsorbs methylene blue but not succinic acid, whereas graphite adsorbs succinic acid but not methylene blue.

Numerous studies made by France[10] and his coworkers lead to the conclusion that the adsorption of a given dye by a growing crystal is a specific process, depending on the structure of the crystal as well as on the shape, presence, and position of polar groups in the dye molecule. Adsorption may occur when only one substituent group of the dye can fit into the ionic planes of a crystal face. France and Wolfe,[10] in a study of the adsorption of thirty isomeric monoazo dyes by crystals of sodium nitrate, sodium bromate, and sodium chlorate, found twenty-five of the dyes to be adsorbed by sodium bromate, six by sodium nitrate, and two by sodium chlorate. Although each of the dyes was adsorbed on at least one type of crystal, only one of the thirty dyes was adsorbed by all three crystals.

Rigterink and France,[10] using potassium sulfate crystals, found the basic structure of the dye influenced the adsorption. Dyes with the nucleus:

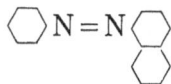

are more likely to be adsorbed than those having the nucleus:

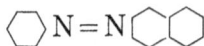

$$\bigcirc N = N \bigcirc$$

In the case of dyes having the first type of nucleus, Rigterink and France also found that those produced from β-naphtholsulfonic acids are more adsorbable than dyes produced from a-naphtholsulfonic acids. When an NH_2 group is substituted for an OH group, the situation is exactly reversed; the dyes formed from α-naphthylaminesulfonic acids are more adsorbable than those formed from β-naphthylaminesulfonic acids.

Although much light has been shed on adsorption by crystals through the extensive studies of France, Buckley, and others, we do not have sufficient knowledge as to just what conditions must be fulfilled to enable an impurity to fit into the crystal structure. As stated by France, "No simple rule can as yet be made whereby one can correctly predict just what foreign materials will be adsorbed by a given crystal."

The adsorption of dyes or other impurities can produce a definite disturbance of the crystal surface as is indicated by the fact that the crystal habit is often altered during further growth.

Surface Forces and Chemical Change

So far we have considered that chemical adsorption involves definite bonds between the adsorbent and the adsorbate. However, adsorption accompanied by a chemical change can occur as a result of purely surface forces from which any known specific bonding is absent.

Deutsch[11] found that aqueous solutions of indicator dyes acquire a different color when vigorously shaken with an immiscible liquid such as benzene. Thus, the color of bromothymol blue at pH 7.4 changes from blue to yellow when treated in this manner. The change in color takes place at the interface of the emulsified droplets, and the original color is restored when the liquids separate. In another study, it was observed that a colorless solution of rhodamine in benzene turns a deep red when vigorously shaken to form an emulsion with water. When the phases later separate, the color almost completely disappears. The phenomenon is not limited to liquid liquid ‖ interfaces, but also occurs at an air ‖ liquid interface; a brownish-yellow solution of thymolsulfonaphthalein (*p*H 2.8) when shaken with air develops a reddish-violet foam.

The reactions of indicators are far from simple and it is not easy to determine just what chemical reactions accompany these changes in color. Some changes are ascribed to the formation and adsorption of undissociated molecules at the interface, but this explanation is not adequate to cover all cases.

The behavior of these dyes is entirely reversible. The change in color does not persist after the phases separate, but can be restored an indefinite number of times by subsequent shaking. However, with some substances, the change is irreversible. Ramsden[12] found that egg albumin and similar substances can be rendered permanently insoluble by shaking their aqueous solution to produce a foam.

Surface reactions of a similar nature no doubt occur at solid surfaces, and here the effect is a function also of the shape of the rigid structure. That various chemical and physical phenomena arise from the spatial arrangement of atoms within molecules has long been accepted. It is perhaps less well recognized that visible properties of shape can modify the chemical and physical behavior of substances. Vapor pressures are higher on convex than on concave surfaces. Cassel[13] reports that soot deposited on a rough surface burns more readily than that deposited on a smooth surface. Copper mirrors deposited electrolytically are not active catalytically as are the rough surfaces produced by the thermal decomposition of organic copper salts. Crystalline structure influences the properties of explosives: Cullen[14] states that detonating mixtures must contain crystals of definite size and shape. Activated carbon, produced as it is from a variety of substances under all sorts of conditions, must have untold possibilities in the way of different forms of surface structure.

It is apparent that surface reactions involve phenomena of great complexity, and it is not surprising to find workers who, in the words of Bancroft, "accept adsorption and let it go at that."

7. SURFACE AREA

Until about 1940, efforts to examine many aspects of adsorption were handicapped by a lack of knowledge as to the total area of the surface of an adsorbent.

The porous nature of carbon excludes measurement by microscopical examination which would give only the external surface. Consequently, indirect methods are used and these include measurements based either on adsorption isotherms or on thermal effects. The various assumptions involved in using these indirect methods are not always recognized. In the adsorption method, the carbon is blanketed with a known quantity of adsorbed gas or solute—the probable covering power of which is then calculated.

The accuracy with which this can be expressed as square meters of surface depends on the exactness with which the number of adsorbed layers can be determined, and on the closeness with which one can approximate the average area covered by each adsorbed molecule. Beyond this, the definition of the area so measured presents some difficulty. Langmuir has pointed out that there must be spaces of all sizes and shapes on a charcoal surface. On a plane surface, a single adsorbed molecule may be attached to a single carbon atom; whereas, in crevices and pores the same size molecule may cover several carbon atoms. Much depends on the size and orientation of an adsorbed molecule. Thus, an aliphatic acid molecule that lies flat on the surface will cover more carbon atoms than when attached at one end. Therefore, it is to be anticipated that the measured surface will depend on the structure of the adsorbate.[1,2]

Determinations based on adsorption from solution are definitely untrustworthy because the adsorption of many solutes is quite specific. This method measures only the restricted area to which the molecules of solute can be attached, and the size of this area may depend on several conditions, e.g., pH. Then too, adsorption from solution determinations usually neglect the area occupied by adsorbed solvent molecules.

Much more confidence can be placed on values calculated from the adsorption of gases. Of the various methods that have been suggested, the one developed by Brunauer, Emmett, and Teller,[15,16] (BET) has won general approval. The basis of this method is to plot an adsorption isotherm under conditions that provide adsorption through purely physical forces; these conditions are met by adsorbing a gas such as nitrogen at low temperatures. The adsorption isotherm is plotted to yield a straight line in which the slope and the intercept give the amount of gas required to form a monolayer. Knowing the probable area occupied by each molecule, one can calculate the probable area of the adsorbent.

Bokhoven and van der Meijden[17] express a preference for the use of the benzene adsorption isotherm, obtained at 20° or 30°C, for the evaluation of their activated carbons. The use of H_2O serves the same purpose and the data obtained will also provide a basis for the calculation of the total surface (BET method) but they believe their benzene adsorption method more practical and far less time consuming.

Bartell[18] developed a method of calculating the surface area based on the heat of immersion which gives results in substantial agreement with many based on adsorption isotherms.

The total surface area does not necessarily mirror the utility of a carbon. Frequently for a given situation a type of carbon having relatively small area of surface will be more effective than another carbon with much greater total surface. In many applications. only part of the surface participates in the adsorption and the rest of the surface is not occupied by the substance being adsorbed.

8. PORE STRUCTURE

Many instances in which a carbon seems unable to adsorb the full amount that could be accommodated are traced to a screening action. A large part of the carbon surface is furnished by the walls of the pores, and such surface is accessible only to molecules that are small enough to enter.

TABLE   13:1

MINIMUM PORE DIAMETER FOR ADSORBATES*

Adsorbate	Minimum Pore, Diameter, Å
Iodine	10
Potassium permanganate	10
Methylene Blue	15
Erythrosine Red	19
Molasses	circa 28

* From *Basic Concepts of Adsorption*, by Richard J. Grant, June 1, 1961; reprinted with permission of the Pittsburgh Chemical Company.

Our knowledge of the internal structure of activated carbon is clouded and imperfect, but it appears to include a connecting network of irregularly shaped gaps and crevices in the carbon crystallite agglomerates. However, for the purpose of calculating sizes of the apertures, they are assumed to be cylindrical. The smaller size apertures—*micropores*—are measured by the manner in which the amount of an adsorbed gas changes during desorption. The classification into separate sizes, *pore size distribution,* is calculated from the data by application of the Kelvin equation. Pore size distributions thus calculated find confirmation in studies of the adsorption of different size molecules (Table 13:1). Micropores provide most of the surface area on which adsorption occurs.

Sizes of the larger apertures, *macropores,* are measured by the pressure required to fill them with mercury. Macropores provide passageways through which adsorbable molecules can more readily reach interior micropores.

Pore size distribution is tangible, and it is moreover one of the more controllable properties in the manufacture of activated carbon; in fact, some report an ability to tailor-make a desired pore size distribution. No one can gainsay that an understanding of this feature has contributed to the manufacturing art, and if pore diameter were the sole decisive factor in usefulness of carbon the task of the manufacturer would be greatly simplified.

There are many experiences with adsorption, however, that cannot be adequately explained on the basis of pore size distribution. Nekrasov[25] and Ginsberg[26] studied systems in which propionic acid was less adsorbed than either the smaller acetic or the larger valeric acid molecule. A similar difficulty applies to data by Linner and Gortner[27] covering the adsorption of aqueous solutions of dicarboxylic acids at high concentrations. They observed that acids with an odd number of carbon atoms were less adsorbable than the adjacent even members of the series. Griffin, Richardson, and Robertson[28] found a similar trend for solutions in alcohol. Still further, different substituent groups, even though of similar size, differ greatly in their influence on adsorbability.

In studies of the adsorption of dyes of similar molecules size, carbon *A* will selectively adsorb certain dyes, whereas other dyes will be preferentially adsorbed by carbon *B*.

Some such enigmas are interpreted as involving shape relations. As we know little of the shape and form of the pores in a given carbon, we are consequently faced with differentiating a screening action due to pore size, and the task of fitting odd-shaped molecules into pores of unknown shape. One such situation is presented by the data in Table 13:2 covering an experiment in which a char was air-activated for a total time of 45 minutes, samples being taken at regular intervals to determine changes in adsorptive power.[29]

TABLE   13:2

RELATION OF SURFACE AREA TO SPECIFIC
ADSORPTIVE POWER

Time of activation[a] minutes	Surface area sq M/g carbon	Adsorption[b]	
		Phenol	Aniline Blue
10	402	0.09	0.05
20	600	0.15	0.11
30	815	0.15	0.14
40	990	0.15	0.20
45	1065	0.15	0.28

[a] Air-activation of acid-washed black-ash at 550° C.
[b] Grams adsorbed per gram carbon at an equilibrium concentration in solution of 0.10 g/liter.

As the activation proceeded and the surface area increased. the carbon acquired increased adsorptive power for aniline blue. The rate of increase accelerated during the final stage of the activation. When, for example, the surface area reached 815 *sq m/g* carbon, a gram of carbon adsorbed 0.14g of aniline blue; but the adsorption doubled to 0.28g when the surface area reached 1065 *sq m/g*. In other words, the final 250 *sq m/g* account for as much adsorption of aniline blue as was accomplished by the entire 815 *sq m/g* of surface formed initially. This final spurt is explained by the fact that during the early period of activation, many pores are too small to admit the large molecules of aniline blue, but in the final stretch many pores become sufficiently enlarged to admit a greater number of these molecules.

The foregoing interpretation, however, gives no clue to explain the pattern of phenol adsorption as disclosed in the same experiment. During the initial activation, adsorption of phenol increased gradually to 0.15 *g/g* carbon when the surface area had reached 600 *sq m/g*; but beyond this point any further enlargement of the surface made no further increase in the amount of phenol adsorbed. From this we might conclude that all surface created above 600 sq m/g carbon had no adsorptive power for phenol.

It is possible that the Dubinin concept of volume filling (13:5) may provide an acceptable interpretation. It would be along the following lines: The initial activation stage enlarges the pores sufficiently to admit phenol molecules which are then firmly held by the attractive forces radiating inward from all sides. When the pores become wider with further activation, the phenol molecules lose contact with some sides of the pores, and the attachment becomes weaker. As a result some phenol molecules will be desorbed, and fewer molecules are held on each unit of surface. That loss offsets the gain from the increased surface area.

But there are many situations for which the pore concept does not offer a reasonable interpretation, and this leads to a consideration that diverse adsorptive qualities can exist on separate areas of the surface of carbon.

9. QUALITATIVE CHARACTERISTICS OF SURFACES

The hypothesis of separate surface areas with specific and selective adsorptive powers is based primarily on circumstantial evidence. Direct proof is not yet available. And so, to convey some perception of the hypothesis, let us picture the erosion of carbon during activation as an etching process in which simple patterns are etched initially on the surface. Then as the activation continues, the initial patterns are modulated into other patterns to provide progressive alterations in specific adsorptive powers and capacities.

That specific affinities can reside in separate areas is indicated in other studies, one of which is the way certain adsorptive powers can be selectively blacked-out.[31] An illustration is found in an experiment in which potassium permanganate was adsorbed on a carbon which was then soaked in a sulfite solution and finally washed with water to remove any residual permanganate and reaction products. After this treatment the carbon had less surface area but adsorbed almost as much methylene blue as the original carbon did (Figure 13:2). In contrast, the ability to adsorb ponceau R or molasses color was greatly diminished.

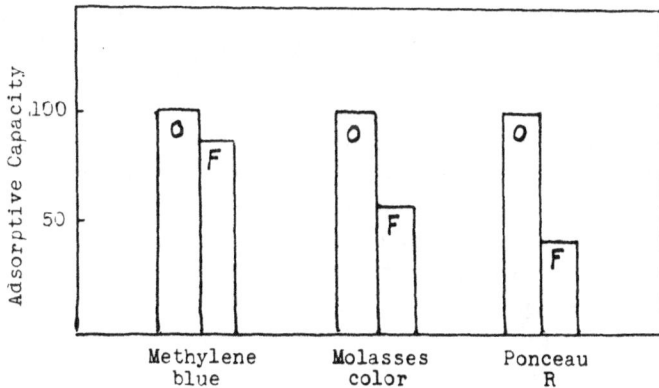

Fig. 13:2   Effect of pretreatment with potassium permanganate on specific
             adsorptive capacity
             O, original adsorptive capacity of Carbon *M*
             F, adsorptive capacity after treatment with permanganate

The concept that selective adsorption can be caused by qualitative characteristics of the surface is supported by studies of adsorption on plane surfaces: Lead nitrate crystals adsorb methylene blue on the 110 faces, whereas a different set of faces does not adsorb this dye but will adsorb picric acid. When placed in a solution containing both dyes, lead nitrate crystals will develop a blue color on one set of faces and a yellow color on others.[32]

Active Centers

It is of interest to interpret qualitative variations of activated carbons in the light of the concept of active centers, as developed by workers in the field of contact catalysis. This theory considers that every adsorptive power does not exist on all portions of a surface: instead, many specific powers are localized in separate subdivisions or patches called the *active centers*.* In these centers,

*The concept of selective adsorption on active centers is not in opposition to that of a screening action by pores. Each has its own sphere and both concepts contribute to an over-all understanding of selective adsorption.

the surface atoms are arranged in patterns having specific affinities. We can picture that each method of activation etches a characteristic finger print. The theory of active centers also furnishes a plausible explanation of the influence of prior history on subsequent stages of activation. Each stage, beginning with the initial carbonization, not only has an immediate effect but the pattern then set carries forward to modulate the action of subsequent stages of activation. The action is analogous to a chemical process that involves successive stages. Thus, in the production of organic derivatives of benzene much depends on whether the benzene is initially nitrated, sulfonated, or subjected to some other action.

We must not overlook that much of the attractiveness of the active-center hypothesis is in the convenience of describing heterogeneous adsorptive behavior in terms of space relations. This must not lead us into thinking that the concept can supply a detailed map of the surface of a carbon. Rather, we should view the active-center hypothesis as a mental invention which enables us to coordinate aspects of behavior that would otherwise be difficult to gather into an organized pattern. With this approach, let us examine characteristics of active centers.

Evidence that different types of active centers coexist on any one carbon surface is provided by various studies.

Different active centers can be alike in some respects and yet be dissimilar in others. Two types of surface which adsorb similar amounts of a given dye under one set of experimental conditions may exhibit divergent adsorptive powers when the conditions are altered. This phenomenon is frequently found when comparing the adsorptive behavior of different carbons: Thus, carbons B and E gave similar adsorption of molasses color at 25°C, but elevating the temperature to 90°C increased the adsorption by carbon B and left E unchanged. Again, carbons A and H adsorbed equal quantities of malachite green from an aqueous solution; whereas from alcohol, carbon H adsorbed only half the dye adsorbed by A (Table 13:3). Evidently, the active centers which adsorb malachite green on carbon H have certain properties unlike those on carbon A. Catalytic activity may be associated with adsorptive power on some carbons and not on others; Zylbertal,[33] in examining two carbons that adsorbed uric acid, found that only one of them was able to catalyze the oxidation.

Even on any one carbon, there may be different species of centers capable of adsorbing a given substance. Chaplin[34] found that

TABLE 13:3

INFLUENCE OF SOLVENT ON SPECIFIC
ADSORPTIVE POWER*

Carbon Code	Methylene Blue[a]		Malachite Green[a]	
	In water	In alcohol	In water	In alcohol
A	0.84	0.26	1.07	0.113
H	0.73	0.07	1.10	0.050

[a] Millimoles adsorbed per gram carbon at an equilibrium concentration of 0.1 millimoles per liter.
* Data from *Ind. Eng. Chem.* 37:645 (1945); reprinted by permission of the American Chemical Society.

adsorbed CO_2 could be partly displaced from the carbon by HCN, but complete displacement could not be effected regardless of how much HCN was added. A similar partial displacement of the CO_2 occurred when CCl_4 was added. When both CCl_4 and HCN were added, however, the CO_2 was completely displaced and this was true regardless of the order in which the HCN and CCl_2 were added. From this, we may deduce that CO_2 can be adsorbed by two different types of centers, one of which is also able to adsorb CCl_4, and the other HCN.

This brings out that, although each active center exhibits selectivity as to the kind of substances that can be adsorbed, the selectivity is not limited to the adsorption of only one molecular species. More than one key may fit each lock. Some centers seem to have adsorptive power for a variety of substances if one can judge by the many compounds that are able to displace one another's adsorption.

The many variations found in the adsorptive patterns of different carbons might suggest a very great variety of diverse centers. But the variations could also arise from varied permutations and combinations of a few species. To this, we do not know the answer. The situation is comparable to a hypothetical task of analyzing a mixture of carbohydrates if we were without means of identifying individual carbohydrates.

10. ISOTHERMS AND SURFACE VARIABLES

Isotherms offer a method for detecting qualitative differences as well as quantitative differences between activated carbons. The path of an isotherm is a function of all components of a system; and a change in any component frequently alters the course of an isotherm. Thus, the path of the isotherm, for the system *carbon A- benzoic acid- water*, will be altered if another acid is substituted for benzoic, or if another solvent replaces water. This suggests that when different carbons do not provide similar types of isotherms, the difference arises from dissimilar qualities of adsorptive power.

Such qualitative differences are frequently found when comparing carbons prepared by dissimilar methods of activation. This is illustrated by the isotherms in Figure 13:3 for the adsorp-

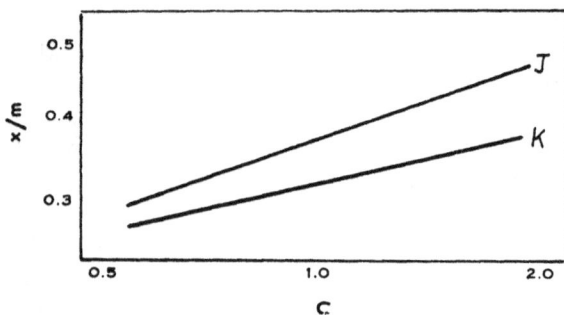

Fig. 13:3   Adsorption of iodine from aqueous solution
(Carbon J, zinc chloride process; Carbon K, calcium chloride process)
C, grams iodine/liter
X/M, grams iodine adsorbed/gram of carbon

tion of iodine from an aqueous solution. The slope of the isotherm for iodine with a carbon activated by the calcium chloride process differs from that of the carbon prepared by the zinc chloride process.

In contrast are the isotherms (Figure 13:4) of carbons R and S prepared by steam-activation at 800°C. These isotherms have similar slopes indicating a similar quality of surface. Although these two carbons probably are alike in quality of surface, they have differences in the level of the isotherms; carbon S at the higher

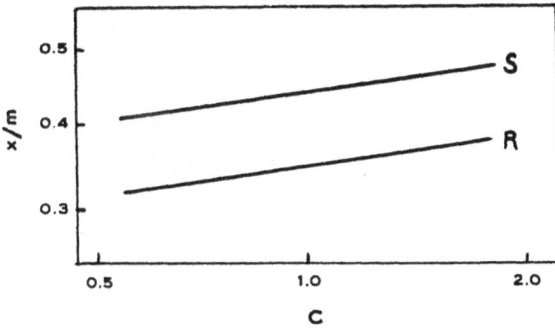

Fig. 13:4   Adsorption of iodine from aqueous solution
(Carbon R, activated by steam, 30 minutes; Carbon S, activated
by steam, 50 minutes)
C, grams iodine/liter
X/M, grams iodine adsorbed/gram carbon

level has greater adsorptive power. This can be traced to the fact
that carbon S was activated for a longer period of time during
which an additional quantity of active surfade could be formed.

Although similar methods of activation give similar isothermal
slopes, the converse does not necessarily follow, and one should
not conclude that two carbons were produced by the same method
because the isothermal slopes are found to be parallel under a
single set of conditions. Thus, the isotherms for molasses color
with carbons L and M were identical at 25°C, but were dissimilar
when the adsorption was conducted at 95°C (Table 13:4). Here,

TABLE   13:4

ADSORPTION OF MOLASSES COLOR BY CARBON
(Influence of Temperature on Adsorption Exponent)

Carbon Code	Adsorbent Exponent	
	25° C	95° C
L	0.21	0.23
M	0.21	0.30

although two carbons have a similar slope at 25°C, the same type of surface is not present on both carbons, as is revealed by the different response to temperature. From a practical point of view, the different response to temperature emphasizes the fact that industrial studies of adsorptive power should be conducted at the temperature to be employed in the process.

11. HYDROLYTIC ADSORPTION

A chemical equilibrium can be displaced when one of the components of the reaction is strongly adsorbable.

In studying the adsorption of dyes and other organic salts, early investigators observed instances in which the ions were not adsorbed in equal amounts. Thus with a basic dye such as methylene blue, the dye (cation) was adsorbed but the chloride ion remained in solution as a salt or as hydrochloric acid. Many such adsorptions were found to be an exchange adsorption in which the activated carbon released hydrogen or mineral cations to the solution in exchange for the adsorbed dye ion.

Studies by Bartell[25] and Miller[36] have shown that the adsorption of dyes and other organic salts can occur through the adsorptive forces of carbon acting in conjunction with the aqueous phase. By working with very pure carbon they eliminated the possibility of exchange adsorption and were able to demonstrate that unequal adsorption of ions can be caused by hydrolytic action.* The study was extended by Miller[36] and many conclusions in this section are drawn from his data.

The adsorption of methylene blue has been thoroughly studied. This dye, when dissolved in water, undergoes hydrolysis:

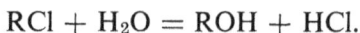

$$RCl + H_2O = ROH + HCl.$$

The solution contains equilibrium concentrations of methylene blue chloride, methylene blue hydroxide, and hydrochloric acid, together with ions of hydrogen, chloride, and dye. Difficulties in analytical procedures prevent quantitative measurements of the amount adsorbed in each form, but it is possible to gain a reason-

---

*This does not exclude the possibility that water is adsorbed as H and OH ions. The principles of the Donnan membrane has been applied to hydrolytic adsorption.

ably clear picture of the over-all mechanism. Small portions of carbon sufficient for partial decolorizing of the solution cause an increase in the amount of free hydrochloric acid. This is interpreted as indicating that at least part of the dye is adsorbed as ROH, shown on the right-hand side of the equation (R is dye radical); this upsets the equilibrium which is restored by further hydrolysis of dye, a process that simultaneously produces more free HCl. Actually, some HCl is adsorbed but the amount adsorbed does not equal that set free by the hydrolysis.

By adding sufficient carbon, the color and the HCl can be completely adsorbed. This raises a question as to the manner in which the latter portion of dye is adsorbed: Is it adsorbed as RCl, or as ROH with separate adsorption of the HCl? Miller studied the adsorption using many different proportions of dye and carbon. He found that the adsorption of dye color and chloride anion followed dissimilar paths, indicating independent adsorption of dye and acid. Separate adsorption is indicated by the fact that when sufficient carbon is used to give complete adsorption, the color is adsorbed more rapidly than the HCl. Still further, some HCl can be washed from the carbon without extracting any color.

Hydrolytic adsorption similar to that observed with methylene blue occurs with other basic dyes, with salts of alkaloids, and, in fact, with all salts from which carbon can preferetially adsorb the basic constituent. Such adsorptions are characterized by an increase in the free acidity in the solution. The extent of the increase depends on the relative adsorption of dye and acid, and this varies according to the particular type of carbon used.

With salts formed from strongly adsorbable acids, adsorption causes the solution to become more alkaline, but fundamentally the mechanism is similar to that found for methylene blue. Consider potassium benzoate. In aqueous solution, this hydrolyzes:

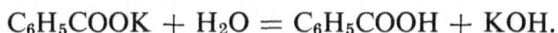

$$C_6H_5COOK + H_2O = C_6H_5COOH + KOH.$$

The solution contains equilibrium concentrations of potassium benzoate, benzoic acid, and potassium hydroxide. Of these, the benzoic acid is most strongly adsorbed and when thus removed, further hydrolysis occurs. The increase in the concentration of potassium hydroxide causes the solution to become more alkaline. The hydrolytic nature of this adsorption is confirmed by extracting the carbon with benzene; much of the *benzoate* that disappears from the solution during the adsorption can be extracted from the carbon as benzoic acid. The amount so recovered is chemically equivalent to the potassium hydroxide set free during the adsorption.

The fundamental principle involved in hydrolytic adsorption rests on the ability of hydrolysis to form an adsorbable acid or base—an effect that can be augmented by suitable adjustments of *p*H. With salts in which the acid component is the adsorbable portion, adsorption is increased by lowering the *p*H. Thus, with potassium benzoate the addition of a mineral acid increases the adsorption of benzoate by converting this into benzoic acid, without the need of hydrolysis mechanism. Similarly, when the basic component is more adsorbable, as in basic dyes, aniline hydrochloride, etc., the adsorption is increased by raising the *p*H with a strong base.

Usually, such adjustment of *p*H need be carried only to a point where conversion to the nonionized acid or base is approximately complete; adjustment beyond that point seldom confers additional benefits. In some adsorptions, it is detrimental to carry the *p*H adjustment too far.

Many inorganic salts are adsorbed hydrolytically. This behavior is typical with salts of metals that form insoluble hydroxides; in fact, the insolubility of the hydroxide is often correlated with the extent of the adsorption. Consider the action of carbon on a solution of ferrous sulfate; iron is adsorbed and the *p*H of the solution falls. A similar phenomenon occurs in the absence of carbon, only in this case, ferrous hydroxide slowly leaves the solution as a precipitate. The parallel suggests that the adsorption is actually an accelerated precipitation. This assumption is supported by the fact that when acid is added to a ferrous sulfate solution, it has an adverse effect on the removal of iron, whether by carbon or by precipitation through normal hydrolysis.

The parallel is not complete because the reaction with carbon results in greater removal of iron and can occur at a lower *p*H than the unaided hydrolysis. It should not be overlooked that carbon is able to adsorb slight quantities of HCl; this would upset the equilibrium conditions and assist the precipitation of $Fe(OH)_2$.

This seems to occur with solutions of lead salts in which the normal hydrolysis does not cause lead hydroxide to precipitate:

$$Pb(NO_3)_2 + 2H_2O = Pb(OH)_2 + 2HNO_3.$$

Carbon removes lead from such a solution, and the pH of the solution is simultaneously raised. The increase in pH suggests that carbon, by absorbing HNO_3, causes the hydrolysis to proceed to the point where $Pb(OH)_2$ precipitates and is removed mechanically by the carbon. This, however, is not the complete explanation because some lead is absorbed from a solution containing sufficient acid to maintain the natural pH of a lead nitrate solution.

That carbon does have an attraction for metallic hydroxides is illustrated by the ability of some carbons to adsorb measurable quantities of the very soluble potassium hydroxide. Kruyt and De Kadt[37] found that activation in air at 400°C produced a carbon capable of adsorbing potassium hydroxide, and they ascribed this to the presence of acidic constituents formed on the surface of the carbon. If, however, a carbon thus prepared is re-heated to 900°C and cooled in the absence of air, it ceases to adsorb alkali and instead will adsorb mineral acids.

12. SURFACE OXIDES[1,38,39]

Observing that the concentration of oxygen is a factor in the adsorption of acids and bases, Shilov[40] was led to study the behavior of carbon exposed to oxygen at different temperatures and pressures. His studies suggested the existence of surface oxides which he termed A, B, and C—oxides A and B being pictured as basic, and C as acidic. The adsorption of acids is considered to involve a reaction with basic oxides, whereas the adsorption of bases is traced to the acid oxide. The mechanism is not essentially different from that described by Miller[41] who found that a carbon that had adsorbed a water-insoluble dye was able to take up considerable quantities of alkali, a property not possessed by the original carbon.

In general, acid oxides predominate in carbons prepared in moist air at 300° to 500°C, and basic oxides in those prepared at 800° to 900°C in air, steam, or carbon dioxide. The transition temperature is in doubt because the acidic and basic properties show a gradual change from one type to another. Amphoteric properties are found in carbons prepared between 500° and 800°C.

The characteristics of the oxide produced at one temperature are altered when a carbon is reheated at a different temperature. A carbon prepared at 400°C, with base-adsorbing properties, when subsequently heated to 900°C becomes acid-adsorbing. If this carbon is again reheated in air at 400°C, base-adsorption reappears. This last change also occurs slowly at room temperature in the presence of moist oxygen. Thus it appears that the surface oxides are reversible from one form to another although the effect is not entirely independent of previous treatment.

The surface covered by surface oxides varies in different carbons. Lowry and Morgan[42] found 1.71 % fixed oxygen in one carbon and 3.75% in another. McKie[43] reports a ratio of 1 oxygen atom to 140 surface carbon atoms. Strickland-Constable[44] suggests that 1 surface carbon atom in every 25 may attach 1 oxygen atom. Most of these determinations were made on carbons prepared at high temperatures or which had been subjected to low-temperature activation of relatively short periods. Oxygen concentrations of 18 % are reported for carbons exposed to oxygen at 400°C.

13. ADSORPTION ASSOCIATED WITH CHEMICAL CHANGE

Somewhat similar to hydrolytic adsorption is *splitting adsorption* in which various complex compounds are disrupted by carbon which selectively adsorbs certain constituent parts. This behavior has been found with complex platinum, palladium, and osmium compounds, also with ammino compounds containing copper, cobalt, manganese, etc.—the metallic group being adsorbed.

Various substances undergo reduction upon adsorption:[1] mercuric salts are reduced to the mercurous state, and ferric salts to ferrous salts. Salts of some metals, *e.g.*, silver and gold, undergo partial or complete reduction to metal when adsorbed; the extent of the reduction is a function of the electrochemical nature, being more pronounced with the less electropositive metals. Another contributing factor is the chemical nature of the associated anion. It is also to be noted that some anions, *e.g.*, permanaganates and chromates, undergo reduction during adsorption. The reducing action is somewhat specific: One carbon may have greater reducing power for ferric chloride whereas another carbon may be more effective for permanganates.

Many of these substances are no longer soluble after such a chemical change and this raises a question as to the mechanism, as it could follow one of several paths:

1. The substance may be first adsorbed and then undergo reduction.
2. The reduction and adsorption may constitute integral parts of an over-all phenomenon.
3. The reduction may occur first and convert the substance into an insoluble form; in which case, the role of the carbon would be a purely mechanical one of sweeping a solid precipitate from the liquid.

Probably each of these mechanisms plays a part in one adsorption or another. Bolam and Phillips,[45] in a study of the adsorption of silver nitrate, found the reaction to occur in two steps: a) rapid adsorption of the silver ion; b) slow reduction to metallic silver. Only part of the adsorbed silver is reduced, the ratio of adsorption to reduction depending on the carbon used and on the time. Prolonged contact increases the amount of silver that is reduced, but does not appreciably alter the amount adsorbed. From all this, it appears that the adsorption of silver nitrate proceeds through mechanisms 1 and perhaps 2.

In other adsorptions, the reduction occurs during a momentary contact or adsorption, and the question of more permanent adsorption depends on the solubility of the product that is formed. Ferric chloride dissolved in distilled water, when treated with carbon, is reduced and adsorbed hydrolytically as ferrous hydroxide. If the solution is acidified sufficiently, the adsorption is neglibible and the newly formed ferrous ions appear in the solution.

14. THE ADSORBED LAYER[1]

We have but little knowledge of the status prevailing in the adsorbed layer, especially in liquid systems. Even when only a few components are present and the substances involved have a simple molecular structure, clues as to the nature of the adsorbed layer usually admit varied interpretations. Fortunately in industrial applications, information as to the status of the adsorbed substance is seldom needed except when catalysis or subsequent elution is involved.

The adsorbed layer is seldom of sufficient thickness to produce interference of X-rays, hence this tool sheds little light on the status of the adsorbed substance. Studies of optical and magnetic properties reveal but little.

The frequent references made to the adsorbed layer as a two-dimensional gas or liquid film are apt to lead to the thought that the adsorbed layer is only a slightly modified form of the bulk gas or liquid. Even in the case of purely physical adsorption, the forces involved are comparable to those producing a change in state. Consequently, the differences between the properties of an adsorbed liquid film and those of the bulk liquid are comparable to the differences found between a liquid phase and the corresponding solid or gas phase. Obviously, adsorption accompanied by chemical or quasi-chemical action would produce even more profound changes.

When a molecule is adsorbed, certain degrees of freedom may be frozen out or suppressed. A study of the heat capacity of matter in the adsorbed state should give some indication as to whether the adsorbed molecules can move freely, vibrate, rotate, or show still more restricted motion. Coolidge[46] points out that measurements of specific heat in the adsorbed state require precautions to provide a constant adsorbed quantity through a change in temperature. He found the specific heat of adsorbed water exceeds that of water vapor; more strictly speaking, the heat capacity of the charcoal-water system exceeds that of the separate components. Porter and Swain[47] found that the heat capacity of water vapor adsorbed on charcoal to be about the same as tht of liquid water, whereas argon or hydrogen has lower specific heat in the adsorbed state than in the solid state.

Monolayers and Multilayers

Much thought has centered around the question of the probable thickness of the adsorbed layer.[1,2] The potential theory suggests that the adsorbed molecules form a miniature atmosphere around the solid. Langmuir considered that a single layer is formed by the adsorptive forces of the solid surface, although he recognized that the first layer could adsorb other molecules to form a second layer. As the forces holding the second layer would be weaker, the molecules would evaporate more rapidly and thus the second layer would contain fewer molecules than the first. He felt it would be improbable for more than three layers to form.

Other theories have since been developed to show mechanisms which allow the formation of multilayers. A multimolecular theory, proposed by Brunauer, Emmett, Teller[48] in 1938 and by Brunauer, Deming, Deming, and Teller[49] in 1940, assumes that only the first adsorbed layer is attracted strongly by the surface, the second layer being adsorbed essentially by the first adsorbed layer. The adsorptive forces are assumed to result from the totality of forces usually designated as van der Waals forces, among which are the *dispersion forces* of London.[2]

Solutes

Adsorption from solution is often discussed as though only the solute were adsorbed. But as is well known, the solvent is also adsorbed—often appreciably so. It is possible that the solute and solvent are adsorbed—not as separate entities—but instead as a solution in which the concentration differs from that of the bulk liquid.

Ions

Inasmuch as factors that repress ionization also lead to increased adsorption, it is natural to deduce that solutes are preferentially adsorbed as undissociated molecules, and for the most part this is probably so. However, separate lines of evidence indicate the presence of at least some ions in the adsorbed layer. One indication is furnished by the migration of carbon particles in an electrical field. Other evidence is found in the ability of adsorbed acids to invert sucrose, but it is reported that this inversion does not occur with small traces of adsorbed acids. In this connection. Miller and Bandemer[50] found that small concentrations of adsorbed hydrochloric acid are removed with difficulty by extraction with sodium hydroxide, a matter of significance in view of the avidity with which these substances normally react.

Effect of Size of Adsorbed Molecules

Except where a screening action of the pores intervenes, it will frequently be found that larger molecules are more adsorbable than smaller molecules of similar chemical nature. Although various factors such as solubility often play a part, the behavior is in line with the Langmuir concept that adsorption is a function of the time lag between the initial attachment and the subsequent desorption of a molecule. Small molecules, which may become attached at a single point, can be desorbed as soon as this bond is broken. Larger molecules can be adsorbed initially by becoming attached through a single atom, the binding being strengthened when other points of contact are made. In the subsequent desorption, however, the larger molecule will not be released until all points of attachment are broken simultaneously, and this will happen less often then the breaking of a single bond holding a smaller molecule.

15.   ORIENTATION

Although the strong attachment of larger molecules to a surface is evidence of multiple points of attachment, sometimes the bonds are localized to a few atoms or groups of atoms. In some adsorptions, parts of the adsorbed molecule are free, and able to extend into the environment. An illustration is found in the ability of adsorbed invertase to invert a sucrose solution.[51] Evidently the active portion of the enzyme extends into the liquid phase.

The influence of substituent groups on adsorption is often viewed as furnishing clues as to the orientation. Thus, if the introduction of a particular group into a molecule enhances the adsorption, that group is presumed to be attached to the surface. Conversely, if the adsorption is diminished, the added group is presumed to be oriented away from the surface of the carbon.

Deductions so reached can be questioned on several grounds. The effect of a substituent group on adsorption is often associated with changes in other properties and this can lead to seeming inconsistencies. A polar group such as OH will diminish the adsorption of a solute from an aqueous solution, but the same polar group will enhance the adsorption of a vapor. The inconsistency can be traced to the influence of a polar group on other properties: A polar group makes a vapor more condensable and therefore more adsorbable; but the same polar group will make a solute more soluble in water and cause the solute to be pulled away from the surface of the carbon.

Moreover, an endeavor to correlate adsorbability with orientation must consider that the effect of introducing a group into a molecular structure is not localized to the position occupied by that group, but may extend to affect the properties of other portions of the molecule. The introduction of a polar group into a benzene ring enhances the chemical reactivity at other positions in the ring; likewise a chlorine atom attached to the hydrocarbon end of an aliphatic acid can greatly alter the properties of a carboxyl group at the other end of the molecule. Similar effects on adsorbability seem entirely possible.

Wettability is viewed as furnishing clues to the orientation of the adsorbed layer. Carbon is *hydrophobic* (adverse to water); that is, it is wet more readily by organic solvents than by water. From this it has been deduced that when organic substances are adsorbed from water, the hydrocarbon part of the molecule is attached to the surface of the carbon, and any polar groups extend into the liquid phase. In general, this deduction is mainly correct; but it may need partial modification, because all carbons are not completely hydrophobic. When some carbons are placed in a mixture of benzene and water, a portion of the carbon will gather at the interface indicating that some areas of the carbon surface have partial *hydrophilic* (attracted to water) characteristics. That the attraction of carbon is not restricted to hydrocarbon groupings is shown by the ability of carbon to adsorb substances that have no carbon atoms, e.g., halogens and oxides of nitrogen.

The nature of the carbon surface suggests that many varied orientations can occur, with polar groups being held at hydrophilic centers, and hydrocarbons at other centers. It is to be mentioned that the relative proportion of hydrophobic and hydrophilic characteristics varies from one carbon to another depending on method of manufacture.

16.   ELECTROPHORESIS

When an electric current is passed through a suspension of a finely divided solid in water, the solid particles tend to migrate to one of the electrodes.[1] This behavior is ascribed to an electric double layer which exists at boundaries between phases and is called electrophoresis. In simple cases, the electric double layer

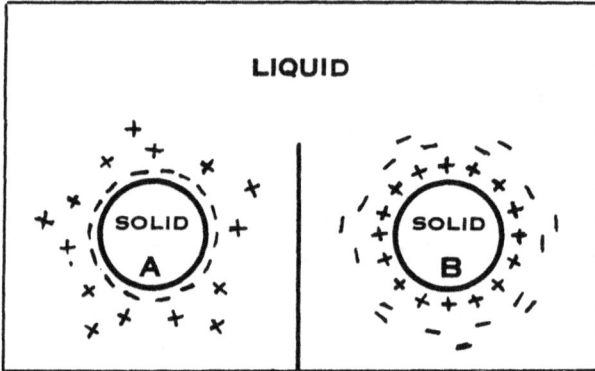

Fig. 13:5   Double layer of ions surrounding a solid suspended in a liquid

may be regarded as a condenser whose potential is called electro-kinetic potential. Even on large particles the electric double layer consists of ions. Consequently electrophoresis and ion migration in an electric field are similar phenomena. The double layer may form through self-dissociation, as in the case of a protein which can split off ions, or it may form through the preferential adsorption of certain ions from the solution.

When an electrolyte, such as hydrochloric acid, is adsorbed on a solid surface. the molecules may be attached in one of several ways. The chloride ions can be preferentially attached close to the surface of the solid, leaving the hydrogen ions to extend into the liquid phase (*A* in Figure 13:5). Under the influence of an electric field, the negative chloride ions move to the positive pole dragging along the attached solid particles, and an equivalent number of the loosely held positive hydrogen ions moves to the negative pole. The migration would be reversed with a solid surface that preferentially attaches the positive hydrogen ions and allows the chloride ions to extend into the liquid phase (*B* in Figure 13:5) then the solid particles would be dragged to the negative pole.

Most activated carbons, when suspended in distilled water, migrate to the positive pole, indicating that they carry a negative charge. The velocity of the migration and thus the strength of the charge depends on the method of preparing the carbon. Olin,[52] and Bennister and King,[53] found that an elevation in the temperature at which carbon is prepared produces an increase in the electrophoretic velocity. This reaches a maximum in samples activate at 850°C, and then decreases if still higher temperatures are used. Buzagh,[54] and also Bennister and King, observed that the electrophoretic behavior is influenced by the concentration of the carbon suspension.

It is reported that carbons with a positive charge in distilled water can be prepared by activation at temperatures above 950°C in the presence of hydrogen or carbon dioxide, but this has not been confirmed by all workers.[1] Failure to obtain a carbon with a positive charge has been attributed to a failure to observe certain precautions; for instance, the carbon after heating must be cooled in complete absence of air to prevent formation of oxides characteristic of lower temperatures.

The electrokinetic behavior of carbon does not appear to be altered by the adsorption of weak organic acids or bases; but dyes and strongly ionized acids, bases, and salts do have an influence. The negative charge on carbon is strengthened by small concentrations of multivalent anions, e.g., SO_4, $Fe(CN)_6$, and weakened by multivalent cations, e.g., Al, Ti. The influence of the hydrogen ion has been studied in some detail.[1,52,55] When hydrochloric acid is added to a suspension of carbon, the electrophoretic velocity decreases until, at a pH between 2 and 4, the migration ceases. This is the *isoelectric* point. If the solution is made still more acid, the particles acquire a positive charge and migrate to the negative pole. The hydrogen ion concentration at which these changes occur varies according to the carbon used and also depends on the nature and concentration of other ions that may be present.

Several explanations have been offered as to the origin of the electric charge on carbon. As the charge is a function of the temperature at which the carbon is activated, this suggests a relation to surface oxides which could yield ions by a reaction with water. The charge also could arise from mineral constituents, and in this connection, Roychoudhury and Mukerjee[56,57] report that a carbon having a positive charge can be prepared by thorough washing to remove all electrolytes.

17. HEAT OF ADSORPTION

Molecules in condensing from a gas to a liquid come to a state of comparative rest and lose kinetic and latent energy, the energy being liberated as heat of liquefaction or condensation. Similarly, heat is liberated when a vapor condenses on a solid surface.[1,2] Stronger forces are involved in condensation on a solid because this can occur under conditions that do not allow normal liquefaction to take place. Consequently, one might expect a greater quantity of heat to be liberated in adsorption than in normal liquefaction, and this is found to be so except in a few cases, *e.g.*, mercury or water on charcoal.[58]

When a definite quantity of a gas is admitted to a sample of activated carbon in a calorimeter, the total heat liberated, called the integral heat of adsorption, is usually reported as calories per mole (or milliliters) of adsorbed gas. Comparisons of integral heats of different substances should be based on corresponding conditions of measurement; thus Lamb and Coolidge[59] report the integral heat evolved when 1 mole of a vapor is adsorbed by 500 grams of charcoal (Table 13:5)

All portions of an adsorbed gas do not liberate equal amounts of heat. therefore the integral heat is the average heat for the total

TABLE   13:5

INTEGRAL HEAT OF ADSORPTION*

Compound	Q^a calories	hm^b calories
C_2H_5Cl	6,220	12,000
CS_2	6,830	12,500
CH_3OH	9,330	13,100
C_2H_5Br	6,850	13,900
C_2H_5I	7,810	14,000
$CHCl_3$	8,000	14,500
$HCOOC_2H_5$	8,380	14,500
C_6H_6	7,810	14,700
C_2H_5OH	10,650	15,000
CCl_4	8,000	15,300
$(C_2H_5)_2O$	6,900	15,500

[a] Q = molecular heat of vaporization.
[b] hm = heat of adsorption, 1 mole of gas per 500 grams of carbon.
* Data from A. B. Lamb and A. S. Coolidge, *J. Am. Chem. Soc.* 42:1146 (1920); reprinted by permission of the American Chemical Society.

gas adsorbed in the experiment. In order to know the amount of heat corresponding to each portion of adsorbed gas, it is necessary to determine the *differential heat* of adsorption. This can be approximated by introducing the gas in very small increments into a calorimeter containing the carbon and measuring the heat evolved by each increment of gas adsorbed. Differential heats are reported as calories per mole of adsorbed gas.

Significant information on the course of an adsorption is revealed by a plot showing the change in the differential heat as increasing amounts of gas are adsorbed.[2] Usually the initial portions of adsorbed gas have a greater differential heat than subsequent portions. The decrease is often small, particularly with compounds having a comparatively high boiling point, such as those in Table 13:5 As expressed by Lamb and Coolidge, such adsorptions show very little fatigue.

With some substances, a high initial differential heat is followed by a sharp decline in the adsorption of later portions. Keyes and Marshall[60] found that the initial portion of chlorine adsorbed by carbon liberates 31,900 calories per mole, and that this decreased to 24,600 for subsequent additions. An extreme example is provided by the adsorption of oxygen on carbon at $0°C$.[1,2,60] The initial portions evolve from 72,000 to 129,000 calories per mole, depending on experimental conditions and probably also on the type of carbon used. As additional portions of oxygen are adsorbed, the differential heat falls rapidly to values of approximately 4000 calories per mole.

Various theories have been offered to explain such changes in the heat of adsorption.[1,2] One view is that a monolayer is first formed in which the molecules are held by stronger forces than in subsequent layers. Other workers ascribe variations to the heterogeneous nature of the surface. The molecules adsorbed initially may enter capillaries and crevices; being held on several sides they should liberate more heat. Another suggestion is that the earlier portions are adsorbed by the more active surface atoms and are held by forces of a different nature than later portions. The last explanation seems to apply to the adsorption of oxygen on carbon at temperatures of $0°C$ and over. Some of the later portions of adsorbed oxygen can be removed as such by applying a vacuum, whereas the oxygen adsorbed initially can be removed only by elevating the temperature. Then, it does not come off as oxygen, but as carbon dioxide and carbon monoxide.

Nature of the Heat of Adsorption

Phenomena related to heat of adsorption have engaged the interest of a number of investigators. Many have noted a relation between the heat of adsorption and heat of liquefaction. Lamb and Coolidge[59] suggested considering the heat of adsorption as consisting of two components: the heat of liquefaction and a remainder which they termed the *net heat of adsorption*. They found that this net heat of adsorption—expressed as calories per milliliter of liquid—was practically identical for all eleven vapors used in their study. This had led many to view such adsorptions as purely physical, and explanations based on various physical properties such as compressibility of the liquid have been advanced. In commenting on this, McBain[62] has pointed out that stress cannot be laid on a particular success in explaining data according to one special hypothesis without considering the whole subject of chemical and physical interaction. He found a similar numerical relation between many heats of reaction in which there was a direct chemical action between the molecules of the reactants.

Too much must not be read into the comparison frequently made between the heat of liquefaction and the heat of adsorption. The solid adsorbent plays an integral part in the phenomenon, and the heat of adsorption should be considered as a property of the entire system. That the attachment in the adsorbed state differs from that between molecules in a liquid is indicated in various ways. The heat of liquefaction is the same for all portions of condensed vapor, whereas the heat of adsorption varies with the amount adsorbed. Then, too, the greatest amount of heat is liberated by the initial portions of adsorbed vapor and, at this stage, the adsorbed molecules are not present in sufficient quantities to form a continuous liquid film. Work by Bangham and Razouk[63] casts doubts on whether the adsorbed layer should be called a liquid film even when a carbon is saturated with a vapor. They found that supersaturated methanol vapor does not condense to the bulk liquid on charcoal; this indicates that the adsorbed methanol does not provide nuclei for condensation and thus is not in a state comparable with the bulk liquid.

Heat of Wetting

The heat evolved when 1 gram of power is immersed in a liquid is known as the *heat of wetting* or the *heat of immersion*.[64] Data on heat of wetting are not expressed in terms of the adsorbate as is done with heats of gas adsorption, because it is not feasible to measure the amount of liquid adsorbed. Consequently, data on heat of wetting are reported as calories per unit mass of adsorbent (Table 13:6)

It is generally assumed that a high heat of wetting is associated with great adsorbability. From this, it might be concluded that a solute will be adsorbed less readily from a solvent having a high heat of wetting. Usually this is so, but exceptions are reported. Bartell and Fu[65] found that bromonaphthalene has a comparatively low heat of wetting with carbon, but is strongly adsorbed from nitrobenzene which has a much higher heat of wetting.

Bartell points out that the determining factor in adsorption is the free surface-energy change. This is not measured by the heat of wetting which instead gives the decrease in total surface energy. He states that the adhesion tension, *i.e.*, the work required to separate a liquid from a solid phase measures the free energy change and thus the relative adsorbability.

TABLE 13:6

HEATS OF WETTING OF CHARCOAL*

Liquid	cal/g Charcoal
Water	8.4
Nitrobenzene	27.1
Benzene	21.0
Carbon disulfide	29.5
Carbon tetrachloride	20.0
Petroleum ether	23.8

* From J. L. Culbertson and L. L. Winter, *J. Am. Chem. Soc.* 59:308 (1937); reprinted by permission of the American Chemical Society.

REFERENCES

1. Deitz, V. R., *Bibliography of Solid Adsorbents*, United States Cane Sugar Refiners and Bone Char Manufacturers and the National Bureau of Standards, Washington, D.C., 1944, pp. 375–431.
   Deitz, V. R., *ibid.*, 1943 to 1953, National Bureau Standards, Circular 56:566, Washington, D.C., 1956.
2. Adam, N. K., *The Physics and Chemistry of Surfaces*, Oxford University Press, London, 1941.
   Adamson, A. W., *Physical Chemistry of Surfaces*, Interscience Publishers Inc., New York, 1960.
   Alexander, J. (Editor), *Colloid Chemistry*, 6 volumes; Chemical Catalog Co. and Reinhold Publishing Corporation, New York, 1926–1946.
   Brunauer, S., *The Adsorption of Gases and Vapors, Physical Adsorption*, Princeton University Press, Princeton, 1943.
   von Buzagh, A., *Colloid Systems*, Technical Press Ltd., London, 1937.
   Freundlich, H., *Colloid and Capillary Chemistry*, E. P. Dutton and Co., New York, 1922; Methuen & Co., Ltd., London, 1926. McBain, J. W., *The Sorption of Gases and Vapors by Solids*, G. Routledge and Sons, Ltd. London, 1931.
   Weiser, H. B., *Colloid Chemistry*, John Wiley, New York, 1949.
3. Zsigmondy, R. Z., *Anorg. Chem.*, 71:356 (1911).
4. Polányi, M., *Verhandl. deut. physik. Ges.*, 18:55 (1916).
5. Eucken, A., *Verhandl. deut. physik. Ges.*, 16:345 (1914).
6. Berényi, L., *Z. physik. Chem.*, 94:628 (1920).
7. Langmuir, I., *J. Am. Chem. Soc.*, 38:2221 (1916); 39:1848 (1917); 40:1361 (1918).
8. Taylor, H. S., *J. Am. Chem. Soc.*, 53:578 (1931).
9. Bragg, W. H., and Bragg, W. L., *X-rays and Crystal Structure*, Geo. Bell and Sons, London, 1924.
10. France, W. G., *Colloid Chemistry*, edited by J. Alexander, Vol. 5, Chapter 21, Reinhold Publishing Co., New York, 1944; includes many references to work of others.
    France, W. G., and Wolfe, K. M., *J. Phys. Chem.*, 45:395 (1941).

Rigterink, M. D., and France, W. G., *J. Phys. Chem.*, 42: 1079 (1938).

11. Deutsch, D. Z., *physiol Chem.*, 136:353 (1928).
12. Ramsden, W., *Proc. Roy. Soc.*, 72:156 (1903).
13. Cassel, H. M., *J. Am. Chem. Soc.*, 58:1309 (1936).
14. Cullen, *Trans. Inst. Chem. Eng.*, 17:14 (1939).
15. Brunauer, S., Emmett, P. H., and Teller, E., *J. Am. Chem. Soc.*, 60:309 (1938).
    Emmett, P. H., *J. Am. Chem. Soc.*, 68:1784 (1946).
16. Emmett, P. H., and Brunauer, S., *J. Am. Chem. Soc.*, 59:1553 (1937).
17. Personal communication. W. C. Bokhoven and C. van der Meijden, of N.V. Norit-Vereeniging Verkoop Centrale, Amsterdam, Holland.
18. Bartell, F. E., and Fu, Y., *Colloid Symposium Monograph*, 7:135 (1930).
19. Grant, R. J., *Basic Concepts of Adsorption*, Pittsburgh Chemical Co., 1961.
20. Emmett, P. H., *Chem. Rev.*, 43:69 (1948).
21. Juhola, *et al., J. Amer. Chem. Soc.*, 71:2069 (1949); 74:61 (1952).
22. Dubinin, M. M., *Chem. Rev.*, 60:235 (1960); *Proceedings of 4th International Symposium on Reactivity of Solids.*, J. H. De Boer, Editor Elsevier Pub. Co., pp. 643–648 (1961).
23. Wiig, E. O., and Smith, Stanton B., *J. Phys. & Colloid Chem.* 55:27 (1951).
24. Gleysteen, L. F., and Scheffler, G. H., *Proceedings Fourth Conference on Carbon*, Pergamon Press, New York (1960).
25. Nekrasov, B., *Z. physik. Chem.*, 136:18 (1928).
26. Ginzburg, D. Z., *Khim, Referat. Zhur.*, 4, No. 2:20 (1941).
27. Linner, E. R., and Gortner, R. A., *J. Phys. Chem.*, 39:35 (1935).
28. Griffin, K. M., Richardson, H. L., and Robertson, P. W., *J. Chem. Soc.*, 2705 (1928).
29. Unpublished data presented at meeting of American Chemical Society, at Atlantic City in 1952.
30. Unpublished data.
31. Hassler, J. W., and McMinn, Wm. E., *Ind. Eng. Chem.*, 37:645 (1945).
32. Adams, N. K., *The Physics and Chemistry of Surfaces*, 3rd Edition, p. 250, Oxford University Press, London, 1941.

33. Zylbertal, Z., *Biochem. Z.*, 236:131 (1931).
34. Chaplin, R., *Trans. Faraday Soc.*, 30:249 (1934).
35. Bartell, F. E., and Miler, E. J., *J. Am. Chem. Soc.*, 44:1866 (1922); 45;1106 (1923); *J. Phys. Chem.*, 28:992 (1924).
36. Miller, E. J., *J. Am. Chem. Soc.*, 46:1150 (1924); 47:1270 (1925); *J. Phys. Chem.*, 30:1162 (1926); 31:1197 (1927); *Mich. Agr. Exp. Sta. Tech. Bull.*, 73 (1925).
37. Kruyt, H. R., and De Kadt, G. S., *Kolloid-Z.*, 47:44 (1929).
38. Steenberg, B., *Adsorption and Exchange of Ions on Activated Charcoal*, Almquist and Wiksells, Uppsala, 1944.
39. Garten, V. A., and Weiss, D. E., *Reviews of Pure and Applied Chemistry*, Vol. 7, June, 1957.
40. Shilov, N., Shatunovska, H., and Chmutov, K., *Z. physik. Chem.*, A 149:211 (1930); A 150:31 (1930).
41. Miller, E. J., *Phys. Chem.* 36:2967 (1932).
42. Lowry, H. H., and Morgan, G. A., *J. Am. Chem. Soc.*, 42:1408 (1920).
43. McKie, D., *J. Chem. Soc.* 2870 (1928).
44. Strickland-Constable, R. F., *Trans. Faraday Soc.*, 34:1074, 1374 (1938).
45. Bolam, T. R., and Phillips, W. A., *Trans. Faraday Soc.*, 31:1443 (1935).
46. Coolidge, A. S., *J. Am. Chem. Soc.*, 49:708 (1927).
47. Porter, J. L., and Swain, R. C., *J. Am. Chem. Soc.*, 55:2792 (1933).
48. Brunauer, S., Emmett, P. H., and Teller, E., *J. Am. Chem. Soc.*, 60:309 (1938).
49. Brunauer, S., Deming, L. S., Deming, W. E., and Teller, E., *J. Am. Chem. Soc.*, 62:1723 (1940).
50. Miller, E. J., and Bandemer, S. L., *J. Am. Chem. Soc.*, 49:1686 (1927).
51. Miller, E. J., and Bandemer, S. L., *J. Phys. Chem.*, 34:2666 (1930).
52. Olin, H. L., Lykins, J. D., and Munro, W. P., *Ind. Eng. Chem.* 27:690 (1935).
53. Bennister, H. L., and King, A., *J. Chem. Soc.*, 991 (1938).
54. von Buzagh, A., *Kolloid-Z.*, 48:33 (1929).
55. Bennister, H. L., and King, A., *J. Chem. Soc.*, 1888 (1938).
56. Roychoudhury, S. P., and Mukherjee, J. N., *Kolloid-Z.*, 57:302 (1931).

57. Roychoudhury, S. P., Nandi, S. K., and Banerjee, J. K., *J. Indian Chem. Soc.,* 13:410 (1936).
58. Coolidge, A. S., *J. Am. Chem. Soc.,* 49:708, 1949 (1927).
59. Lamb, A. B., and Coolidge, A. S., *J. Am. Chem. Soc.,* 42:1146 (1920).
60. Keyes, F. G., and Marshall, M. J., *J. Am. Chem. Soc.,* 49:156 (1927).
61. Pearce, J. N., and Reed, G. H., *J. Phys. Chem.,* 35:905 (1931): 39:293 (1935).
62. McBain, J. W.; *The Sorption of Gases and Vapors by Solids.* G. Routledge and Sons, Ltd., London, 1932, pp. 410, 429.
63. Bangham, D. H., and Razouk, R. I., *Trans. Faraday Soc.,* 33:1459, 1463 (1937).
64. Culbertson, J. L., and Winter, L. L., *J. Am. Chem. Soc.,* 59:308 (1937).
65. Bartell, F. E., and Fu, Y., *J. Phys. Chem.,* 33:1758 (1929): *Colloid Symposium Monograph,* 7:135 (1930).
66. See chapter 20, section 13 under Polanyi and Dubinin.

14

Contact Catalysis

Activated carbon is used as a contact catalyst in various reactions of isomerization, polymerization, oxidation, and halogenation. It also finds use as a carrier for other catalysts, perhaps to a greater extent than generally realized.

Any description of a catalyst must be surrounded by so many qualifications that it becomes difficult to give any precise definition.[1,2] In very general terms, a catalyst may be described as a substance which alters the speed of a chemical reaction without itself disappearing as such to appear as an ingredient of the reaction products. No apparent net change occurs in the chemical composition of a catalyst, but it does not always escape changes in surface characteristics caused by the mechanism of the reaction or as a result of poisoning. These changes, when they occur, cause a decay in catalytic power and limit the effective life.

The length of life necessary to justify the industrial use of a catalyst depends on technological and economic circumstances surrounding each specific process. In many cases, the weight of products formed is only fifty to one-hundred times that of the catalyst. At the other extreme is the use of platinum gauze in the oxidation of ammonia where, with suitable precautions, there is no apparent limit to the amount of product that can be formed.

Catalysts can provide many and varied benefits. The ability to accelerate a reaction reduces the amount of processing equipment otherwise necessary. Some reactions which normally proceed at an immeasurably slow rate can be accelerated sufficiently by a catalyst to become a worth-while industrial process. Catalysts make the preparation of certain products a practical operation and they provide a possibility for preparing them from less costly materials.

Catalysts often eliminate the need of accessory chemicals which apart from their cost may also result in the formation of by-products having little or no value.

Catalysts frequently enable a reaction to proceed at lower temperatures; the consequent savings in heat, power, and reduced depreciation of equipment are important in processes that normally involve high temperatures and pressures. In some processes, a catalyst eliminates the need of using other forms of energy, *e.g.,* actinic or electrical. Applebey[3] states that the energy necessary in industrial practice to fix a given amount of nitrogen by the use of a catalyst is about one quarter of the energy required to fix the same amount electrically.

A suitable catalyst can often direct a reaction to produce the minimum of by-products; this arises from the fact that catalysis can be negative; that is, a reaction can be suppressed by a catalyst. Often a catalyst accelerates certain reactions and inhibits others; therefore, one can alter the course of a reaction by an appropriate catalyst. Thus, compound *A* may normally react to produce *B* as a major product with minor quantities of *C*. In the presence of an appropriate catalyst, the reaction producing *C* may be inhibited and consequently *B* can be produced in a purer form and with less waste. On the other hand, if *C* is the desired product, it is necessary to search for a catalyst that accelerates that reaction and simultaneously retards the reaction producing *B*. In general, catalysts only serve to decrease the time required for a system to reach equilibrium, but there are cases in which the final concentrations attained with a catalyst are not the same as would be anticipated if the role of the catalyst were ignored.[4,5]

1. THEORIES OF CONTACT CATALYSIS

The industrial applications of catalytic processes have been generally in advance of theoretical knowledge of the mechanisms involved. Most of the catalytic processes developed during the 19th century and the early decades of the 20th century were the results of *cut-and-try* methods inspired by what Hilditch[6] has called "chemical intuition." Even today, many industrial catalytic processes are still guided by empiricism.

For many years, the general principles underlying the mechanism of catalytic processes appeared shrouded in mystery, but in recent decades, numerous studies by many investigators have given a clearer insight of the probable mechanism of many catalytic reactions. It can hardly be said, however, that an ultimate explanation has been found. The present state of our knowledge is well expressed in a statement made by Adkins:[5]

> There are no physical or chemical measurements which can be made *a priori* to distinguish the good from the poor catalyst. After a particular substance prepared in a certain way has been found to be the best catalyst available, the fact can usually be explained.

At a meeting of one of the learned societies, an "elder statesman" got up and discouraged further discussion by observing that he had heard innumerable theories of catalysis in his life, but never one which he could not disprove. In relating the incident, Hinshelwood[7] made the comment:

> One may feel that he, or those who tried to satisfy him, were seeking the impossible. There is no theory of catalysis. The only question is whether we understand catalytic phenomena well enough to arrange them into a picture of which we like the pattern.

Much of the difficulty disappears if we adopt a pragmatic approach and regard these theories not as objective reality but as mental highways on which we can move to a fuller understanding. And, just as the choice of a highway by a motorist depends upon where he wants to go, so also we will do well to select the theory that provides the most help in the particular problem at hand.

Some of the various theories will be briefly outlined—partly because they provide a convenient way to describe catalytic behavior in a systematic manner, partly because the theories of catalytic surfaces shed light on the nature of adsorption by activated carbon.

Active Areas

Catalytic activity is often concentrated on a relatively small portion of the surface of a catalyst; only rarely is the entire surface active.[2] This conclusion has developed from poisoning studies, the amount of poison required to completely suppress a reaction often being far less than needed to cover the entire surface. Pease and

Stewart[8] found that carbon monoxide in quantities sufficient to cover only 1 % of the surface of a copper catalyst suppressed 90 % of the catalytic power; in other words, 90 % of the catalytic power was concentrated in 1 % of the surface. Poisoning studies by Almquist and Black[9] showed that in a pure iron catalyst, only one atom in two thousand was active.

Recognition that only part of the surface need be active has dispelled much of the mystery of *poisoning;* this is, the ability of small traces of certain impurities to destroy the efficiency of a catalyst. Today, poisons are regarded as substances that are preferentially adsorbed by the active portions of the surface; when the active area is small, it can be completely blanketed by small amounts of a poison. The active area is regarded as consisting of a number of patches distributed more or less uniformly over the surface.[2,10] The active patches are also termed *active points,* or more often *active centers,* and much study has been devoted to learn the probable nature of these areas.

Taylor[11] suggests that surfaces contain spots in which crystallization is incomplete, and that activity may reside in individual atoms existing in varying degrees of unsaturation, ranging from atoms completely saturated by neighboring atoms to those which are attached to the surface by only a single valence, *i.e.,* only a single step removed from a gaseous atom.

Edges and corners of crystals have been suggested as the seats of activity because atoms at such positions are in a more unsaturated state.[12] This view finds support in the fact that many edges and corners are present in the rough surfaces characteristic of catalysts. Metallic mirrors lack the activity of the rough surfaces that are formed through the rapid decomposition of organometallic salts.[13] Fire-polished glass will not catalyze the recombination of atomic hydrogen, but a fractured surface will initiate the reaction.[14]

Smekal[15] points out that although modern concepts of structure explain many properties of crystals, certain properties are not dependent on the space lattice but are related to *imperfections* that normally occur in all crystals. It is suggested that faults such as surface cracks and interfaces between minute crystals can act as active centers.

Many aspects of catalytic phenomena are clarified by the concept of a heterogeneous surface over which active centers are distributed, but the concept in itself does not describe the mechanism of catalysis.

Concentration Theory

At one time it was believed that the acceleration of a reaction by an adsorbent resulted from an increased concentration of the reactants at the surface—a mass-action effect.[2] Today this influence seems relatively unimportant. It is now known that adsorption often accelerates a reaction to a far greater degree than can be explained by increased concentration alone. Furthermore, adsorption does not always accelerate, and in some cases it may even diminish the rate of reaction.[16]

The concentration theory completely fails to explain the selective nature of catalysis. Why, for example, does formic acid decompose into hydrogen and carbon dioxide with a zinc oxide catalyst, whereas with titanium oxide, it breaks down to carbon monoxide and water? Or, to quote another example, why do carbon monoxide and hydrogen form methane in the presence of nickel, whereas quantitative yields of methanol are produced with a zinc chromite catalyst.[6]

Such effects suggest an influence arising from the chemical properties of the catalyst. Specific catalytic effects depend on properties, of the surface atoms, and these properties in turn, are a function of the chemical nature of the catalytic substance. This is shown by the fact that the same type of active surface is not developed on copper as on platinum, and both differ from the active surface that can be formed on carbon. Properties of surface atoms that are important in catalysis appear to be linked to specific adsorptive powers, thus hydrogenating and dehydrogenating catalysts adsorb hydrogen, whereas oxidizing catalysts adsorb oxygen.

Intermediate-Compound Theory

Sabatier[17] noted that the vigor of the reaction induced by a catalyst is not necessarily related to the amount of reactants adsorbed; thus, although platinum adsorbs more hydrogen than does nickel, yet nickel is often the better hydrogenation catalyst. From this, he concluded that activity depends on the ability of the catalyst to attach the substance in a very reactive form. In the case of nickel and hydrogen, the activity is presumed to depend on the formation of a temporary hydride, NiH_2 or Ni_2H_2, from which the hydrogen is subsequently released to a suitable acceptor such as ethylene—the catalyst surface being left in its primitive form, capable of reproducing the cycle.

The reverse process of dehydrogenation proceeds through a similar mechanism. Hydrides can be formed by taking hydrogen from various substances which are thus dehydrogenated; ammonia gas is rapidly decomposed into its elements when passed through a bed of red-hot copper or iron turnings.

This theory, known as the *intermediate-compound theory*,[2] has been very fruitful in many research studies. Sabatier and his co-workers, who made many important contributions to the industrial utilization of catalysts, were guided by this theory. Although the theory still receives support, it now occupies a somewhat subordinate position. In most catalytic reactions, it has not been possible to establish the separate existence of the intermediate compound stage, and in the minds of many, this places a question mark on the validity of the theory. However, it is well recognized that the formation of stable compounds is not conducive to high activity. Elements that form stable hydrides, *e.g.,* chlorine or oxygen, do not show the hydrogenating powers found in nickel or platinum, and the stable sodium chloride is useless as a carrier of chlorine. Therefore, one would expect that the ability of intermediate compounds to participate in a further reaction would be associated with instability, and this very instability would make it difficult to show their separate existence.

Some have sought to cover the difficulty of isolating the intermediate stage by adopting the designation *indefinite intermediate-compounds* or *association complexes*. Closely akin to this concept is the *activated adsorption* theory which introduces energy relationships. Bancroft[2] points out that the definition of intermediate compounds has become so elastic that it is just as convenient to rewrite many interpretations of the intermediate compound theory in terms of adsorption. As he further points out, the distinction between intermediate compounds and adsorption is psychological rather than chemical.

Activated Adsorption

The theory of activated adsorption is founded on a general concept proposed by Arrhenius to explain the influence of temperature on the rate of chemical reactions.[1,2] Arrhenius suggested that every system contains *normal* and *active* molecules, but only the active molecules participate in a chemical reaction. There are diverse views as to the mechanism of activation of molecules, but it is generally agreed that the essential difference between normal and active molecules is that the active molecules have a higher content of energy.[2,19] The distinction is relative because the amount of energy needed for a molecule to be considered active depends on the reaction that is involved. In order that a molecule of A may undergo chemical transformation into a different molecular species B, it must have at least a certain minimum content of energy—the so-called critical energy for reaction. Usually this is higher than the average energy, and the additional energy over and above this average, which a molecule A must acquire in order to react, is known as the *energy* of *activation*.

Taylor[20] observed that temperature changes often produce fundamental differences in the adsorptive behavior, and suggested that an elevation in temperature can produce an activated condition so that molecules become adsorbed in a special state. This special state may lead to a strong chemisorption, as is the case with oxygen molecules adsorbed as surface oxides on carbon. These molecules do not enter into further oxidation reactions. In other cases, adsorbed molecules in the special state are able to participate in further reactions and at lower temperatures than in the absence of a catalyst.

Various mechanisms have been suggested to explain why less energy is needed for molecules to react when in the adsorbed state. Hinshelwood[7] gives an analogy to illustrate one possibility: When a safe is opened by a key, little energy is needed, but precise geometrical conditions must be fulfilled; when it is blown open, much more energy is used, but the exact geometry of the lock ceases to be of importance. Similarly, the change of monoclinic to rhombic sulfur can be accomplished in several ways, each with different energy requirements. One way is to vaporize monoclinic sulfur and condense the vapor on a portion of the rhombic form at a suitable temperature. Another method is to cool a mixture of monoclinic and rhombic sulfur below the transition temperature. In this latter case the change is accomplished by atoms or molecules of sulfur at the interface becoming detached from one phase to be oriented into the new structure without ever being free; this change can be initiated by far less energy than needed to vaporize sulfur. Based on these analogies, it can be pictured that the transformation of molecule A into species B could be accelerated by adsorption in an ordered fashion to provide nuclei for the growth of the new species. In a reaction $AB + XY = AY + BX$, a catalyst, by causing strains and suitable orientation, may enable the atoms to be exchanged without ever being free.

Strains due to adsorption may cause some adsorbed molecules to be disrupted into atoms or ions.[2] Polanyi[21] has pointed out that if molecules AB and CD were adsorbed as atoms, A, B, C, D, then when these atoms are desorbed they would have an opportunity to form new molecular combinations. Such a mechanism would hinge on the ease with which the atoms are desorbed; the binding to the surface should not be strong enough to prevent a rapid rate of desorption. This may be a clue to the phenomenon so often observed, that catalytic activity seldom coincides with the temperature of maximum adsorption. The amount of adsorption is often greater at a temperature lower than that at which catalytic activity becomes appreciable. According to Langmuir,[22] the measured amount of adsorption depends not only on the number of molecules that become attached to the surface, but also on the lapse of time between adsorption and desorption. Thus, the reduced adsorption, as measured at elevated temperatures, may be due not to a smaller number of molecules participating in adsorption, but rather to a shorter staying time in the adsorbed state. An increased pace at which changes of state occur would increase the opportunities for reaction.

It has also been suggested that the disruption of molecules into atoms could initiate a path on which a transformation from A to D could proceed through a series of steps $A \rightarrow B \rightarrow C \rightarrow D$, with each step requring only a fraction of the energy needed to go directly from A to D.[23] Moreover, any energy released at each intermediate stage could help activate the subsequent stage.

When adsorbed molecules are disrupted into atoms, the adsorbed atoms being bonded to the surface must lack some properties associated with free atoms. Burke[24] has suggested that stretched molecules may be more reactive than those disrupted into atoms. The distortion by stretching could change the distribution of energy within the adsorbed molecules, and the energy may be momentarily concentrated in some particular bond enhancing the activity at that point. Moreover, some of the energy

of adsorption normally released as heat may be momentarily
stored in potential or vibrational forms in the adsorbed molecules,
during which time an opportunity may occur for reaction with
other molecules.[25]

A molecule does not exist continually in one form but resonates
between different forms, *e.g.,* carbon monoxide can exist in the
forms:[26]

$$C^- \equiv O^+, \quad C = O, \quad C^+ - O^-$$

A molecule persists for a longer time in the form that has the least
energy, but during briefer intervals it will assume other forms.
Adsorption may provide energy to lengthen the time during which
a molecule remains at a higher energy level.

The influence of time may also be expressed by saying that
when two molecules *AB* and *CD* are capable of exchanging
partners, the exchange is more apt to occur when they are in a
more energetic state and in close proximity, with the partners
suitably oriented toward each other. The chances are not great
that all the required favorable conditions will exist simultaneously
during momentary contacts between molecules in the gas phase.
However, when two molecules are adsorbed side by side there is
greater probability that all favorable conditions will occur at some
time during their stay.[27]

Although the discussed mechanisms indicate various ways in
which a catalyst could enable a reaction to proceed at lower temperatures with less activation, they do not explain why the course
of a reaction is not the same on all catalysts.

Orientation[2]

Some students of contact catalysis suggest that an adsorbed
molecule is anchored to the surface at only one point of attachment, leaving the balance of the adsorbed molecule free to extend into the solution or gas phase where it can be bombarded
by molecules of the other reactant. Bancroft[18] has speculated on
how different orientations could cause different reactions; *e.g.,*
the oriented adsorption of ethyl acetate could account for three
different reactions depending on whether the molecule is attached
through the ethyl, the methyl, or the carboxyl group. The orientation theory is also capable of explaining cases in which a reaction
is suppressed by a catalyst. Here it is assumed that the reacting

group is held so as to be shielded from contact with molecules of the other reactant.

Specific Activity

The theory of orientation, and indeed all theories that trace a relation between the kind of products formed and the nature of a catalyst, ultimately rest on an assumption that differences exist in the quality of solid surfaces.

As molecular configuration influences the chemical properties of molecules in the vapor and liquid state, it is reasonable to believe that the specific behavior of active centers on solids could arise from definite arrangements of surface atoms, a concept proposed by Adkins[28] in 1922. He and his co-workers found that the ratio of competing reaction products formed by a catalyst could be modified by changes in the preparation of the catalyst. One typical study employed two titanium oxides, one of which was prepared by the hydrolysis of tetraethyl orthotitanate, and the other from tetrabutyl orthotitanate. When these two catalysts were used separately against ethanol, it was found that they differed in the ratio of ethylene and ethane produced. Adkins[29] suggested that spatial properties, existing in the original source materials, survived to some degree in the finished catalyst. The space relations of several adsorbing centers, acting on a single molecule, could influence the manner in which the molecule is distorted. As expressed by Adkins,[29]

Differences in spacing of the centers of activity might produce different molecular fragments, just as the size and shape of fragments of a piece of paper would be in part determined by the relationship in space of the two hands which tore the original sheet.

Burke[24] and Balandin[30] have developed the concept of the influence of the geometry of the surface. Balandin considers adsorption to be accompanied by catalysis only when the active centers are distributed at suitable distances to form groups which he terms *multiplets*. According to this theory, the arrangement of active ceters within the multiplets affects the specificity of the reaction that is induced. This can be illustrated by the presumed behavior of ethyl alcohol with a catalyst. If we assume the hydroxyl group together with one hydrogen atom of the alcohol molecule to be held on one spot, and the balance of the molecule on another spot, then, when the molecule is torn apart, the hydrogen and hydroxyl on the one spot will unite to form water and the remaining portion of the molecule will form ethylene. If at a different multiplet, one active spot adsorbs two hydrogen atoms, leaving the balance of the molecule on an adjacent patch, this would favor dehydrogenation and result in the formation of an aldehyde. Similarly, the exchange of partners between AB and CD would be facilitated when A and D, or B and C, are adsorbed on the same centers of a multiplet. The absence of suitably arranged multiplets would by this theory explain cases in which adsorption occurs without catalysis.

Active Centers

It is apparent that the hypothesis of active centers is extensively employed to interpret catalysis, but it should be stated that the concept has been questioned. Direct evidence of the existence of active centers is somewhat lacking. X-ray studies of an active catalyst always seem to give the usual diffraction pattern of the fully crystalline material. This, however, is not considered a disproof of the existence of active centers, or that their structure is identical with the normal space lattice. As is pointed out, surface layers are not of sufficient thickness to give an X-ray pattern of their own. Electron diffraction is reported to indicate a difference between catalytically active and inactive platinum, and also to distinguish between certain supported metallic catalysts.

The indirect evidence of the existence of active centers is reached from so many different approaches as to seem almost conclusive. Nevertheless, we must not overlook that some of the preference for interpreting adsorption and catalysis in the light of active centers may arise from the ease of visualizing a concept expressed in space relations. We must keep in mind that mere geometric arrangements in themselves are insufficient to produce the changes that occur, and the action of active centers must be associated with a deeper disturbance of energy relations.

Supports

Inert refractory supports have long been used as economizers to extend the surface of active catalysts;[1,2] thus asbestos has been used to support platinum to catalyze the oxidation of the SO_2. The original purpose of supports seems to have been to extend the surface of expensive catalysts. Atoms that are buried in the bulk phase, when a catalyst acts as its own support, are brought to the active surface when distributed over another, cheaper, support.

Inert supports also provide other advantages. Some metals are most active when reduced from the oxide or salt at the lowest possible temperature, and a suitable support enables such a minimum temperature to be utilized. Then too, there is a tendency for a catalyst to recrystallize, forming a more stable but less active surface at temperatures far below the melting point—a behavior known as *sintering*. When distributed over an inert support, the molecules of catalyst have less opportunity to cohere to one another and sintering is thus reduced. Some supports prolong the life of the catalyst by suppressing the susceptibility to poisoning, and a supporting material is often selected with this in mind. The nature of this action is obscure although in some cases, it appears to be caused by preferential adsorption of the poison on the support.

Supports that do not directly participate in a catalytic reaction can increase the rate of catalysis indirectly by providing additional areas on which reacting molecules can be captured. Molecules adsorbed by physical forces on inactive areas may retain sufficient mobility to enable them to migrate over the surface and rapidly reach an active center.[31] The active centers are thus supplied with reacting molecules more rapidly than could occur by direct hits from a vapor or solution phase.

Promoters

Much more positive effects of supports are frequently found; the activity of the catalyst may be multiplied many times. This behavior springs from the ability of certain pairs of substances to enhance each other's activity and produce effects far greater than can result from the use of either alone. The phenomenon is known as *promotion,* and a substance that accelerates the activity of another is termed a promoter.[1,2] Such behavior has led to the use and addition of substances other than those needed to support the catalyst. Frequently, only small quantities are needed to produce great effects. Medsforth[32] found that hydrogenating activity of nickel for carbon monoxide or dioxide is increased nearly twelve-fold by the addition of suitable oxides, only 5% of cerium oxide being needed.

Many theories have been offered to explain the action of promoters, but as is true of much catalytic behavior, we are a long way from understanding the precise mechanism provided by each promoter. Many of the theories are not mutually exclusive and several mechanisms may occur simultaneously.

An interesting theory of promoter action is based on the belief that interfaces are the seat of catalytic activity—the active centers.[1,2] It is apparent that new boundaries will be provided and existing ones altered when one solid is incorporated with another, and the enhanced activity may reside in specific configurations produced at boundaries of the two solids. That boundaries between surfaces can be the seat of activity is supported by the fact that a decomposition or combination may be difficult to start on a solid surface consisting entirely of a single component. Overburnt lime does not hydrate readily because water does not combine except where hydration has already begun.[22] Faraday found that hydrated salts do not commence to effloresce unless the surface is scratched. The action of manganese dioxide on the decomposition of potassium chlorate is attributed to specific boundary influences. Pease and Taylor[33] found that the reduction of copper oxide by hydrogen occurs almost entirely at the interface between copper and copper oxide.

Boundary influence can be an important factor, yet, as Bancroft[2] has pointed out, it fails to explain the cases in which a promoter has a maximum influence at concentrations far below that which gives a maximum extent of interface.

Another suggestion covering the action of promoters is that one of the reactants is activated on the catalyst and the other on the promoter. In other reactions, the promoter may provide needed links in a chain reaction, some steps occurring on the catalyst and others on the promoter.

Promoters can provide more suitable geometric arrangements of the active centers. The ability of very small concentrations of a promoter to produce large effects has been attributed to the promoter molecules completing a necessary configuration, just as a single piece can complete an elaborate jig-saw puzzle. This concept fits into the Burke-Balandin multiplet theory, the promoter furnishing active atoms to complete suitable multiplets.

In this connection, it is of interest to note an influence that promoters can have on the course of a reaction. Adam[27] reports that pure ferric oxide converts carbon dioxide and hydrogen into methane and liquid hydrocarbons; but when a trace of sulfur is present in the oxide, methanol is formed instead.

Beneficial Poisoning

Bancroft[18] has called attention to the fact that the action of poisons is often similar to that of promoters. Rosenmund,[34] using a palladium catalyst, found that benzaldehyde is not formed by treating a solution of benzoyl chloride in benzene with hydrogen, unless a trace of sulfurized quinoline is present. The hydrogenation of benzaldehyde gives greater yield of the alcohol with smaller yield of other products when a trace of quinoline is present. The term poisoning implies a detrimental effect, but it is apparent that poisoning can be helpful under certain conditions. Poisoning may be beneficial by blanketing undesired active centers and thus enabling the catalyst to concentrate on the desired reaction. Poisoning may also cause a reaction to stop at a desired intermediate stage of a chain of reactions.

Hilditch[6] points out cases in which poisoning is anesthetic rather than definitely *toxic,* for under suitable conditions, the poison is displaced from the catalyst which then resumes its normal activity. In such instances, beneficial poisoning may be provided by maintaining a definite concentration of the poison in the reactants. It is claimed that the presence of small quantities of oxygen in hydrogen aids various hydrogenation processes by minimizing the formation of undesired by-products.

2. ACTIVATED CARBON AS A CATALYST

Many reactions are catalyzed by activated carbon.[36,37] The catalytic power of an activated carbon is often quite specific. The specific properties depend on the method of manufacture. Two carbons, each made by a different process, may be similar in adsorptive properties and yet be dissimilar in specific catalytic power. One of them may be more effective for certain reactions whereas the other carbon may prove more appropriate for other reactions.

Activated carbon finds many applications as a carrier for other catalysts. Apart from providing a large surface, various other benefits have been observed to attend the use of activated carbon as a carrier for other catalysts. One value is that the catalyst can be prepared at low temperatures—a feature generally desirable. Thus nickel salts deposited on activated carbon can be formed into an efficient catalyst at 350° to 400°C; whereas temperatures of 420° to 450°C are required when kieselguhr is used as a support, and 550°C or higher with silica gel.[6]

Other benefits are illustrated in the use of activated carbon as a carrier for aluminum chloride. Various difficulties attend the use of unsupported aluminum chloride as a catalyst. Sticky intermediate products are formed in certain reactions which coat the aluminum chloride granules, causing them to adhere to one another and form large agglomerates with little activity. Moreover, aluminum chloride tends to sublime at 180°C under normal pressure, a feature which restricts its usefulness. Studies by Ipatieff and Pines[38] indicate that both difficulties are overcome by mixing the aluminum chloride with a chloride of another metal, e.g., zinc chloride, and adsorbing the mixture on activated carbon. A catalyst thus prepared shows little or no tendency to agglomerate. Then too, the adsorptive power of the carbon retards the volatilization of the aluminum chloride.

In other cases, the carbon acts as a promoter and increases the rate of the reaction, or enables it to proceed at lower temperatures. Working with zinc oxide deposited on charcoal, Adaurov[39] observed that the carrier may alter the course of the reaction.

The promoting action is not the same for inactive as for activated carbon, and dissimilar effects are found with different types of activated carbons. Sabalitschka and Moses[40] studied the influence of the carrier on a palladium catalyst employed for the hydrogenation of maleic and fumaric acids. They found that the activity of the catalyst depends on the extent to which the palladium compound is adsorbed prior to reduction. With carriers of low adsorbing power, a more efficient catalyst can be prepared by using the readily adsorbable palladium hydroxide rather than the less adsorbable chloride. With good adsorbents, equally efficient catalysts can be prepared with either of the palladium compounds.

Adadurov[41] prepared catalysts by adsorbing lead nitrate on carbon, then drying, and heating to 360°C. He found that when these catalysts were employed for the decomposition of formic acid, the course of the reaction depended on the conditions under which the lead nitrate had been originally adsorbed. Formic acid was decomposed into carbon dioxide and hydrogen when the catalyst was prepared by adsorbing the lead salt from an 0.01 N solution; whereas appreciable amounts of carbon monoxide were formed with a catalyst prepared with a 0.05 N solution of lead nitrate.

The behavior of a catalyst can be modified by changing the conditions. Johne and Weden[42] found that ferrocyanide is almost completely oxidize to ferricyanide in the presence of air and carbon in a slightly acid medium; whereas in a strongly alkaline medium, the reverse reaction occurs.

The hydrogen-ion concentration is a factor in other reactions; a high pH promotes the catalytic oxidation of uric acid[43] but inhibits the hydrolysis of aspartic acid.[44] Hartnung[45] found that the type of reaction products formed in the catalytic reduction of nitriles depends on whether or not hydrogen chloride is present in the ethanol used as a solvent.

Catalysis in the liquid phase can be influenced by the nature of the solvent used, and the effect may be quite specific.[46] Bente and Walton[47] found that the oxidation of hydroquinone, in the presence of carbon, proceeds less readily in organic solvents than in water; dioxane suppresses the oxidation almost completely. Taylor and Coughtrey[48] observed that the stereoisomeric changes of sulfoxides which took place readily with tetraline as a solvent did not occur when benzene was so employed.

Temperature and pressure can alter the course of a reaction as would be expected from thermodynamic considerations. The influence of temperature is shown in the reaction between chlorine and ethylene, which, in the presence of carbon, yields dichloroethylene at 120° to 125°C; whereas higher temperatures favor the formation of hexachloroethane.[49]

Many studies have been made of specific applications of activated carbon as a catalyst and also as a promoter or carrier for other catalysts. Of these, a few will be briefly described.

Reactions Involving Halogens[50]

Originally, phosgene was produced by the action of sunlight on chlorine and carbon monoxide, but it has since been found that the reaction proceeds readily in the absence of light when a mixture of the gases is passed over activated carbon.[6,36] One pound of activated carbon can catalyze the production of 1 ton of phosgene before reactivation of the carbon becomes necessary.

Sulfuryl chloride is produced by the reaction of sulfur dioxide with chlorine in the presence of activated carbon.[6] Formerly, the reaction was conducted in the vapor phase, but now it is carried out in the liquid phase.[50] Activated carbon is suspended in liquid sulfuryl chloride through which equal volumes of sulfur dioxide and chlorine are passed.[51] One pound of the more active types of carbon will catalyze the production of 250 pounds of sulfuryl chloride per hour.

An efficient and safe method of preparing hydrogen chloride consists of passing an equimolecular mixture of hydrogen and chlorine through a bed of granular activated carbon.[36] Activated carbon also catalyzes the formation of hydrogen chloride from steam and chlorine.[52]

Newman[53] and his coworkers report that bromine and steam react at 600°C to form hydrobromic acid when activated carbon is present. The reaction temperature can be lowered to 500°C when iron oxide is mixed with the carbon. The concentration of the resulting acid depends on the proportion of steam that is used, and fuming acid (60% HBr) can be made readily by this method. The oxygen liberated by the reaction gradually oxidizes the carbon to carbon dioxide.

The preparation of chlorinated organic compounds involves reactions that seldom go to completion and some free chlorine is present in the end products. The free chlorine may be removed by passing the product through a bed of activated carbon saturated with water. The carbon acts as a catalyst to promote the reaction between chlorine and water, and the hydrochloric acid that is formed drains away with the water at the base of the bed.

The formation of vinyl chloride from acetylene and hydrogen chloride is catalyzed by activated carbon impregnated with mercuric and mercurous chlorides.[54,55] The catalyst is prepared by saturating activated carbon with mercury vapor at 200°C and then adding hydrogen chloride to form mercuric and mercurous chlorides.

Dechlorination of Water Supplies

The removal of chlorine from water involves both adsorption and catalysis.[56,57,58] Presumably the chlorine is initially adsorbed and then reacts:

$$2Cl_2 + 2H_2O \rightleftharpoons 4HCl + O_2$$

The dechlorinating power of carbon can be influenced by the presence of other constituents in the water supply. Ammonia will form chloramines that are less readily removed by carbon. The catalytic removal of chlorine is accelerated at a low pH and retarded at a high pH.

Instances are reported in which the dechlorinating action will temporarily diminish and then recover if the carbon is given a brief rest period. Under normal circumstances this fatigue does not develop; in fact, it occurs only when rather high concentrations of chlorine are present.

Bokhoven and van der Meijden[58] report that for an average quality of granular or pelletized carbon about 3 inches of a carbon layer will reduce the concentration of chlorine by 50%; here the principle of the *halving value* is involved. The height of the carbon column should be calculated in accordance with this principle of halving value, particularly when a low final chlorine concentration is required. In addition, a reserve layer is recommended with a view to a possible build-up of organic compounds in the carbon bed which could reduce the catalytic properties of the carbon.

Isomerization

Cowan[59,60,61] and his coworkers have developed a process for the isomerization of nonconjugated oils to conjugated forms. The process consists of bringing a vegetable oil—cottonseed, linseed, or soybean oil—in contact with a nickel-carbon catalyst at 170°C. An inert atmosphere is maintained during the reaction, which requires 6 to 8 hours for completion at a catalyst concentration of 7.5%. When 20% of catalyst is present, only 20 to 30 minutes are required for isomerization. A sample of catalyst can be used to isomerize five to ten batches of oil before its activity is lost. The nickel-carbon catalyst does not split ester linkages, and no further chemical treatment of the isomerized oil is required.

Cowan and his coworkers observed that certain types of activated carbon are much more suitable than others for the preparation of the catalyst.

Hydrogenation

Schuster[62] reports that activated carbons possess hydrogenating power for adsorbed olefines whereas inactive charcoal acquires this property only when hydrogenating metals are deposited. Kailan and Stüber[63] found that wood char is better than animal char as a support for nickel in the hydrogenation of oleic acid.

Reactions Involving Sulfur Compounds

Activated carbon will adsorb considerable quantities of hydrogen sulfide, but much better removal can be accomplished by employing the carbon to catalyze the oxidation of hydrogen sulfide to sulfur.[64] In a process developed by Engelhardt,[65] the reaction is catalyzed by adding a small amount of ammonia to the gas. The process has been adopted in some operations, but generally, iron oxide or various liquid systems are employed. These methods, although effective for removing hydrogen sulfide, are not suitable for removing organic sulfur compounds. The organic sulfides can be broken down by certain catalysts at temperatures of 250° to 400°C, but a more practical method of removal is to pass the gas through a bed of granular activated carbon.[66] In order to accomplish efficient and complete removal, the gas should be relatively dry because moisture interferes with the adsorption of sulfur compounds. Giller and Winkler[67] find that the efficiency is increased by introducing small quantities of air and ammonia into the water gas prior to the treatment with carbon; presumably the effect is similar to that reported by Engelhardt for hydrogen sulfide.

The carbon can be regenerated for re-use by removing the adsorbed sulfur compounds. When oxidizing conditions exist during the adsorption, a large portion of the sulfur compounds is oxidized to sulfur which can be extracted with a suitable solvent,[68] e.g., ammonium sulfide. An alternative method of removing the sulfur compounds and thus regenerating the carbon has been developed by Lander, Sinnatt, King, and Bakes.[69] Steam, at a temperature of 250° to 350°C, is passed through the carbon bed in a direction counter to that of the gas during the adsorption stage. The steam should contain 2.5% air and 1% ammonia. The sulfur compounds are carried off by the steam and can be recovered by condensation. Even under the best conditions, the regeneration is not complete. The carbon gradually loses efficiency with re-use and is eventually replaced with fresh carbon.

Olin, Scarth, and Starkweather[70] report that the removal of hydrogen sulfide from fuel gas by iron oxide is promoted by activated carbon. The extent to which the action is promoted varies with the type of carbon used.

In recent years the use of carbon impregnated with iron or copper (or both) is receiving increasing attention for the removal of traces of sulfur compounds from gas streams. According to one view, the impregnated carbon catalyzes the conversion of hydrogen sulfide to elemental sulfur. Other studies indicate that the removal is not due to catalytic action but is, rather, a chemical reaction involving the formation of metallic sulfides plus the adsorption of organic sulfides by the carbon.

The impregnated carbon can be regenerated by air and steam under appropriate conditions. Information on this is available from the supplier.

Reactions Involving Oxygen

Activated carbon catalyzes the oxidation of ferrous sulfate to the ferric form. In a procedure described by Schumacher and Heise,[71] air is bubbled through an acidulated aqueous solution of ferrous sulfate in which activated carbon is present. Burdick[72] found that the oxidation of nitric oxide is increased five-hundred fold in the presence of activated carbon; moisture decreases the activity.

Janes[73] describes a process for the electrolysis of aqueous solutions of caustic alkali for producing oxygen without the simultaneous liberation of hydrogen. This is accomplished by a cathode of activated carbon designed so that air adsorbed from the atmosphere reacts with the nascent hydrogen to form water.*

*Some hydrogen peroxide is also formed, the oxygen being released when it migrates to the anode.

The over-all drop in voltage through the cell is thus reduced and this results in decreased consumption of electrical energy. No water is consumed by the process, the oxygen liberated at the anode being chemically equivalent to that taken up at the cathode. Basically, the process is equivalent to extracting oxygen from air at the cathode and delivering it in a pure form at the anode.

Berl[74] describes the use of an activated carbon cathode to prepare hydrogen peroxide by an electrochemical process. A porous pipe, covered with activated carbon, is used as the cathode in an electrolyte solution, the cathode being separated from the anode by a diaphram. Air or oxygen is forced through the cathode pipe; in the presence of the activated carbon, the oxygen reacts with the cathodically evolved hydrogen to form hydrogen peroxide. The operations is improved by impregnating the carbon with paraffin, but only a very thin coating should be applied to avoid impairing the activity of the carbon. A preferred method of impregnation is to dip the carbon into a thin solution of paraffin in a volatile solvent and subsequently evaporate the solvent. The process is also adaptable to the production of perborates.

Activated carbon catalyzes the decomposition of hydrogen peroxide,* and for this certain types of carbon are more effective than others. The reaction may be utilized when it is desired to remove hydrogen peroxide from a solution without introducing other chemical contamination. The reaction is also employed in certain applications to effect greater decolorization of a product. Skumbardis[75] reports that the simultaneous addition of hydrogen peroxide and activated carbon to a raw sugar solution effects a more rapid and more complete decolorization than is obtained by either hydrogen peroxide or carbon acting alone.

Martus and Becker[76] utilize the oxygen-adsorptive power of activated carbon to depolarize the cathode of an alkaline primary cell. A cathode, made of activated carbon, is designed so that part of it is in contact with the atmosphere. Air adsorbed from the atmosphere diffuses through the carbon and depolarizes the cathode. Martus and Becker report that activated carbon, when very wet with water, has less adsorptive power for oxygen. They overcame this behavior by impregnating the carbon with a slight amount of oil.

---

*In general, a substance that catalyzes a reaction in one direction will also accelerate the reverse reaction. The direction taken depends on the point of thermodynamic equilibrium.

Depolarization by Air in Dry Cells

Types of activated carbon which have electrical conductivity can be used instead of manganese dioxide as a depolarizer in dry-cell batteries.[77] Powdered activated carbon is treated with ammonium chloride and pressed into a porous cylindrical shape that is placed in a zinc container forming the outer electrode. If provision is made for a sufficient supply of air, the effect of the activated carbon cylinder is similar to that of manganese dioxide. The potential of the activated carbon battery is 0.1 volt lower than that of the manganese dioxide type, but the voltage is steadier and is practically constant throughout the life of the battery.

This type of dry cell has been widely adopted in Europe where a specially developed carbon is employed in the manufacture.[78] The carbon cell is especially suitable where it is necessary to furnish low continuous outputs of energy for long periods of time. The shelf life of these cells is long because there is practically no self-discharge. Consequently these cells are extremely efficient as sources of current over long periods in remote places such as telephone and telegraph block stations, and electric installations for wire fencing.

Miscellaneous

Indirect effects of carbon are sometimes confused with direct catalytic activity, *e.g.*, some reactions are accelerated by activated carbon as a result of the adsorption of inhibitors. An example is found in the recovery of iodine from iodides present in petroleum salt brines. Nitrous acid is employed to oxidize the iodides to iodine—a reaction that may be retarded by the presence of inhibitors in the brine. Treatment of the brine with an activated carbon removes the inhibitors and enables the oxidation reaction to proceed.

Activated carbon causes some reactions to proceed beyond the point of normal equilibrium, and this property is utilized in some applications. Thus, under normal conditions, it is not possible to completely neutralize the fatty acids in edible oils, because the reaction

$$RCOOH + NaOH = RCOONa + H_2O$$

is reversible. When activated carbon is present, however, the RCOONa (soap) is adsorbed, and the reaction proceeds to completion.

Summary

This chapter endeavors to present a perspective of the use of activated carbon in heterogeneous catalysis. The applications selected for description are but a fragment of the vast number that are reported in patents and scientific journals; all of which are described in CHEMICAL ABSTRACTS, where specific applications can be readily located indexed under *Carbon, Active* and/or *Charcoal Active*

REFERENCES

1. Frankenburg, W. G., Komarewsky, V. I., and Rideal, E. K. (Editors), *Advances in Catalysis and Related Subjects*, Volumes 1 to 7, Academic Press Inc., New York, 1948 to 1960.
2. Reports of the Committee on Contact Catalysis:
   Bancroft, W. D., *Ind. Eng. Chem.*, 14:326, 444, 545, 642 (1922); *J. Phys. Chem.*, 27:801 (1923).
   Taylor, H. S., *J. Phys. Chem.*, 28:897 (1924); 30:145 (1926).
   Reid, E. E., *J. Phys. Chem.*, 31:1121 (1927).
   Burke, R. E., *J. Phys. Chem.*, 32:1601 (1928).
   Armstrong, E. F., and Hilditch, T. P., *J. Phys. Chem.*, 33:1441 (1929).
   Frazer, J. C. W., *J. Phys. Chem.*, 34:2129 (1930). Twelfth Report of the Committee on Catalysis, John Wiley and Sons, New York, 1940.
3. Applebey, M. P., *Proc. Roy. Soc.*, A127:240 (1930).
4. Freundlich, H., *J. Chem. Soc.*, 164 (1930).
5. Adkins, H., *Ind. Eng. Chem.*, 32:1189 (1940).
6. Hilditch, T. P., and Hall, C. C., *Catalytic Processes in Applied Chemistry*, D. van Nostrand Inc., New York, 1937.
7. Hinshelwood, C. N., *J. Chem. Soc.*, 1203 (1939).
8. Pease, R. N., and Stewart, L., *J. Am. Chem. Soc.*, 47:1235 (1925).

9. Almquist, J. A., and Black, C. A., *J. Am. Chem. Soc.*, 48: 2820 (1926).
10. Beebe, R. A., *J. Phys. Chem.*, 30:1538 (1926).
11. Taylor, H. S., *Proc. Roy. Soc.* (London), 108A:105 (1925).
12. Schwab, G. M., and Pietsch, E., *Z. Elektrochem.*, 35:573 (1929).
13. Constable, F. H., *Proc. Roy. Soc.*, A110:283 (1926).
14. Wood, R. W., *Proc. Roy. Soc.* (London), A102:1 (1922).
15. Smekal, A., *Z. Elektrochem.*, 35:567 (1929); *Z. Physik*, 55:289 (1929).
16. Kruyt, H. R., and Duin, C. F. van, *Rec. trav. chim.*, 40:249 (1921).
17. Sabatier, P., *La catalyse en chemie organique*, Van Nostrand Co., New York, 1920; *Chem. and Ind.*, 46:681, 702 (1927); *Ind. Eng. Chem.*, 18:1005 (1926).
18. Bancroft, W. D., *Applied Colloid Chemistry*, McGraw-Hill Book Co., New York, 1932.
19. Lewis, G. N., and Smith, D. F., *J. Am. Chem. Soc.*, 47:1508 (1925).
20. Taylor, H. S., *J. Am. Chem. Soc.*, 53:578 (1931).
21. Polányi, M., *J. Soc. Chem. Ind.*, 54:123T (1935).
22. Langmuir, I., *J. Am. Chem. Soc.*, 38:2221 (1916).
23. Taylor, H. S., *Proc. Roy. Soc.* (London), 113A:77 (1926).
24. Burke, R. E., *J. Phys. Chem.*, 30:1134 (1926); 32:1601 (1928).
25. Hinshelwood, C. N., *Proc. Roy. Soc.*, A127:240 (1930).
26. Kimball, G. W., "Twelfth Catalysis Report," Chapter 1, John Wiley and Sons, New York, 1940.
27. Adam, N. K., *Physics and Chemistry of Surfaces*, 2nd Edition, Oxford University Press, New York, 1938, pp. 277, 240.
28. Adkins, H., *J. Am. Chem. Soc.*, 44:2175 (1922).
29. Adkins, H., and Millington, P. E., *J. Am. Chem. Soc.*, 51: 2449 (1929).
30. Balandin, A. A., *J. Gen. Chem.* (U.S.S.R.), 2:166 (1932); *Z. Physik. Chem.*, B2:289 (1929).
31. Volmer, M., *Trans. Faraday Soc.*, 28:359 (1932).
32. Medsforth, S., *J. Chem. Soc.*, 123:1452 (1923).
33. Pease, R. N., and Taylor, H. S., *J. Am. Chem. Soc.*, 43:2179 (1921).
34. Rosenmund, K. W., and Zetsche, F., *Ber.*, 54:425 (1921).
35. Rosenmund, K. W., and Jordan, G., *Ber.*, 58 (B):160 (1925).

36. Deitz, V. R., *Bibliography of Solid Adsorbents*, United States Cane Sugar Refiners and Bone Char Manufacturers and National Bureau of Standards, Washington, D.C., 1944. Abstracts of scientific literature on the use of active carbon as a catalyst, pp. 652–688.
Supplementary volume published in 1956.
37. Hassler, J. W., *Active Carbon, the Modern Purifier*, Industrial Chemical Sales Division, West Virginia Pulp & Paper Company, New York, 1941. References to the use of active carbon as a catalyst, pp. 111–13.
38. Ipatieff, V. N., and Pines, H., U.S. Patent 2,327,188.
39. Adadurov, I. E., and Krainii, P. Ya., *J. Phys. Chem.* (U.S.S.R.), 5:1132 (1934).
40. Sabalitschka, T., and Moses, W., *Ber.*, 60B:786 (1927).
41. Adadurov, I. E., *J. Phys. Chem.* (U.S.S.R.), 5:1139 (1934).
42. Johne, F., and Weden, H., *Biochem. Z.*, 273:147 (1934).
43. Truszkowski, R., *Biochem. J.*, 24:1349 (1930).
44. Wunderly, K., *Helv. Chim. Acta.*, 16:515 (1933).
45. Hartung, W. H., *J. Am. Chem. Soc.*, 50:3370 (1928).
46. Adkins, H., *Ind. Eng. Chem.*, 32:1189 (1940).
47. Bente, P. F., and Walton, J. H., *J. Phys. Chem.*, 47:133 (1943).
48. Taylor, T. W. J., and Coughtrey, W. C. J., *J. Chem. Soc.*, 974 (1935).
49. Alekseevskii, E. V., *J. Russ. Phys. Chem. Soc.*, 55:403 (1924).
50. Pope, W. J., *Rec. trav. chim.*, 42:939 (1923); British Patent 122,516.
51. McKee, R. H., and Salls, C. M., *Ind. Eng. Chem.*, 16:279, 351 (1924); Canadian Patent 251,586.
52. Ruff, V. T., *J. Chem. Ind.* (U.S.S.R.), 13:348 (1936).
53. Neumann, B., Steuer, W., and Domke, R., *Z. angew. Chem.*, 39:374 (1926).
54. Harshaw Chemical Company, Cleveland, Ohio.
55. Girdler Catalysts, Lousville, Ky.
56. Parrish, William P.: Personal communication.
57. Communication, Ted Barnebey of Barnebey-Cheney, Columbus, O.
58. Bokhoven, W. C., and Meijden, Chr. van der: Personal communication, United Norit Sales Corporation, Amsterdam, Holland.

59. Radlove, S. B., Teeter, M. H., and Cowan, J. C., *Northern Regional Research Lab. Report*, October 1, 1945.
60. Radlove, S. B., Teeter, H. M., Bond, W. H., Cowan, J. C., and Kass, J. P., *Ind. Eng. Chem.*, 38:997 (1946).
61. Falkenburg, L. B., Schwab, A. W., Cowan, J. C., and Teeter, H. M., *Ind. Eng. Chem.*, 38:1002 (1946).
62. Schuster, C., *Trans. Faraday Soc.*, 28:406 (1932).
63. Kailan, A., and Stüber, O., *Monatsh.*, 62:90 (1932).
64. Committee on Gas Manufacture, 1928 Report, *Gas J.*, 184:526 (1928).
65. Engelhardt, A., *Z. angew. Chem.*, 34:293 (1921).
66. Bakes, W. E., King, J. G., and Sinnatt, F. S., *Fuel Research Board (England, Tech. Paper No. 31* (1931).
67. Giller, F., and Winkler, F., U.S. Patent 2,168,933.
68. Engelhardt, A., *Gas-u. Wasserfach.*, 71:290 (1928).
69. Lander, C. H., Sinnatt, F. S., King, J. G., and Bakes, W. E., British Patent 337,348.
70. Olin, H. L., Scarth, V., and Starkweather, W. L., *Gas Age-Record*, 70:561 (1932).
71. Schumacher, E. A., and Heise, G. W., U.S. Patent 2,365,729.
72. Burdick, C. L., *J. Am. Chem. Soc.*, 44:244 (1922).
73. Janes, M., U.S. Patent 2,390,591.
74. Berl, E., *Trans. Faraday Soc.*, 34:1040 (1938); British Patent 389,244.
75. Skumburdis, K., *Kolloid Z.*, 55:156 (1931).
76. Martus, M. L., and Becker, E. H., U.S. Patent 1,673,198 and 2,120,618.
77. Blumenthal, L. M., *U.S. Department of Commerce Fiat Final Report* No. 1181 (June, 1947).
78. Literature furnished by United Norit Sales Corporation Ltd., Amsterdam, Holland.

PART V

BIOCHEMICAL PROPERTIES

15

Biochemical Aspects of Activated Carbon

Adsorbents, such as activated carbon, find many uses in biochemical studies. They are employed in the preparation of biological products; they are used as medicaments; and they provide a tool for the study of biological processes.[1,2,3,4]

Surface chemistry has an essential role in biological processes. Investigators have observed a close relation between adsorption phenomena and various aspects of biological behavior. Many vital processes are interpreted as manifestations of the interchange of energy between the phases of heterogeneous systems which constitute living cells.

1. BIOCHEMISTRY

Surrounding the protoplasm of living cells is a thin membrane with a double surface, one facing the outer world and the other enclosing the protoplasmic system. The membrane is not a mere separating wall; being semipermeable, it provides an avenue for the discharge of waste materials and also enables the cell to select from the outer world raw materials needed for its processes. At these membranes, surface forces synthesize relatively simple compounds into complex biological factors essential to life and growth, but the mechanism of those syntheses is but dimly understood. Freundlich[5] has pointed out that nature—operating within a small temperature range and utilizing a wide variety of interfaces—can produce complex compounds whose preparation in the laboratory taxes the ingenuity of the trained chemist.

Studies of biological systems have revealed data of fundamental importance to students of surface chemistry. Perhaps one of the clearest pictures of active centers on surfaces is provided by Quastel's[6] work on adsorption by bacterial cell walls. More often, however, the approach is reversed and simple systems are searched for clues to the mechanism operating in complex living systems.

Certain forms of catalytic behavior found in biological processes are simulated by inorganic adsorbents.[1,7] Carbon can act as a catalase, oxidase, peroxidase, dehydrogenase, etc. The formation of urea from ammonium carbonate solutions at the temperature of the human body in the presence of activated carbon parallels the action of various animal tissues.

Poisoning has been traced to adsorption phenomena. Substances that poison biological systems often are well adsorbed on carbon and other adsorbents. The hemolytic activity of alkaline soaps parallels their adsorbability on activated carbon and this is also true of the bactericidal activity of phenyl-substituted acids.[9] Narcotics show a similar relation.[10] Poisoning in biological systems often can be traced to the adsorption of substances on cell walls. In some instances it ends the flow of materials in and out of the cell; in others, adsorption of the poison on active centers prevents the synthesis of vital substances.

Warburg,[11] who suggested that cell respiration is catalyzed by heavy metals contained in living cells, observed that chemicals that poison such action also inhibit similar catalytic properties in charcoal. Thus, hydrocyanic acid poisons living cells and also the catalytic properties of carbon. Both effects are ascribed to adsorption on active centers that contain iron.

Larson[12] and coworkers, seeking to learn why peptone inhibited the action of sulfapyridine on pneumococcus, observed that the ability of carbon to adsorb sulfapyridine was reduced when peptone was present. This suggests that peptone may prevent the adsorption of sulfapyridine by bacterial surfaces.

Although data from simple systems provide clues to the mechanisms involved in biological behavior, there are definite limitations. It is obvious that the multiplicity of factors involved in biological processes makes it difficult to reach sure conclusions from the evidence provided by simple nonliving systems.

The ability to prepare substances in a relatively pure form by the use of adsorbents often makes it possible to distinguish effects of impurities or of companion substances present in the natural product from effects of the product itself. Chick and Roscoe,[13] and also Kinnersly and Peters,[14] employed adsorption to differentiate similar vitamins. By selective adsorption and desorption, Willstätter[15] and his coworkers isolated individual enzymes in a relatively pure form and thus advanced our knowledge of these substances. They found that although the specificity of an enzyme remained unchanged, many properties ascribed to enzymes in their natural state depended on companion substances. These companion substances affect the stability of the enzyme, the action of poisons, and the influence of temperature and pH on the activity of the enzyme. Thus, lipase from human prancease is most active in a slightly alkaline medium, whereas lipase from human gastric juice is most active at a pH of 5 to 6. When gastric lipase is treated with an adsorbent (kaolin), the enzyme develops maximum activity at pH 8—similarly to the pancreatic lipase. Apparently, the adsorbent removes a companion compound which inhibits the activity at the higher pH. Willstätter produced purified enzymes with many-fold their natural activity, a matter of importance to industries that employ enzymes to replace chemical methods.

Willstätter worked mainly with mineral adsorbents, but the techniques he employed have been applied to carbon by other workers. A number of enzymes are appreciably adsorbed by carbon if suitable conditions are provided,[1,2,16] but a subsequent elution has not always been accomplished with any great degree of success. Pepsin has been eluted with phosphate[17] and rennin with saponin,[18] but invertase has resisted various methods of elution that have been tried.[19]

Enzymes vary in the extent to which they retain activity in the adsorbed state.[20] Free and adsorbed uricase appear to be equally active on dissolved uric acid, but adsorbed uric acid does not react.[21] In the case of amylase, Sabalilschka[22] found that the activity depended on the pH at which the enzyme was adsorbed. The activity of adsorbed invertase is affected by the conditions under which it is adsorbed. Miller[19] found that small concentrations of adsorbed invertase did not invert sucrose, whereas there was appreciable activity when larger amounts of invertase were adsorbed. Invertase adsorbed from purified dialyzed solutions had less activity than when adsorbed from unpurified solutions. Miller also found that acetic acid—adsorbed either before or simultaneously with the enzyme—enhanced the adsorption; moreover, the adsorbed enzyme had greater activity. This effect, however, did not occur when the acid was added after the adsorption of the enzyme.

The utility of carbon for the purification of a biological product may depend on the stability of the product while in the adsorbed state. A number of biological products are stable when adsorbed and this is true even of substances that are susceptible to destructive changes. Regna[23] reports that penicillin adsorbed on activated carbon is stable for at least 6 hours. Jensen and De-Lawder[24] found that adsorption on carbon did not alter the activity of crystalline insulin. In some cases, the stability is enhanced in the adsorbed state. Thus, under ordinary conditions, the adrenal cortex hormone is rapidly destroyed when the adrenal glands are removed from the animal body, but when adsorbed on activated carbon, the hormone is stabilized and the product retains its activity indefinitely at room temperature.[25]

There are certain compounds in which oxidation and other chemical changes are catalyzed by activated carbon. Amino acids are an outstanding example.[1,26] These acids frequently undergo aminolysis; that is, the NH_2 group is replaced by OH to give the hydroxy acid. In the presence of air, oxidation also occurs, forming an aldehyde with one less carbon atom; or the oxidation may be complete giving carbon dioxide, water, and ammonia. The course of the reaction depends on experimental conditions; hydrolysis, for instance, is retarded by a high pH and accelerated by a low pH.[27]

All amino acids are not affected to the same extent. Bergel and Bolz[28] found that leucine is oxidized ten times faster than alanine. Fürth[29] reports that histidine and proline are relatively stable, and Wunderly[30] finds that the deaminating effect is specific for the true amino grouping.

Some carbons have greater catalytic activity than others on amino acids.[31] Warburg's[32] studies indicate that the activity is promoted by the presence of iron in certain forms. He reports that by poisoning an activated carbon with cyanide, the oxidizing effect can be minimized without affecting the adsorption.

Fox and Levy[33] point out that, in some cases, activated carbon can be employed to accomplish a mild oxidation unaccompanied by undesirable by-products; thus ascorbic acid is oxidized to dehydroascorbic acid. Schwob[34] found that uracil is completely oxidized by hydrogen peroxide in the absence of carbon, but in the presence of carbon, the action can be controlled to produce intermediate compounds of biological importance. Frèrejacque[35] reports that at room temperature, uric acid in potassium hydroxide solution is oxidized to allantoin, and potassium oxonate in the presence of activated carbon. At 0°C oxonamide is produced.

2. PHARMACOLOGY

Activated carbon, in the form of a dry powder enclosed in a bandage, is applied as a dressing for ulcers, suppurating or gangrenous wounds, and other malodorous secretions. Although used principally to eliminate fetid odors, some antiseptic value is also claimed. Schöbesch[36] finds carbon of value for prophylactic treatment of yperite wounds. Nahmmacher[37] recommends the intrauterine application of charcoal for various forms of endo-metritis, especially in infected abortions. A surgical dressing, prepared from regenerated cellulose and 10 to 15% of activated carbon, is reported to stimulate secretions in wounds; moreover, the dressing absorbs secretions that are not absorbed by cellulose alone.[38] The dressing has been used advantageously in combination with sulfonamides to treat wounds that develop pus; the sulfon-amides, when used alone, are inactive when pus is secreted.

Plaster bandages and casts applied over wounds often develop an odorous suppuration. Ryan[39] reports that the disagreeable odors that develop in such cases can be eliminated by incorporating 5 to 12% of activated carbon in the plaster. The carbon used should be free from salts that would alter the setting properties of the plaster. Seddon and Florey[40] report that offensive odors are prevented by covering the plaster cast with a cloth impregnated with activated carbon. Best results are obtained when the limb is completely enclosed in a bag made from impregnated cloth. The bag should fit over the cast like a loose sleeve or stocking. In hospitals, odors from wounds can be controlled by withdrawing air from beneath the blankets and passing it through a canister of activated carbon.[41]

Adsorbents have long been employed to treat various intestinal disturbances. In 1712, Father Deutrecolle, a Jesuit missionary in

China, described the use of clay to treat diarrhea.[42] Braafladt reports that the use of kaolin together with a hypertonic salt solution reduced the mortality from cholera from 60 to 3 % during the Balkan war of 1910.[42] McRobert advocated kaolin to treat acute bacterial food poisonings encountered by the British army in India.[42] Favorable effects have been reported to result from the use of charcoal in infections such as dysentery, cholera, typhoid, and food poisoning.[1]

Some investigators report that benefits arise from the adsorbent adhering to the membranes of the intestinal tract and forming a protective coat. The elimination of gas is probably the result of the adsorption of substances that cause the formation of gas rather than to direct adsorption of the gas. Activated carbon, when wet, has very little adsorbing power for the gases found in the alimentary tract.[43] The beneficial results have been ascribed to removal of bacteria and toxins. Carbon of itself has no disinfecting power, but when certain metals, e.g., silver, are deposited on the surface of carbon, they confer disinfecting properties.[44,45] It is to be noted that the disinfecting power of soluble disinfectants can be greatly reduced by carbon if the active agent is adsorbable. Eisler[46] found that when phenol or mercuric chloride, adsorbed on carbon, is used for disinfection, about a hundred times as much disinfectant is required as when used in the free state.

Limited data are found in the literature on the removal of bacteria by carbon in vitro.[47,48] In many cases, the removal is ascribed to the bacteria being mechanically trapped when the liquid is filtered. The selective removal of certain types of bacteria indicates that more than a mechanical action is involved. Gunnison and Marshall[47] found that carbon adsorbed Lactobaccilus acidophilus, but not Escherichia coli or Clostridium welchii. Salus[49] found typhoid bacteria to be more readily adsorbed than Escherichia coli and less readily than cocci. Oksentyan,[50] in a study of lactic acid bacteria, found that adsorption did not affect the reproduction of the bacteria but lowered their physiological functions. Lasseur[51] observed Pseudomonas chlororaphis to become affixed at one pole and Bacillus megatherium frequently came to rest a short distance from the carbon granule. Of the bacteria studied by Gunnison and Marshall. these investigators were not able to demonstrate microscopically an affinity between adsorbents and bacteria.

Although bacteria are colloidal systems, their large size makes it questionable whether their removal can be regarded as true adsorption. One suggestion is that carbon, by adsorbing protective colloids, may leave the bacteria in a condition that permits them to clump and be mechanically trapped by the carbon. Studies by Buzagh,[52] dealing with conditions under which large particles can adhere to one another, are of interest in this connection.

Evidence on the adsorption of toxins and viruses is more positive, but the clinical value of the data is limited by the fact that they are derived mainly from experiments conducted *in vitro*. The studies include a number of viruses, venoms, toxins, antitoxins, and agglutinins.[1,2] The extent of each adsorption is influenced by experimental conditions and by the type of carbon used. In some cases, the activity disappears in the adsorbed state, whereas in others, it is reduced.[53] Poppe and Busch[54] found that at *p*H 6.5 to 8.4, the foot-and-mouth virus is so strongly adsorbed that neither the adsorbent nor the supernatant fluid is infectious to guinea pigs. Cordier[55] reports that the virus is not destroyed but the activity is reduced; small doses of adsorbed virus immunize the animal whereas large doses cause ulcers. An experience of Seibert[56] is worth mentioning. She found that injections of a purified derivative from old tuberculin did not stimulate the production of antibodies, but when it was adsorbed on carbon and injected in that form, it became antigenic; that is, it incited the production of antibodies. Such an increase in activity is in contrast to the decrease often found with adsorbed substances.

In some cases, advantages are provided by administering medicaments adsorbed on activated carbon.[57,58] The gradual release of the substance maintains a more uniform effective concentration. Goiffon[59] found that secondary reactions resulting from eserine are avoided by administering the drug on charcoal. He also reports the use of charcoal containing adsorbed hydroxyamino-phenylarsonic acid for the treatment of amebic and parasitic infections. Grollman[25,60] reports success in administering adrenal cortex hormone adsorbed on charcoal—the hormone being eluted by the gastro-intestinal fluids. Sklow[61] reports that the effect of certain hormones is prolonged when adsorbed on carbon. He found that when testosterone adsorbed on carbon is implanted subcutaneously, the action lasts five times longer than when administered in oil. The activity of adsorbed progesterone was prolonged to an even greater extent.

Activated carbon sometimes is incorporated with nauseating drugs to make them less objectionable. The procedure is practical except in the case of strongly adsorbable drugs that are not released in the gastro-intestinal tract.

Antidote

Holt[62] and Holz have prepared an informative review of the use of activated carbon as an effective emergency antidote for many ingested poisons. They stress the importance of selecting a suitable type of activated carbon and emphasize that the carbon should be administered promptly after the poison has been swallowed. They recommend that the carbon be used in the form of a fine powder that is stirred into water to the consistency of a thick soup; this can be drunk or used for lavaging. Holt and Holz state that a bottle of activated carbon on every medicine shelf would go a long way to combat serious poisonings in the home.

The successful administration of carbon in acute intestinal infections and as an antidote has led to its use for chronic disorders of the gastro-intestinal system. Although activated carbon is in itself entirely nontoxic and passes out of the digestive system unchanged except for the adsorbed substances it acquires, pharmacologists caution against regular and continued use. It is pointed out that the adsorptive power of carbon can deplete vitamins and enzymes from the system.[1,62]

The required dose of carbon varies according to the condition to be corrected.[62,63,64,65] For ordinary gastric irritations, 0.5 gram may be sufficient, whereas 5 grams or more may be required for acute infections. Still larger quantities are necessary when used as an antidote for a poison. Much depends on the site of the poison or infection. Thus, larger doses are necessary when action is desired in the lower intestine because much of the adsorptive power of the carbon is exhausted by the time this region is reached.

Various procedures have been developed to evaluate activated carbons for medicinal purposes.[1] The adsorption of substances such as Methylene Blue, strychnine, antipyrene, mercuric chloride is suggested for evaluation. In addition to measurements of adsorptive power, the U.S. Pharmacopeia includes tests to insure the absence of heavy metals, sulfides, and cyanides.[65]

3. INFLUENCE OF CARBON IN BIOLOGICAL PROCESSES

Brief mention can be made of cases in which activated carbon has been used to study changes in biological functions.

Gottlieb and Ludwig[66] studied the adsorption of dyes from serum. They found that carbon adsorbs much less dye from serum of patients with liver disease than from normal serum. The reaction is suggested for use as a test for liver function, and also for the study of changes in blood proteins in disease of the liver.

Grollman, Shumacker, and Howard[67] adsorbed estrogenic substances from urine and fed the carbon-estrogen combination to experimental animals. They found that the estrogenic substances are eluted from the carbon by the gastro-intestinal fluids and produce effective results. By this method, one can easily study variations in the estrogenic effects of urine, and thus follow changes in the concentration of estrogen in urine under various physiological conditions.

Wolff, Hawkins, and Giles,[68] in determining the nicotine content in the blood of smokers, observed that blood contained certain substances (termed *nicotine blank material*) which give a color with the test reagent similar to that produced by nicotine. They found that activated carbon will adsorb nicotine but not the blank material. This led them to an indirect approach in which the apparent nicotine concentration was determined before and after adsorption with activated carbon. The content of nicotine is assumed to be the difference between the two determinations.

Fermentation

Various effects are possible when activated carbon is present during fermentation.[1] Apart from any direct catalytic action, indirect effects may occur. Carbon may accelerate biological activity by adsorbing poisons, or may retard it by adsorbing nutrients necessary for the growth of microorganisms. Reactions may also be retarded by the adsorption of enzymes. In some cases, the course of the reaction is altered and different products are formed.

Amati[69] noted that carbon retarded the fermentation of sugars, but several other workers have reported an acceleration of biological activity.[1-70] Lampe,[71] working with molasses, reported that 0.1 % of activated carbon greatly accelerated the fermentation. He also found that formation of aldehyde was reduced sufficiently to make it unnecessary to use an aldehyde column in the subsequent distillation. In contrast, Abderhalden's[72] studies indicate that appreciable quantities of aldehydes develop when fermentation is conducted in the presence of activated carbon. He also notes that the formation of glycerol during the fermentation of dextrose by yeast is enhanced by carbon, if air is excluded from the fermentation vessel. Tomoda[73] found that activated carbon added to a sugar mash increased the yield of fusel oil.

4. BIOCHEMICALS

At the time the first edition of this text was prepared, many envisioned that large markets would result from studies of adsorption—desorption processes in the manufacture of biochemicals. The markets have not developed but the potential still exists. Activated carbon is extensively employed for the decolorizing and deodorizing of many biochemicals but we lack information as to specific applications.

Biochemical substances usually are present in only small concentrations in the source materials and to be prepared in usable form, they must be concentrated and also separated from a variety of other ingredients. The properties of these substances make them amenable to purification by adsorption processes, but the exact procedure to be employed varies from one case to another. In a few cases, it is possible to selectively adsorb the desired biochemical substance and leave most of the impurities in solution. The carbon cake is then eluted with a solvent that preferentially extracts the desired compound. Generally, the procedure is much more involved as it is seldom possible to obtain an adsorbent that will strongly adsorb the desired substance and simultaneously exclude the many other compounds present in the source materials.

Elution presents similar difficulties. In some cases, companion substances in the source materials act as coadsorbates and aid the adsorption and elution, whereas in other cases the companion substances interfere and should be removed by a prior precipitation or extraction. Frequently, several stages of adsorption are

desirable. In some processes, the selection of suitable conditions
for the first stage will allow the bulk of the impurities to be
selectively adsorbed, leaving the desired substance in a less contaminated condition in the filtrate. The conditions are then readjusted to permit the adsorption of the desired compound on
another adsorbent from which it is subsequently eluted.

The variety of treatments that can be employed is well illustrated
by the many different processes that are described in the scientific
literature. We lack information, however, on the extent to which
specific processes are employed on a commercial scale.

Penicillin

The carbon process was very important in the early development of manufacturing penicillin but has since been succeeded by
other methods. A description of the process is included both
because it is of historical interest and because it illustrates techniques useful in other applications.

A suitable broth is sterilized and inoculated with a seed culture
of *Penicillium notatum* and then aerated with sterile air for 70 to
80 hours at 24°C. At the end of the fermentation, the broth is
filtered. The clear brown broth (penicillin concentration about 30
parts per million) is usually chilled to the lowest possible working
temperature during the subsequent processing to avoid destruction
of the penicillin. Sufficient carbon is added to the filtered broth to
adsorb the penicillin; the mixture is stirred for 10 minutes and then
filtered. The carbon cake is eluted with a suitable solvent. Originally amyl acetate was used, but later this was replaced by 80%
acetone. The elution is conducted preferably by removing the
carbon cake from the filter, stirring with the eluting agent for
about 20 minutes, and then filtering.

When acetone is used as an eluant, it is separated from the
filtrate by distillation in a vacuum or by extraction with a solvent
that is not miscible with water. The resulting aqueous concentrate
of penicillin is chilled to 0°C, acidified to *p*H 2.0, and extracted
with an organic solvent. Penicillin, a fairly strong acid, is very
soluble in a number of organic solvents, *e.g.*, ethyl acetate, alcohol,
ether, amyl acetate, cycloxhexanone, dioxane, chloroform.

The organic solvent containing the penicillin is stirred with a
dilute sodium bicarbonate solution. The penicillin salt which is

formed goes into the aqueous phase and this is filtered through a Seitz-type biological plate-and-frame filter. The relatively concentrated solution is transferred to vials and dehydrated by freezing and high vacuum.

In some operations, the penicillin is treated at an intermediate stage with activated carbon under conditions that enable color and other impurities to be adsorbed.

Streptomycin

Although other methods are now employed to manufacture streptomycin, the carbon process as originally developed is described because of features that are of value in other applications.[1]

Sterilized mash is inoculated with a pure culture of *Streptomyces griseus* and aerated under aseptic conditions for several days at 25° to 30°C. The broth is filtered and activated carbon is added to the clear filtrate to adsorb the streptomycin. The carbon dosage must be closely controlled as an insufficient amount results in incomplete adsorption, and an excess of carbon reduces the yield by elution. The carbon is separated by filtration and washed with alcohol to remove loosely held impurities, after which the streptomycin is eluted from the carbon with acidified methanol. The eluate is neutralized and evaporated at reduced pressure under carefully controlled conditions to prevent decomposition of the streptomycin. At this point, the streptomycin is 25 to 30% pure and a number of subsequent steps are required to obtain a final product free from all toxic impurities. In some operations, a subsequent purification with activated carbon is provided under conditions in which the impurities are adsorbed and not the streptomycin.

Vitamins

a) Vitamin-bearing Oils:

The conventional method of decolorizing and deodorizing edible oils with adsorbents at an elevated temperature is of little value for vitamin-bearing oils because, under such conditions, considerable amounts of vitamin A are destroyed by oxidation. Buxton[74] describes a method in which the vitamin potency is preserved. Activated carbon is mixed with a solvent such as heptane or ethylene dichloride, and nitrogen is bubbled through the suspension for about 5 minutes to deaerate the carbon. The vitamin-bearing oil is then added, and the mixture stirred for 30 minutes at room temperature in an atmosphere of nitrogen, after which the mixture is filtered. Vitamin A is not adsorbed under these conditions and practically all of it is found in the filtrate. The solvent is removed by evaporation at reduced pressure and the remaining oil is greatly improved in color and odor. Some anti-oxidants are removed by the carbon and these should be replaced.

In a modification of the foregoing process, the carbon is mixed with water and heated to over 50°C. The deaerated carbon slurry is added to a vitamin-containing oil, and the whole mass is stirred at a slightly elevated temperature under reduced pressure until taste and odor impurities are adsorbed. The heat applied has a dual purpose in that it drives off the moisture and accelerates the rate at which the impurities are adsorbed.

Hennessy[75] found that when fish oils are hydrogenated in the presence of activated carbon, the odor is greatly improved and the vitamin potency is not impaired.

b) Vitamin A:

De[76] found that vitamin A can be separated from carotene by adsorbing the latter on activated carbon. Marcussen[77] describes a process for separating vitamins A and D. A concentrate prepared from fish-liver oil is dissolved in heptane and percolated through a column of activated carbon. When additional pure heptane is passed through the column, the vitamin D is eluted first and the vitamin A subsequently.

In a procedure employed by Holmes, Cassidy, Manly, and Hartzler,[78] vitamin A is concentrated by fractionation in Tsweet adsorption columns, using a column of oxygen-free carbon and a column of a special type of magnesia.

c) Vitamin C:

Fox and Levy[79] found that activated carbon converts ascorbic acid into the reversibly oxidized dehydroascorbic acid, the extent of the conversion depending on the type of carbon used. Very little irreversible oxidation occurs. Watanabe[80] observed that the optimum pH for the reversible oxidation ranged from 3.2 to 4.4, depending to some extent on the type of carbon used. Dehydroascorbic acid can be regenerated to ascorbic acid by hydrogen sulfide. Indovina[81] found that ascorbic acid is adsorbed without decomposition when the treatment is conducted in an atmosphere of nitrogen. Kuhn and Gerhard[82] observed that the oxidation of ascorbic acid is diminished by impregnating the carbon with a reducing agent such as sodium sulfide or sodium thiosulfate.

d) Vitamin D:

Funk and Dubin,[83] using autolyzed yeast, adsorbed vitamin D with fullers' earth or activated carbon. Acetic acid was employed to elute the vitamin from the activated carbon; but for fullers' earth, barium hydroxide was more effective.

Vitamin G (Riboflavin):

Booher[84] employed adsorbents to prepare a vitamin G concentrate from whey. In one procedure, the vitamin in the whey is adsorbed on fullers' earth at low temperature and then eluted with hot water. The vitamin in the eluate from the fullers' earth is further concentrated by adsorption on activated carbon from which it is subsequently desorbed with an ethanolbenzene mixture, the latter being separated from the vitamin concentrate by evaporation.

Vitamin H:

Booher[84] describes the use of activated carbon to obtain a 60- to 90-fold increased concentration of vitamin H* from rice polishings. The success of the process depends on certain preliminary steps of purification. An acidulated aqueous extract of rice polishings is mixed with fullers' earth, which adsorbs vitamin B^1. The neutralized filtrate is desiccated and the dry solids are extracted with absolute ethyl alcohol. The extract is diluted with water and treated with activated carbon to adsorb vitamin H which is subsequently eluted with n-butyl alcohol or other suitable solvent.

Antineuritic Factor:

Kinnersly and Peters[86] studied methods for concentrating thiamine (torulin) an antineuritic factor present in yeast. In one procedure, neutral lead acetate is added to an aqueous yeast extract to precipitate certain impurities. The filtrate is treated with

*Later identified as a combination of pyridoxine and pantothenic acid.

barium hydroxide to precipitate gums—the excess barium hydroxide being removed with sulfuric acid. Other impurities are removed from the solution with mercuric sulfate. Activated carbon is added to the filtrate to adsorb the active principle which is subsequently extracted with 0.1 N hydrochloric acid in 50% alcohol.

Kinnersly and Peters[86] point out that many variable factors are involved and that a procedure suitable for recovering thiamine from one raw material may not be applicable when a different source material is used. The adsorption of thiamine depends on the presence of a coadsorbate in the solution; consequently, the preliminary treatments should be such as to remove interfering substances without affecting the coadsorbate. The pH at which the adsorption should be conducted depends on the preliminary treatment. For example, when a previous precipitation with mercuric sulfate is provided, the optimum pH is 7.0; but when this step is omitted, the adsorption should be conducted at pH 5.0 to 6.0.

Folic Acid:

Mitchell, Snell, and Williams[87] describe a method for concentrating folic acid from spinach. The process includes three successive stages of adsorption–desorption on activated carbon after which the solution is further purified by various chemical reagents. Folic acid, when adsorbed from crude solutions, can be eluted with hot ammonia, but the elution is not successful when the folic acid is adsorbed from purified solutions. Because of this, it becomes necessary to modify the procedure during the second and third stages of the adsorption–desorption process. Frieden, Mitchell, and Williams[88] suggested that impurities adsorbed from a crude solution cause the folic acid to be held less tenaciously. They studied the action of other substances and found that when activated carbon is pretreated with aniline, folic acid adsorbed from a purified solution can be eluted without difficulty.

Liver Extract:

Kyer[89] prepared a concentrated liver extract by precipitating the proteins with lead acetate and then adsorbing the active principle on carbon. The use of the lead salt had certain disadvantages; in a later method, calcium chloride and sodium carbonate were added to form a precipitate of calcium carbonate which removed the protein as did lead acetate. The filtrate adjusted to pH 5 is treated with activated carbon, and the filtrate from this is treated with an additional quantity of carbon. The carbon cakes are extracted with hot 50% alcohol and the eluate is concentrated *in vacuo*. Tyrosine also is adsorbed and eluted with the active substance, although, with some carbons, the adsorption of tyrosine is incomplete.

Precipitate Factor:

Elvehjem[90] and coworkers found a dietary factor in liver, yeast, and milk, which they called the *alcohol-ether precipitate* factor. This factor can be adsorbed by activated carbon, but difficulty has been experienced in eluting the active substance. However, when carbon that contained the adsorbed factor was fed to vitamin-deficient animals, good growth was obtained. This suggests that elution *in vitro* could be accomplished under appropriate conditions.

L. Casei Factor:

Stokstad, Hutchings, and Subba Row[91] studied the isolation of this factor from liver. Liver extract containing 20,000 units of activity per gram was dissolved in water, adjusted to pH 8.5, heated to 80°C, and calcium chloride added to flocculate the precipitate that formed. The *L. casei* factor was adsorbed from the filtrate by activated carbon at pH 3.0. The carbon cake was washed with 60% ethanol to remove inert substances and then it was treated with 0.5 N ammonium hydroxide in 60% ethanol at 70°C to elute the active factor. The filtrate, adjusted to pH 1.3, was percolated through a column of Super Filtrol to adsorb the active factor, and the latter was subsequently eluted with 0.5 N ammonium hydroxide in 60% ethanol.

Hutchings[92] and co-workers isolated the *L. casei* factor from a solution obtained by aerobic fermentation of an unidentified bacterium of genus *Corynebacterium*. The filtered solution was adjusted to pH 3.0 and activated carbon was added. After 30 minutes of contact, the carbon was filtered off, washed with water and then with 50% ethanol, and the washings were discarded. Following this, the carbon was eluted with a solution containing 10% ammonium hydroxide and 50% ethanol. Approximately 65% of the *L. casei* factor was recovered.

Hormones[93]

a) Insulin:

Moloney and Findlay[94] outline a method of purifying insulin. Activated carbon is added to a partially purified solution of insulin adjusted to pH 2.5 with hydrochloric acid. After 12 hours of contact, the mixture is filtered and the carbon cake is washed with water and then extracted with a solution containing 5% acetic acid and 60% ethanol. This extracts certain impurities but does not disturb the insulin. The washed carbon cake is digested for several hours at room temperature with a 12% solution of benzoic acid in 60% ethanol to elute the adsorbed insulin. The ethanol is evaporated from the filtrate and the benzoic acid in the concentrate is extracted with ether. The resulting aqueous solution contains the purified insulin together with traces of dissolved ether. The ether is subsequently removed by evaporation.

Jensen and DeLawder[95] purified commercial insulin by adsorption on activated carbon followed by elution with 90% phenol. The phenol eluate was diluted with water and the insulin that flocked out was washed with absolute alcohol and ether. Insulin thus prepared has an activity about equal to that of crystalline insulin. The activity was not further increased by a subsequent adsorption and elution. This finding is supported by data of Fisher and Scott,[96] and is in contrast to an observation by Dingemanse[97] that a product of greater activity than crystalline insulin can be prepared by this procedure.

b) Oxytocic Hormone:

Methods of concentrating this hormone have been developed by Freeman, Gulland, and Randall.[98] In one procedure, an acid extract of powdered posterior lobe, adjusted to pH 11, is treated with fullers' earth to adsorb impurities. The active principle remaining in solution is adsorbed on activated carbon from which it is subsequently extracted with glacial acetic acid.

c) Adrenal Cortex Hormone:

Grollman and Firor[99,100] found that adrenalectromized animals could be maintained in a normal state of health with adrenal cortical hormone adsorbed on carbon administered orally, the hormone being eluted by the gastrointestinal fluids.

To prepare the hormone, the adrenal glands are extracted with acetone, after which the acetone is removed by distillation *in vacuo* at 35° to 40°C. The remaining aqueous concentrate, after being filtered and adjusted to *p*H 7.0 with sodium hydroxide, is shaken with activated carbon to adsorb the hormone. To get rid of impurities incidentally adsorbed on the carbon, it is washed successively with dilute alkali, hydrochloric acid, and alcohol. The alkali removes products of phenolic decomposition; the acid elutes epinephrine; and the alcohol dissolves certain lipoidal impurities.

When it is desired to administer the hormone uncombined, it is eluted from the charcoal with chloroform. The chloroform is removed by distillation *in vacuo,* leaving a brownish resin that is soluble in aqueous ethyl alcohol.

d) Plant Wound Hormone:

English and Bonner[101] prepared a concentrate of this hormone from bean pods. An alcoholic extract of the pods was evaporated in an atmosphere of nitrogen and the residual syrup was extracted with water. Activated carbon was added to the aqueous extract and the mixture was stirred at room temperature for 1 hour in an atmosphere of nitrogen. After filtration, the active substance was eluted from the carbon with pyridine and purified further by chemical methods.

Removal of Pyrogens

It is necessary that solutions used for parenteral injection should have no harmful effects. Sterility is obviously important. Even though a solution is free from microorganisms in its final state, however, it may be unsuitable because of by-products of bacterial action formed before or during sterilization. Certain bacteria produce thermo-stable substances which are classified as *pyrogens*. Pyrogens, when present in parenteral solutions, will cause a physiological reaction that occurs from 15 minutes to 8 hours after the injection. The reaction is characterized by a rigor and chill accompanied by a sharp rise in temperature and an increase in the pulse rate. This reaction is followed by profuse sweating and a fall in temperature. There may also be nausea, vomiting, headache, and albuminuria. As most patients who are given parenteral injections are already in a weakened condition, such reactions can be dangerous.

As large quantities of pyrogens may be present in ordinary distilled water, specially designed stills are required for the removal of pyrogens by distillation. A number of workers,[1,102] Todd, Gemmell, Brindle, Rigby, Lees, and Levy have found that pyrogens can be removed by powdered activated carbon. Great care is required in the filtration because it must provide complete removal of all solid particles and also maintain aseptic conditions. Hudson[103] has designed a filter unit.

The addition of medicaments, *e.g.*, dextrose, to a carefully prepared water can introduce pyrogens, and as such solutions cannot be purified by distillation, the use of carbon becomes necessary. Certain medicaments are adsorbed by the carbon and then it becomes necessary to add sufficient quantity to compensate for the amount adsorbed.

Granular carbons are not effective and only certain powdered carbons are suitable. In addition to adsorptive power for pyrogens, the carbon must be free of soluble inorganic substances, some of which may produce an opalescence during the subsequent sterilization.

The original publications should be consulted for complete details of the technique for the use of activated carbon in the preparation of pyrogen-free solutions.

This chapter covers only some of the ways in which activated carbon can offer utility in this field. Other possibilities are suggested in References. For a complete review the reader is referred to *Chemical Abstracts* where all items dealing with the subject can be readily located in the index under *Carbon, active;* and *Charcoal, active;* and also to Deitz, *Bibliography of Solid Adsorbents,* which contains a section on biochemical, medical, and pharmaceutical applications. Recent developments are reported in Chapter 20.

REFERENCES

1. Deitz, V. R., *Bibliography of Solid Adsorbents,* United States Cane Sugar Refiners and Bone Char Manufacturers and National Bureau of Standards, Washington, D. C., 1944. Supplementary volume published in 1956.
2. Hassler, J. W., *Active Carbon—the Modern Purifier,* Industrial Chemical Sales Division, West Virginia Pulp & Paper Co., New York, 1941. Classified bibliography of applications to biochemistry, pharmacology, and related subjects, pp. 105–110.
3. Donnon, F. G., *J. Chem. Soc.,* 1387 (1929).
4. Armstrong, E. F., and Hilditch, T. P., *J. Phys. Chem.,* 33: 1441 (1929).
5. Freundlich, H., *J. Chem. Soc.,* 164 (1930).
6. Quastel, J. H., *Biochem. J.,* 20:166 (1926); *Trans. Faraday Soc.,* 26:853 (1930).
7. Schwob, C., Biegner, J. E., Carson, K. J., and Scott, G. V., *J. Am. Chem. Soc.,* 64:2276 (1942).
8. Abderhalden, E., and Buadze, S., *Fermentforschung,* 9:89 (1926).
9. Cavier, R., *Compt. rend.,* 216:255 (1943).
10. Eyster, H. C., *Science,* 96:140 (1942).
11. Warburg, O., *Naturwissenchaften,* 11:159 (1923); *Biochem. Z.,* 136:266 (1923).
12. Larson, W. P., Bieter, R. N., Levine, M., and Hoyt, R. E., *Proc. Soc. Exp. Biol. Med.,* 41:200 (1939).
13. Chick, H., and Roscoe, M. H., *Biochem. J.,* 21:698 (1927).
14. Kinnersley, H. W., and Peters, R. A., *Biochem. J.,* 19:820 (1925).
15. Willstätter, R., *J. Chem. Soc.,* 1359 (1927).
16. Koda, N., *Acta School Med. Univ. Imp. Kioto,* 10:43 (1928).
17. Kikawa, K., *J. Biochem.* (Japan), 6:275 (1926).
18. Moloney, P. J., and Findlay, D. M., *J. Phys. Chem.,* 28:402 (1924).
19. Miller, E. J., and Bandemer, S. L., *J. Phys. Chem.,* 34:2666 (1930).
20. Eyster, H. C., *Plant Physiol.,* 21:68 (1946).
21. Przylecki, S. J., *Acta Biol. Exp. Warsaw,* 1, No. 6:1 (1928).

22. Sabalitschka, T., and Weidlich, R., *Biochem. Z.*, 210:414 (1929).
23. Regna, P., *Trans. Am. Inst. Chem. Eng.*, 40:759 (1944).
24. Jensen, H., and De Lawder, A. M., *J. Biol. Chem.*, 87:701 (1930).
25. Grollman, A., and Firor, *Bull. John Hopkins Hospital*, 57: 281 (1935).
26. Wachtel, J. L., and Cassidy, H. G., *J. Am. Chem. Soc.*, 65: 665 (1943).
27. Baur, E., and Wunderly, K., *Biochem. Z.*, 272:1 (1934).
28. Bergel, F., and Bolz, K., *Z. Physiol. Chem.*, 215:25 (1933).
29. Furth O., and Kaunitz, H., *Bull. soc. chim. biol.*, 12:411 (1930).
30. Wunderly, K., *Helv. Chim. Acta*, 16:1009 (1933).
31. Wunderly, K., *Helv. Chem. Acta*, 17:523 (1934).
32. Warburg, O., and Brefeld, W., *Biochem. Z.*, 145:461 (1924).
33. Fox, F. W., and Levy, L. F., *Biochem. J.*, 30:208 (1936).
34. Schwob, C., *J. Am. Chem. Soc.*, 58:1115 (1936).
35. Frérejacque, M., *Compt. rend.*, 191:949 (1930).
36. Schöbesch, O., *Antigaz* (Bucharest), 12:436 (1938).
37. Nahmmacher, H., *Surg. Gynec, and Obstet.*, 873 (1930).
38. Blumenthal, L. M., *U.S. Department of Commerce Fiat Final Report No. 1181* (June 5, 1947).
39. Ryan, J. F., U.S. Patent 2,402,779.
40. Seddon, H. J., and Florey, H. W., *Lancet*, 6200 (1942).
41. Naval Med. Research Inst. Natl. Naval Med. Center, Bethesda, Md., Report (June 25, 1946).
42. Emery, E. S., *J. Am. Med. Assoc.*, 108:202 (1937).
43. Steenberg, B., *Svensk Farm. Tid.*, 44:549, 565 (1940).
44. Friderich, L., and Godel, A., *Chimie & industrie*, 30:1038 (1933).
45. Harris, P. S., Goetz, A., and Pasadena, R. L. T., *Phys. Rev.*, 60:162 (1941).
46. von Eisler, M., *Biochem. Z.*, 172:154 (1926).
47. Gunnison, J. B., and Marshall, M. S., *J. Bact.*, 33:401 (1937).
48. Bock, J. C., *J. Am. Chem. Soc.*, 42:1564 (1920).
    Kraus, R., and Barbara, B., *Wien. klin. Wochschr.*, 28:810 (1915).
    Stich, C., *Chem. Ztg.*, 67:349 (1943).
49. Salus, G., *Wien. klin. Wochschr.*, 29:846 (1916).

50. Oksent'yan, U. G., *Microbiology* (U.S.S.R.), 9, No. 1:3 (1940).
51. Lasseur, P., Dombray, P., and Palgen, W., *Trav. lab. microbiol. faculté pharm. Nancy*, 7:117 (1934).
52. von Buzagh, A., *Kolloid Z.*, 51:105,230 (1930);52:46 (1930); 85:318 (1938).
53. Pyl, G., and Hobohm, K. O., *Kolloid Z.*, 104:63 (1943).
54. Poppe, K., and Busch, G., *Z. Immunitäts.*, 68:510 (1930).
55. Cordier, G., *Compt. rend.*, 208:1364 (1939).
56. Seibert, F. B., *J. Immunol.*, 28:425 (1935).
57. Sabbatani, L., *Zentr. Biochem. Biophysik.*, 12:470 (1911).
58. N. V. Orgachemia, British Patent 530,455.
59. Goiffon, R., *Semana med.* (Buenos Aires), 37:1133 (1930).
60. Grollman, A., Firor, W. M., and Grollman, E., *J. Biol. Chem.*, 109:189 (1935).
61. Sklow, J., *Endocrinology*, 32:109 (1943).
62. Holt, L. Emmett Jr., and Holz, Peter H., Department of Pediatrics, New York University Medical Center, New York, N.Y.; manuscript received January, 1963.
63. Wood, H. C., and La Wall, C. H., *Dispensatory of United States*, Lippincott, Philadelphia, 1937, p. 288.
64. Hatcher, R. A., *Useful Drugs*, 10th Edition, American Medical Association, Chicago, 1936, p. 75.
65. *U.S. Pharmacopoeia*, 13th Revision, Mack Publishing Co., Easton, Pa., 1947.
66. Gottlieb, H., and Ludwig, H., *Z. klin. Med.*, 131:358 (1937).
67. Grollman, A., Shumacker, H. B., and Howard, E., *J. Pharmacol and Exper. Ther.*, 54:393 (1935).
68. Wolff, W. A., Hawkins, M. A., and Giles, W. E., *J. Biol. Chem.*, 175, No. 2:825 (1948).
69. Amati, A., *Ann. chim. applicata*, 28:487 (1938).
70. Abderhalden, E., *Fermentforschung*, 5:255 (1922).
71. Lampe, B., *Z. Spiritusind*, 54:75, 313 (1931); *Brennerei Ztg.*, 49:6 (1932).
72. Abderhalden, E., *Fermentforschung*, 5:89 (1921). Abderhalden, E., and Stix, W., *Fermentforschung*, 6:345 (1922).
73. Tomoda, Y., *J. Soc. Chem. Ind., Japan*, 42, Suppl. binding, 379 (1939).
74. Buxton, L. O., *Ind. Eng. Chem.*, 34:1486 (1942); British

Patent 535,385; U.S. Patents 2,255,875, 2,328,053.
Buxton, L. O., and Simons, E. J., British Patent 537,403.
75. Hennessy, D. J., U.S. Patent 2,321,913.
76. De, N. K., *Indian J. Med. Research*, 25:17 (1937).
77. Marcussen, E., *Dansk Tidsskr. Farm.*, 13:141 (1939).
78. Holmes, H. N., Cassidy, H., Manly, R. S., and Hartzler, E. R., *J. Am. Chem. Soc.*, 57:1990 (1935).
79. Fox, F. W., and Levy, L. F., *Biochem. J.*, 30:208 (1936).
80. Watanabe, K., *J. Soc. Trop. Agr., Taihoku Imp. Univ.*, 9:369 (1937).
81. Indovina, R., *Atti X⁰ congr. intern. chim.*, 4:586 (1939).
82. Kuhn, A., and Gerhard, H., *Kolloid Z.*, 103:130 (1943).
83. Funk, C., and Dubin, H. E., *J. Biol. Chem.*, 48:437 (1921).
84. Booher, L. E., and Work, L. T., U.S. Patent 2,175,014.
85. Booher, L. E., *J. Biol. Chem.*, 119:223 (1937); U.S. Patent 2,202,307.
86. Kinnersley, H. W., and Peters, R. A., *Biochem. J.*, 21:777 (1927).
87. Mitchell, H. K., Snell, E. S., and Williams, R. J., *J. Am. Chem. Soc.*, 66:267 (1944).
88. Frieden, E. H., Mitchell, H. K., and Williams, R. J., *J. Am. Chem. Soc.*, 66:269 (1944).
89. Kyer, J. L., *Proc. Soc. Exp. Biol. and Med.*, 32:1102 (1935).
Sladek, J., and Kyer, J., *Proc. Soc. Exptl. Biol. Med.*, 39:227 (1938).
90. Elvehjem, C. A., Koehn, C. J. Jr., and Oleson, J. J., *J. Biol. Chem.*, 115:707 (1936).
91. Stokstad, E. L. R., Hutchings, B. L., and SubbaRow, Y., *Annals. N.Y. Acad. Sci.*, 48, Art. 5:261 (1946).
92. Hutchings, B. L., Stokstad, E. L. R., Bohonos, N., Sloane, N., and SubbaRow, Y., *Annals. N.Y. Acad. Sci.*, 48, Art. 5:265 (1946).
93. Studies on hormones not otherwise classified:
Elden, C. A., *J. Biol. Chem.*, 101:1 (1933).
Grollman, A., Shumacker, H. B., and Howard, E., *J. Pharmacol. and Exper. Ther.*, 54:393 (1935).
Katzman, P. A., and Doisey, E. A., *J. Biol. Chem.*, 98:739 (1932).
Peritz, G., and Brahm, C., German Patent 608,414.
Sato, G., *Arch. exptl. Path. Pharmakol.*, 130:323 (1928).

Zondek, H., and Bansi, H. W., *Biochem. Z.*, 195:376 (1928).

Organon, Dutch Patent 48118.

Soc. Anon. pour L'Industrie Chimique, British Patent 244,055.

94. Moloney, P. J., and Findlay, D. M., *J. Phys. Chem.*, 28:402 (1924).

95. Jensen, H., and DeLawder, A. M., *J. Biol. Chem.*, 87:701 (1930).

96. Fisher, A. M., and Scott, D. A., *Trans. Roy. Soc. Canada V.*, 28:75 (1935).

97. Dingemanse, E., and Laqueur, E., *Arch. néerland. physiol.*, 12:259 (1927).

98. Freeman, M., Gulland, J. M., and Randall, S. S., *Biochem. J.*, 29:2211 (1935).

99. Grollman, A., and Firor, W. M., *Bull. Johns Hopkins Hospital*, 57:281 (1935).

100. Grollman, A., Firor, W. M., and Grollman, E., *J. Biol. Chem.*, 108:189 (1935).

101. English, J., and Bonner, J., *J. Biol. Chem.*, 121:791 (1937).

102. Todd, J. P., *Pharm. J.*, 146:258 (1941).

Todd, J. P., and Gemmell, D. H. O., *Lancet*, 112 (1942).

Gemmell, D. H. O., and Todd, J. P., *Pharm. J.*, 154:126 (1945).

Brindle, H., and Rigby, G., *Australian J. Pharm.* (Melbourne), November. 30, 1946, 987; *Pharm. J.*, 157:85 (1946).

Lees, J. C., and Levy, G. A., *British Med. J. Lancet*, 430 (1940).

Howard, F., and Spooner, E. C. R., *Chemistry & Industry* 186 (1946).

103. Hudson, F. A., *Pharm J.*, 131 (1944).

PART VI

LABORATORY PROCEDURES

16

Adsorption of Gases and Vapors
Laboratory Procedures

1. STATIC ADSORPTION METHODS

In the oldest commonly used method for measuring the adsorption of gases, a measured amount of gas is introduced to a known weight of adsorbent. The amount adsorbed is calculated by comparing the resulting gas pressure with that which would be expected if no adsorption occurred. Many varied assemblies have been developed to fit specific situations. Each requires careful attention to diverse details, on which full information is already available in readily accessible publications[2,3,4]. It should be mentioned that few users of activated carbon conduct gas adsorption tests. Many rely on the warranty of the supplier.

In another static method of measuring gas adsorption, the amount of gas adsorbed is measured by the increase in the weight of the adsorbent. For this, a very useful balance was developed by McBain and Bakr[5] Figure (16:1), the essential part of which is a helical spring of silica with a hook at each end, and a bucket containing the adsorbent is suspended from the lower hook. The balance is calibrated by determining the elongation under known loads, which is read by a travelling microscope. The balance is supported within a glass tube containing a thin-walled glass bulb holding the liquid whose vapor is to be adsorbed. After evacuating the system and sealing off the unit, the glass bulb containing the liquid is broken by a magnetic device. The elongation of the balance spring measures the amount adsorbed.

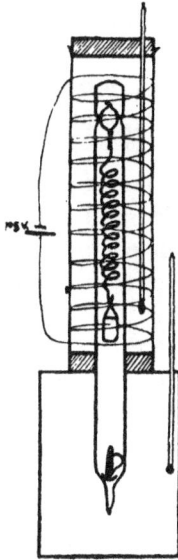

Fig. 16:1 Sorption balance in superposed adjustable thermostats From J. W. McBain and A. M. Bakr, *J, Am. Chem. Soc.*, 48:690 (1926); reprinted by permission of American Chemical Society.

This method has various advantages. Pressure within the apparatus may be adjusted in the case of vapors, by cooling a portion of the glass tubing to various temperatures and having sufficient vapor present to keep some liquid condensed, whereupon its vapor pressure will be the pressure prevailing throughout the apparatus. McBain reports that adsorption and desorption can be observed directly and repeatedly under a whole series of experimental conditions.

Degassing

In static studies of gas adsorption, the carbon usually is *degassed* or *outgassed* before the adsorption.[1,2] To do this, the carbon is heated at a temperature within the range 300° to 900°C, and then cooled; the entire process is conducted under high vacuum. McBain[6] found that degassing alters the isotherm to a type in which nearly all the adsorption occurs at very low pressures. Degassing also increases the rate of adsorption, and equilibrium conditions are attained without the delay often encountered otherwise.

Harned[7] found degassing to be more effective when the carbon is given an intermediate washing-out with the gas to be studied. The procedure involves heating the carbon to over 700°C in a high vacuum, then cooling in a vacuum, after which the carbon is exposed to the gas to be adsorbed, reheated, and finally cooled in a high vacuum. The adsorptive power of the carbon for the gas is then studied, although in some cases, the washing-out followed by heating is repeated. Such treatment increases the velocity of adsorption. Harned considered the effect of washing-out analogous to soaking up gasoline with a sponge already covered with a film of water; practically no gasoline will be taken up until the sponge is washed out a number of times with gasoline.

In evaluating the effect of degassing, one is faced with an uncertainty as to whether the operation may produce changes in the carbon other than removal of preadsorbed gases. Other lines of evidence indicate that qualitative changes in adsorptive power may occur when a carbon is heated above 500°C. More than simple removal of preadsorbed gas may be involved, and a carbon may not have the same intrinsic properties before and after degassing at high temperatures.

Hysteresis

Previously, it has been stated that an equilibrium exists between the concentration in the adsorbed and unadsorbed states. From this one might expect the same equilibrium points to be reached when the process starts from low concentrations and goes to higher concentrations as when the procedure is reversed. Frequently, however, it is found that when carbon containing adsorbed vapor is subjected to diminishing pressure, the desorption isotherm fails to retrace the path of the isotherm which formed when the pressure was increased (Figure 16:2) The carbon then holds more gas for a given pressure on the curve of desorption than on that of adsorption. In such cases, all or part of the desorption isotherm is above the adsorption isotherm, indicating a resistance to desorption.

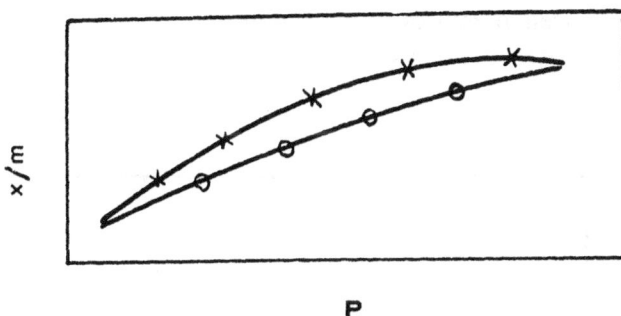

Fig. 16:2   Illustration of Hysteresis
O, adsorption isotherm
X, desorption isotherm
P, pressure of gas at equilibrium

This behavior is known as *hysteresis*. Hysteresis has been found with carbons that were carefully degassed before adsorption and when ample time has been provided to reach equilibrium.

Studies have been made in which a succession of adsorption and desorption stages follow one another; not infrequently each adsorption isotherm duplicates the previous adsorption isotherm and each desorption isotherm retraces the previous desorption isotherm. In such cases, we have no clear guide as to whether the adsorption or desorption isotherm represents the true equilibrium.

The phenomena of hysteresis are reminiscent of behavior such as supersaturation and supercooling. Some students explain hysteresis as a consequence of capillary condensation.[8] Gleysteen and Deitz,[9] in a careful review of the phenomena, suggested an interpretation based on the multimolecular theory of adsorption developed by Brunauer and coworkers.

Discontinuous Isotherms

Although most isotherms appear as continuous curves, there is evidence that some adsorptions are discontinuous, because the isotherm appears as a series of steps[2] (Figure 16:3) Discontinuities are observed only when the separate observations are close together. Burrage[10] in a study of discontinuities determined 350 points between pressures of 0.04 mm and 81 mm. for the adsorption of carbon dioxide.

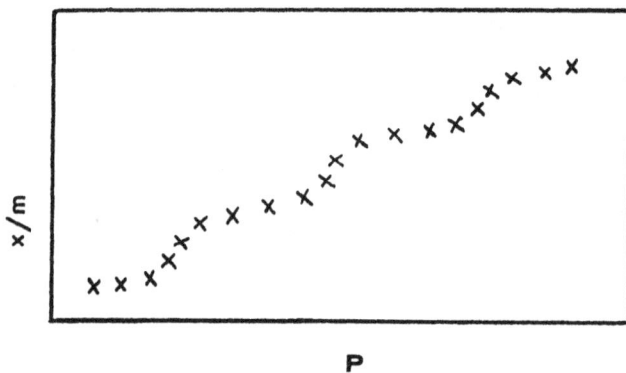

Fig. 16:3   Discontinuous isotherm P, pressure of gas at equilibium

2.   DYNAMIC ADSORPTION METHODS

Static methods of gas adsorption will not necessarily reflect the behavior in industrial operations in which air containing a gas or vapor is passed continuously through a carbon bed. For this, dynamic test methods are designed to parallel industrial processing. A known concentration in air of the gas or vapor under study is passed at a constant rate through a definite volume of carbon. The first appearance at the effluent of a detectable amount of the gas or vapor is the breakpoint. When a greater concentration is acceptable as the endpoint, the test is continued until that endpoint is reached. The service time is the time elapsed from the beginning of the test to the termination.

The first procedures in general use were developed by the Chemical Warfare Service for the selection of gas mask carbon during World War I.[11,12,13] A selected concentration in air of the gas to be studied is passed through a tube containing a measured volume of carbon.

When carbons with greater adsorptive capacity were developed, accelerated test methods were adopted using higher gas concentrations and faster flow rates.[14] As in the case of static gas adsorption tests, most users depend on the warranty of the supplier. For those who wish to conduct the test, full information is readily accessible in available literature.[2,14]

According to Chaney,[11] accelerated dynamic gas adsorption tests serve to give relative values for manufacturing control at a given plant, but they can provide distorted values when applied to carbons of different origin.

Retentivity

Many investigators place value on retentivity data. Retentivity cannot be defined exactly but it refers to the phenomenon that when a char has taken up a vapor to capacity from a relatively strong concentration, a certain proportion of this adsorbed vapor can be removed with ease in a current of air while the remainder is retained more firmly. Data retentivity, as given by different investigators, are comparable only if obtained under corresponding conditions of time, temperature, and velocity of air stream.[15]

Chaney[11] describes an accelerated routine procedure. In this, dried and evacuated carbon is saturated with toluene vapor at 22°C, the weight of adsorbed vapor giving the *saturation value*. The carbon is then evacuated to 2 millimeters of mercury at 100°C, and weighings are taken each half hour. When the last three weighings lie on a straight line, the intercept of this line extended to the Y axis is taken as the retentivity value (Figure 16:4)

In the method developed by Macy,[16] a sample of carbon saturated with a gas is placed in a U-tube of 1.4 centimeters diameter, through which dry air at 25°C is passed at the rate of 1000 milliliters per minute. The amount of gas retained at the end of 2 hours is called the *retentivity*.

Retentivity (an aspect of irreversible adsorption) is desirable for gas mask carbons, but is not desirable for carbons that are to be recycled as in the recovery of solvent vapors.

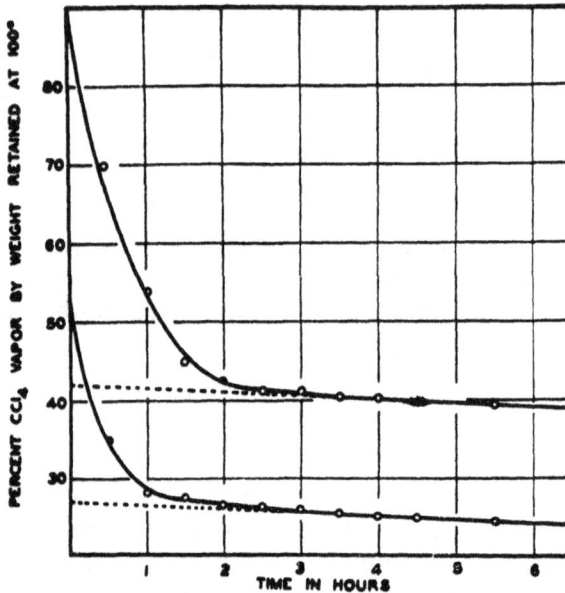

Fig. 16:4   Graphical method for determination of retentivities From H. S.
Stone and R. O Clinton, *Ind. Eng. Chem. Anal. Ed.*, 14:131 (1942);
reprinted by permission of American Chemical Society

3. MISCELLANEOUS METHODS

Heat of Wetting

The heat of wetting is often utilized as a guide to the relative
gas-adsorbing power of carbons, a higher heat of wetting being
associated with greater gas-adsorbing power. Bell and Philip[17]
have observed a relation between the heat of wetting and the
retentivity.

The heat of wetting is the amount of heat liberated when a
measured quantity of a carbon is immersed in the liquid contained
in a well insulated container, *e.g.*, a Dewar flask. The test is rapid,
it can be conducted in very simple apparatus, and it requires less
skill than do the gas-adsorption tests.

An assembly described by Macy[16] is very satisfactory. The apparatus consists of a test tube to serve as a reaction vessel; this is inserted in a somewhat larger tube, the space between the two tubes providing an insulating air jacket.

The test for heat of wetting is generally applied to granular carbons and for these, particle size has little or no influence on the results. Powdered carbons, which do not immerse quickly present difficulties in manipulation. Moisture in the carbon gives a lower value for the heat of wetting. Impurities in the liquid also affect the value obtained.

Heat of Adsorption

Different types of assemblies have been developed to measure the heat of adsorption, each one being designed to fit specific conditions. It would be impractical to present an adequate discussion of the varied techniques that can be used and the interested reader is referred to the original publications.[2]

Chaney Activity Test

Chaney[11] observed a correlation between the retentivity of a carbon and the adsorptive power for iodine, which he termed the *activity*. The activity is obtained by determining the amount of iodine which 1 gram of dried 200-mesh carbon will adsorb from 50 milliliters of 0.2 N iodine-potassium iodide solution. The mixture is agitated for 3 minutes, filtered, and the first portions of the filtrate are discarded. The iodine concentration of an aliquot portion of the filtrate is determined by titration. Highly active carbons for gas-adsorption will remove approximately 90% of the iodine.

Measurement of Odor Concentration

An instrument known as the Scentometer has been developed to measure the magnitude of the concentration of odors at a given site.[18] The principle employed is to measure the extent to which an odor-bearing air must be diluted with odor-free air in order to reach the threshold point. The odor-free air is supplied by the Scentometer, by passing air through activated carbon contained in the instrument. The Scentometer enables field studies to be made to learn the degree of purification required to provide air of suitable quality in the situation under study.

Carbon Tetrachloride Activity and Retentivity

In industrial circles, many suppliers now evaluate carbons for gas and vapor systems by tests based on the adsorption and desorption of carbon tetrachloride. The adsorption stage gives a value designated as the *Carbon Tetrachloride Activity*, and the desorption stage furnishes a value termed the *Carbon Tetrachloride Retentivity*.

The principle involved in the test for activity is: Pure dry air saturated with carbon tetrachloride is passed through a bed of activated carbon until the carbon has taken up all the carbon tetrachloride it can adsorb at the temperature of the test. Conditions of the experimental procedure vary from one supplier to another; therefore the supplier involved should be consulted to learn the details of the conditions to be observed.

The retentivity is measured by passing pure dry air for 6 hours through the carbon from the preceding test, which initially is saturated with carbon tetrachloride. The amount of carbon tetrachloride retained after 6 hours is a measure of the retentivity. The amount retained will be influenced by the conditions of the test and these have not been standardized. Therefore, as in the case of the test for activity, the supplier should be consulted as to the experimental details.

It should be mentioned that test procedures for activated carbon are now under study by a committee of the American Society of testing and Materials.

REFERENCES

1. Brunauer, S., *The Adsorption of Gases and Vapors, Physical Adsorption*, Princeton University Press, Princeton, 1943. McBain, J. W., *The Sorption of Gases and Vapors by Solids*, G. Routledge Sons Ltd., London, 1931.

2. Deitz, V. R., *Bibliography of Solid Adsorbents,* United States Can Sugar Refiners and Bone Char Manufacturers and the National Bureau of Standards, Washington, D.C., 1944. Supplementary volume published in 1956.
3. Deitz, V. R., and Gleysteen, L. F., J. Research Nat. Bur. Standards, 29:191 (1942).
4. Taylor, H. N., and Bonilla, C. F., *Ind. Eng. Chem.,* 39:871 (1947).
5. McBain, J. W., and Bakr, A. M., *J. Am. Chem. Soc.,* 48:690 (1926).
6. McBain, J. W., Lucas, H. P., and Chapman, P. F., *J. Am. Chem. Soc.,* 52:2668 (1930).
7. Harned, H. S., *J. Am. Chem. Soc.,* 42:372 (1920).
8. Cohan, L. H., *J. Am. Chem. Soc.,* 66:98 (1944).
9. Gleysteen, L. F., and Deitz, V. R., *J. Research Natl. Bur. Standards,* 35:285 (1945) (Research Paper No. 1647).
10. Burrage, L. J., *J. Phys. Chem.,* 36:2272 (1932).
11. Chaney, N. K., Ray, A. B., and St. John, A., *Trans. Am. Inst. Chem. Eng.,* 15:309 (1923).
12. Lamb, A. B., Chaney, N. K., and Wilson, R. E., *Ind. Eng. Chem.,* 11:420 (1919).
13. Fieldner, A. C., Overfell, G. G., Taegue, M. C., and Lawrence, J. N., *Ind. Eng. Chem.,* 11:519 (1919).
14. Stone H. W., and Clinton, R. O., *Ind. Eng. Chem.,* Anal. Ed., 14:131 (1942).
    Coull, J., Engel, H. C., and Miller, J., *Ind. Eng. Chem.,* Anal. Ed., 14:459 (1942).
    Berl, E., *Trans. Faradsy Soc.,* 34:1040 (1938).
    Dole, M., and Klotz, I. M., *Ind. Eng. Chem.,* 38:1289 (1946); *Ind. Eng. Chem. Anal. Ed.,* 18:741 (1946).
15. Allmand, A. J., and Burrage, L. J., *J. Soc. Chem. Ind.,* 47:372T (1928).
    Fieldner, A. C., Hall, R. E., and Galloway, A. E., *Bur. Mines Tech. Paper,* No. 479 (1930).
16. Macy, R., *J. Phys. Chem.,* 35:1397 (1931).
17. Bell, S. H., and Philip, J. C., *J. Chem. Soc.* (1934), 1164.
18. Huey, N. A., Broering, L. C., and Gruber, C. W., *Air Pollution Control Assoc. Magazine,* December 1960.
    Barnebey-Cheney, Columbus, Ohio.
    U.S. Public Health Service.

17

Laboratory Adsorption Test Procedures For Liquid Systems

1. INTRODUCTION [1,2]

In industrial laboratories, experimental procedures have three objectives:

  a) to determine if activated carbon will accomplish effective purification

  b) to select the most suitable carbon

  c) to learn optimum operating conditions

In view of the specific character of adsorption, the tests must be conducted on the actual product to be processed in the contemplated process. Moreover, all preliminary tests should be conducted under identical operating conditions.

2. PRELIMINARY PROCEDURE

Separate identical size portions of the liquid product to be studied, together with weighed quantities of carbon are placed in flasks or beakers. The size of the liquid portion will depend on the amount of filtrate that will be required to evaluate improvements in properties of the product. Several dosages of each carbon are used, and should include a dosage considered economically practicable together with 120%, 80%, and 60% of that dosage.*
The reason for using the 120% dosage is that laboratory tests usually require more carbon than is necessary in plant operation to accomplish a desired purification.

*It is often helpful to add 0.5 or 1.0 g filteraid in each test portion.

Each mixture of carbon and liquid (together with a blank—*i.e.*, a liquid portion containing no carbon) is agitated under temperature and other conditions to be employed in a plant operation. The mixing procedure should continue for about an hour. It should be mentioned that granular carbons are pulverized to a powdered form for the preliminary studies.

On completion of the contact period, the mixtures are filtered. Each filtrate is evaluated by appropriate tests. If the data appear favorable, the study is extended to learn the effects of time, also of changes in other conditions such as temperature. That study will be aided by the use of Freundlich isotherms. When the identity and gravimetric concentration of the impurities are not known it is often possible to use some property of the impurity, for example color, to furnish the concentration data. The procedure is described in section 17:6.

3. COUNTERCURRENT APPLICATION

The savings through countercurrent application can be investigated.[3,4] It will be recalled that in this process, the once-used carbon, after being filtered from the finished liquid, is reused to bring another batch of original liquid to an intermediate purity. Less virgin carbon is then needed to convert the intermediate product to the finished form. In order to secure the maximum saving of carbon, it becomes necessary to select an appropriate intermediate purity. Then the amount of virgin carbon needed to provide the desired final purity will furnish semispent carbon with sufficient reserve power to bring a batch of original liquid to the intermediate stage.

The dosage for countercurrent application can be determined by a cut-and-try laboratory study.

The study can be shortened and simplified by the use of charts prepared by Helbig,* one of which appears in Figure 17:1. To use this chart, an adsorption isotherm is developed for the system under study and the value of $1/n$ (the isotherm slope) is determined. Then, knowing the carbon dosage required to give the desired residual concentration of color or other impurities in a single-stage treatment, an inspection of the chart will disclose the percentage of

*Atlas Chemical Industries, Wilmington, Del.

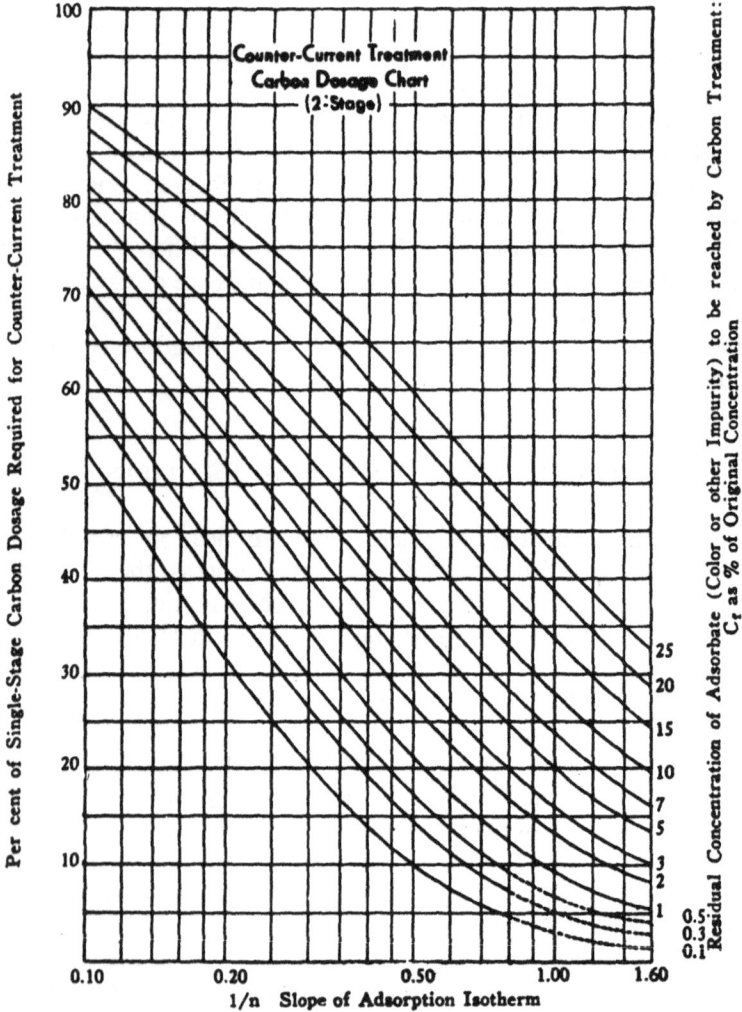

Fig. 17:1 Chart for calculating carbon dosage for two-state counter-current application From W. A. Helbig, in *Colloid Chemistry* (J. Alexander, Ed.), 6:814 (1946); reprinted by permission of Reinhold Publishing Corporation, New York.

that dosage which is required for a two-stage countercurrent treatment.

4. GENERAL ASPECTS TO CONSIDER

In the preparation of many industrial products, the substance to be purified is in different forms at various stages of the process. This can influence the selection of the proper point to apply carbon as can be illustrated by organic acids.

In the preparation of an organic acid, a salt stage may precede the free-acid stage. Here, several factors must be weighed in selecting the proper stage for carbon treatment. Salts of many organic acids hydrolyze with water to give an alkaline reaction which is unfavorable for the adsorption of certain impurities. When the treatment is applied at the acid stage, some acid will be adsorbed, but this seldom presents a serious problem except with very costly acids. A more adverse factor is the tendency of the iron in the carbon to dissolve in acids, an action that is minimized when the pH is elevated. In some operations, these various difficulties are solved by applying carbon at an intermediate stage during the transition of the salt to the acid form. This makes it possible to take advantage of optimum pH conditions.

In processes where carbon can be applied at different points— *e.g.*, before or after a concentration stage—it is desirable to determine at which point carbon is most efficient. In such cases, the comparison should be based on the amount of carbon required to purify a given weight of material (calculated on a dry basis). Let us assume that 2 grams of carbon remove 90% of the color from 1 liter of concentrated liquor containing 500 grams of sugar per liter. We want to know whether less carbon could be used if applied to a weaker liquor (250 grams of sugar per liter) before concentration. To determine this, we find out how much carbon is required to remove 90% of the color from 2 liters of weak liquor (*i.e.*, 500 grams of sugar).

When variations occur in the quality of a raw material, it may be difficult to secure samples of thin and thick liquor that correspond to the same batch of raw material. If so, the thin liquor for testing should be prepared by diluting a portion of the thick liquor. Studies of the effect of concentration are sometimes made by adding the purified solid substance (*e.g.*, pure sugar) to the thin liquor. This method should not be used since the ratio of impurities to solid substance then is not the same in the thin as in the thick liquor.

Any conclusions reached from such a laboratory study must be weighed in the light of other factors. A greater adsorption at the thin stage may be offset when the evaporation stage causes the development of color bodies or other decomposition products that will require supplementary carbon treatment. In other cases, application at the thin stage may be necessary to remove color-formers, eliminate foaming tendencies, and provide other intangible benefits in the subsequent processing.

Industrial applications of carbon often seek to accomplish more than one objective. Thus, in the treatment of corn-sugar liquors, it is desired to remove color, iron, bitter flavor, hydroxy-methyl furfural, and foam-producing substances. Because of specificity in adsorptive power, it is not always possible to find a carbon that excels in all respects and it may be necessary to accept a compromise on some desired features.

In some cases, advantages are provided by mixtures of different types of active carbon. This was found to be true in a process formerly employed to recover iodine from petroleum brines. In that operation, the iodides present in the brine were oxidized to iodine which was adsorbed by active carbon. The oxidation of the iodides was retarded by other substances present in the brine. Certain carbons which adsorbed these inhibitors lacked the needed adsorptive power for iodine, and on the other hand these properties were reversed in other types of carbon. In such cases, it becomes necessary to use a mixture of carbons.

Although the application of active carbon frequently constitutes a necessary step in the preparation of chemically pure products, more often the objective is less far reaching. In a number of applications, carbon is used to improve the quality of a product so that it will be more acceptable to the consumer. This includes the removal of objectionable tastes, odors, and colors. In other cases, the aim is to eliminate processing difficulties such as poor filtration, color-formers, and foaming.

When carbon is used in an industrial process, obviously the dosage should be sufficient to yield a product that meets the desired quality standard, but it is seldom worth while to go much beyond this point. For instance, in the decolorization of a sample of vinegar, 1 gram of carbon per liter removed 90% of the color; doubling the carbon dosage removed only 2% more color, *i.e.,* a total of 92%.

In the selection of a carbon, physical properties, such as the filtration rate and density, cannot be ignored, and last but far from least is the matter of cost. Although an expensive carbon may be essential in certain applications, it does not follow that the price tag is an index of value for all purposes. A carbon may be expensive because of the higher costs required to confer certain specific adsorptive powers, or to secure freedom from soluble ash ingredients. Except where these properties are needed for the success of a process, the purchase of such a carbon may be an extravagance.

5.   EVALUATION OF DIFFERENT ACTIVATED CARBONS

Much work is involved in gathering data for the selection of a carbon for a new application and in view of the many available brands, it is desirable to screen out those grades unlikely to be useful. In this, consultation with the supplier can be helpful. Usually it will suffice to outline the general objective, *i.e.,* whether the treatment is to correct color, odor, foam, or other characteristics that can be broadly defined.

The supplier should be informed if the application would require limits on properties such as are listed in Table 18:1. However such specifications should not be included unless they are actually necessary to the process. Thus if a moderate amount of inorganic content can be tolerated, it is unwise to insist on a low inorganic content that would add to cost and possibly exclude useful grades.

In some situations, circumstances make it practical to predetermine whether batch contact or percolation will be used, otherwise both granular and powdered carbon should be included in the preliminary survey.

When suppliers submit more samples than can be included in a full study, broad spectrum general purpose test methods are often helpful to screen out carbons less likely to be useful. Two such tests are the iodine test and molasses test, see 17:10 To simplify the discussion we will express the findings of these tests as the

I-RE (*Iodine Relative Efficiency*), and *M-RE* (*Molasses Relative Efficiency*). In general, the I-RE will show the relative effectiveness of carbon for situations that involve tastes, odors, and other volatile substances. The M-RE is useful for rough screening when large molecules *e.g.*, color, colloids are to be removed. For general, average situations, it is advisable to use both the *I-RE* and the *M-RE*.

Probably the most dependable method of further screening is by *single point* adsorption of the product to be purified, in which the data are conjoined with a *Relative Efficiency Chart*. The first step is to select a carbon having a good *IRE* and *MRE*. This carbon becomes the *Standard Reference Carbon*. In the study covered by data in Table 17:1, various dosages of the Standard Reference Carbon were applied to the liquor to be decolorized and the color removals were entered to form the *Relative Efficiency Chart* shown in Table 17:1. Then, one gram of each of the carbons to be screened was applied to liquor to be decolorized. Inspection of the color removal guided the screening. Those carbons of which one gram removed 80% or more of color were subjected to further study. Those carbons which gave less than 80% decolorization were proportionately less efficient. Thus, one gram of carbon K removed only 55% of color and therefore had only half the efficiency of the Standard Reference Carbon.

TABLE   17:1

RELATIVE EFFICIENCY CHART FOR
SINGLE POINT SCREENING OF CARBONS

Dosage Standard Carbon *grams per 100 ml*	*Percent Color Removed*
1.0	80
0.9	77
0.8	73
0.7	68
0.6	62
0.5	55

6.   CALCULATION OF ADSORPTION OF IMPURITIES OF UNKNOWN IDENTITY AND/OR GRAVIMETRIC CONCENTRATION

As previously mentioned, the impurities in industrial process liquors are often of unknown identity and/or gravimetric concentration. Their presence is made evident by properties such as color, odor, foam, vicosity, filterability. The characteristics are usually specific to each individual process; and means of measurement are usually known by the technical personnel guiding the operation. Our endeavor will be to describe how such measurements can be integrated into a study of the removal of impurities by adsorption.

The general procedure can be applied to any measurable property. We will use a decolorization study to illustrate how the procedure is employed. When we do not know the gravimetric concentration of the parent color substance, it becomes necessary to establish some arbitrary starting point. For this we define the original liquor as containing 100 units of parent color per 100 ml liquor. Therefore, the visual color of the original shows the presence of 100 color units of parent color substance.

If 50 ml of the original liquor are diluted with 50 ml of water, we know that the concentration of parent color is only half of the original, namely 50 units. Similarly, if 10 ml of the original liquor are diluted with 90 ml water, the color of that mixture represents a concentration of 10 color units per 100 ml. A full set of standards for visual observation of concentration can thus be prepared.

Water has been mentioned for use in making dilutions. However; whenever possible the dilutions should be made using purified original liquor, which is prepared by treating some of the original liquor with sufficient activated carbon to remove all adsorbable impurities. The filtrate is used for making dilutions. Such a step is obviously essential when the product is itself colored or contains impurities that are not adsorbable.

Visual comparisons have a virtue in that they reveal changes in hue that occur in some adsorptions. However, when many tests are involved, the work is expedited by use of a photoelectric color comparator, spectrophotometer, or one of the earlier forms of colorimeters. When the units of the color scale are not linearly proportional to the concentration of the parent color substances, then the scale is calibrated as in Table 17:2.

If a filtrate produced by treating 100 ml of the original solution with 2.0 grams of carbon gives an electrophotometer reading of 80, inspection of Table 17:3 shows that this corresponds to a residual color concentration of 20 units. Since the original solution contains 100 units, it follows that X, the amount adsorbed is 80

TABLE   17:2

CALIBRATION OF COLORIMETER SCALE
TO FURNISH RELATIVE CONCENTRATIONS

ml original liquor diluted to 100 ml	units of parent color substance per 100 ml	Klett Summerson scale reading of diluted solution
100	100	305
50	50	164
40	40	136
30	30	108
20	20	80
10	10	50

units (100 minus 20). The weight of carbon (M) is 2 grams. Therefore, 80/2, equals 40. The same procedure is followed with other dosages of the carbon to provide an isotherm.

The procedure just described can often be employed to measure concentrations of substances causing other unwanted characteristics and thereby develop isotherms for properties such as odor, foam, poor filterability, and poor shelf life. However, in some applications the conversion can be difficult. Therefore, it should be pointed out that although isotherms provide definite advantages in an adsorption study, they are not essential to the successful use of carbon to accomplish a purification.*

Frequently the data are expressed in the units of the instrument employed to measure the property to be corrected. This can be illustrated by an experiment conducted to select a carbon for the decolorization of a process liquor. A finished color of 80 on the scale of the photoelectric colorimeter was desired. The data are in Table 17:3, and are plotted in Figure 17:2. Inspection of the data discloses that a finished color of 80 was furnished by 1.5 grams of carbon A, and by 2.0 grams of carbon B.

TABLE   17:3

DECOLORIZATION OF A PROCESS LIQUOR

| | Carbon | Residual color |
Grade	Dosage g/100 ml	Reading on Colorimeter
A	2.0	65
A	1.5	80
A	1.0	105
B	2.5	67
B	2.0	80
B	1.5	100

*This aspect was previously mentioned in Chapter 4.

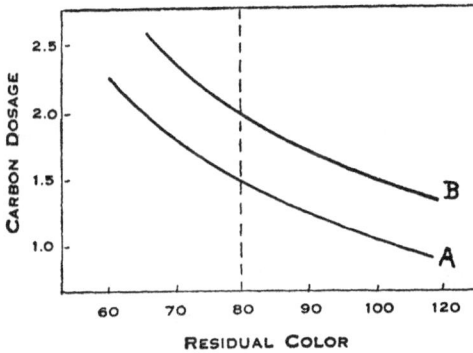

Fig. 17:2 Adsorption of color from an industrial process liquor; residual color expressed as color reading with Klett Summerson electrophotometer (blue filter); carbon dosage expressed as grams/100 ml.

7. MEASUREMENT OF SOME OTHER PROPERTIES DUE TO THE PRESENCE OF IMPURITIES

Foam

Various procedures are in use to measure this property. A convenient method is to shake a measured quantity of liquid under arbitrarily established conditions and measure the height of foam and also the length of time the foam persists. The test is applied to the appropriate dilutions of the original liquid to form a relative efficiency chart, and to the carbon-treated filtrates.

With some products it can be difficult to determine the relation of concentration to height and persistency of foam. In such situations, an isotherm cannot be formed, and therefore the data are plotted as in Figure 17:3.

Filterability

From many liquids, carbon removes substances that retard the flow of the liquid through a filter medium. The extent of the

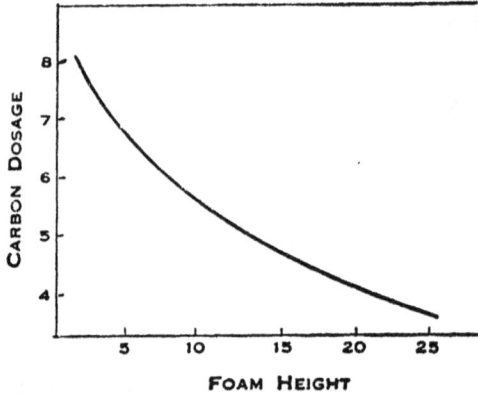

Fig. 17:3   Adsorption of foam-producing substances from an industrial
            liquid
            Carbon dosage expressed as g/liter
            Foam height expressed as ml foam after shaking filtrate 20
            seconds.

removal can be determined by measuring the time required for a
definite volume of liquid to flow through filter paper on a Buechner funnel while maintaining a constant vacuum. The test is
applied to the original liquid and also to the filtrates after treatment with carbon, employing uniform conditions of temperature
and vacuum on all samples. The data are plotted as for foaming
characteristics.

Viscosity

The change in viscosity produced by carbon treatment can be
measured by a viscometer.

Surface Tension

Changes in this property are measured by suitable instruments.

Improved Crystallization

Improvement in the form and appearance of crystals can be observed by visual inspection. Improvement in color is determined by dissolving the crystals in a suitable solvent and measuring the color of the solution. The extent to which a yield of crystals is increased by carbon treatment is difficult to measure accurately in laboratory experiments. Such information should be derived from plant experience.

Stability of Product; Shelf Life

The effect of carbon treatment on product stability can be guaged by the shelf life. Inasmuch as this may require long periods of time, appropriate accelerated tests are often developed by the technical personnel familiar with the characteristics of the product involved.

8. TASTES AND ODORS

Even when the identity of the parent taste and odor substance is known, the concentrations may be so minute as to be difficult to measure by chemical or physical methods. Because of this, dependence must be placed on the senses of taste and smell. This method is generally satisfactory because it is the same as that used by the consumer in evaluating the quality of the product.

The use of the senses of taste and smell are subject to restrictive limitations. First of all they are not precise. Various opinions on the numerical difference in the intensity of odors that an individual can distinguish give a range from 30% to well over 100%. Another limitation inherent in any test based on odor perception is imposed by the rapid onset of fatigue; the sense of smell quickly becomes dulled. Various views exist as to the period of time over which reliable data on odor can be obtained. Much depends on the testing procedure, but even with the most favorable circumstances the maximum time is probably less than an hour. The permissible time is very brief when dealing with odors that have anesthetic effect on odor-perception.

Jar Tests

These are often employed to determine the carbon dose needed to provide the desired improvement in odor. Filtrates from a range of carbon dosages are placed into flasks and each is sniffed to find the dosage that reduces the odor to an acceptable level. All samples must be at an identical temperature during the observation. The temperature for sniffing may be a matter of personal preference, subject to special characteristics of the substance being examined, but to avoid rapid fatigue the temperature should not be higher than 60°C.

Threshold Odor Test[5,6,7,8,9]

The evaluation of odor has been of great importance in the study of potable water, and there the *Spaulding Threshold Test* is extensively used. This test was invented by Charles Spaulding and further developed by Braidech, Laughlin, Feben, Hulbert, and others.[2]

This test is based on a comparison with an odor-free water obtained by passing tap water through a column of activated carbon. The water under test is diluted with odor-free water until the odor is no longer detectable. The last dilution at which an odor is observed is the *threshold odor point*.

Useful data are obtained only by strict adherence to various details and the reader is referred to publications in which these have been given.[1,9]

Usually it is feasible to determine the approximate relative efficiency of two, or even three, carbons within the allowable time for conducting a test on odor, but when more carbons are to be evaluated the odor-perception will probably be dulled before the testing can be completed. Therefore, when many carbons are to be evaluated, the study should be subdivided and conducted in separate sessions, using two carbons at each session.

The separate data are correlated by using a standard carbon for reference at each session. All this can introduce large errors. Therefore, when many carbons are to be compared, the relative power of each one to remove odor is often evaluated by a suitable indirect test.

Experience has shown that the relative power of carbon in the removal of odor runs parallel to the *Iodine Relative Efficiency*—the *I-RE*. Therefore the iodine test will provide data on which to select a carbon for the removal of taste and odor, and it usually involves less uncertainty than an extended series of tests for odor. For those that hesitate to go the whole way, the *I-RE* can be used as a screen to select a smaller number of carbons for a direct odor test.

The phenol test also gives a good index of the relative odor-removal power of carbons, but the test is less convenient to conduct.

When both color and odor are to be removed, only carbons that have decolorizing power need be considered for testing. All decolorizing carbons have some capacity for adsorbing odors, but the reverse does not necessarily follow. A carbon can have large adsorptive power for odors and have negligible decolorizing power.

9. QUALITY CONTROL

In some applications, one single grade of carbon will stand out as furnishing superior purification. More often, several types of carbon will be found satisfactory, and in this case the judgment will be influenced by such factors as rate of filtration, wettability, available technical service, and the price per pound.

After making a decision on the type of carbon to be purchased, the buyer must reach an understanding with the supplier on measures to ensure that the quality of carbon in all deliveries will equal that of the trial lot. Specifications can readily be framed for physical and chemical properties such as ash, density, mesh size. Arrangements for measurement of adsorptive powers will be governed by traits and characteristics of each individual system.

With many industrial products that are treated with carbon, large variations occur in the quality of the source materials; and by comparison, the variations in quality of carbon are negligible. For many situations, shipments of satisfactory carbon can be assured by specifications based on appropriate synthetic tests, *e.g.*, a requirement that deliveries equal the I-RE and/or M-RE of the original sample or shipment.

The foregoing arrangement will guard against irregularities in over-all adsorptive power which would result in some batches being above standard and others below. This procedure, however, will not reveal an imbalance that can occur when all batches of a grade of carbon do not have a consistent assortment of adsorptive affinities; and as a consequence some batches may lack a specific affinity needed for a particular product being processed.

To safeguard against such a situation arising, a first step can be taken after a grade of carbon has been tentatively selected for purchase. For this, a laboratory study is conducted in which the process liquor is treated with different batches of that grade of carbon; this will disclose whether special quality controls are needed to ensure trouble-free shipments.

For products and processes that require rigid control of uniform quality of carbon, the process liquor itself would appear as a logical yardstick for deliveries; but process liquors are seldom the same from day to day, and few have the keeping qualities needed for reproducible data. But even where the problem of instability does not intervene, other considerations rule out any general use of processing liquors for quality control. Obviously, if carried out for many customers, it would create an impossible schedule for the supplier;

This brings us to the matter of quality control as practiced by industrial producers. Manufacturers will use those tests that reveal utility, and the utility is gauged on a pragmatic basis, namely, suppliers select the tests that result in the fewest complaints from customers.

The molasses test, *M-RE* is perhaps one of the most widely used tests, and in some earlier days was the only control test used in the manufacture of many brands of *decolorizing* carbons. At that time many assumed that when a batch of carbon attained a desired *M-RE* all adsorptive powers would be uniformly developed; and based on a general absence of customer complaints such a conclusion seemed to be justified. Incidents in more recent years have led to a modification of that earlier view. It is now recognized that the *M-RE* does not mirror every mutation and deviation in the activation environment; and some deviations that do not affect the *M-RE* can and do alter other adsorptive powers. Consequently, a carbon may have a satisfactory *M-RE* and still be deficient in certain other affinities that can be essential for the purification of a particular product.

Consequently, with shipments to customers who require adsorptive powers not disclosed by the *M-RE* or by any other single test in use, it becomes necessary to include appropriate additional tests in the routine of quality control. The benefit of employing more than a single test is illustrated by the data in Table 17:4. All the carbons having both good *M-RE* and good *I-RE* were uniformly satisfactory for all the applications tested, but each test considered by itself failed to give a reliable picture. Thus carbons with a good *M-RE* but a poor *I-RE* gave good results on some products but not on all; and a high *I-RE* had little significance except when joined with a high *M-RE*.

The ability of different control tests to mirror different types of adsorptive power is a primary consideration in the selection of the tests to be included in a multiple program. It must be kept clearly in mind that when more than a single test is used, each additional test should be such as will catch deviations not detected by the other control tests in use. The selection can be based on a laboratory study in which a variety of synthetic test solutions are reacted with different batches of the grade of carbon under study.

When suitable multiple control tests have been selected, all batches of production going to critical customers should be so tested.

TABLE 17:4

RELATION OF *M-RE* AND *I-RE* TO DECOLORIZING POWER
OF ACTIVATED CARBON ON DIVERSE PROCESS LIQUORS

Carbon batch	M-RE	I-RE	Decolorizing efficiency on coded process liquors					
			AA	BB	CC	DD	EE	FF
1	100	100	Good	Good	Good	Good	Good	Good
2	100	100	Good	Good	Good	Good	Good	Good
3	100	100	Good	Good	Good	Good	Good	Good
4	100	100	Good	Good	Good	Good	Good	Good
5	100	80	Good	Good	Fair	Fair	Good	Fair
6	100	60	Good	Fair	Fair	Poor	Fair	Poor
7	80	60	Fair	Fair	Poor	Poor	Poor	Poor
8	80	100	Fair	Fair	Fair	Fair	Fair	Good
9	60	100	Poor	Poor	Poor	Fair	Fair	Fair

Good represents over 90% efficiency
Fair represents 75 to 90% efficiency
Poor represents less than 75% efficiency

In some manufacturing operations, batches of product that are sub-standard are blended with batches of high quality. When this is done it is important to recognize that all adsorptive affinities may not be brought to the desired standard of quality. This aspect is illustrated in the data in Table 17:5 mixing equal parts of batches #033 and #037 gives an average *M-RE* of 100, which is equal to the standard; but this same blend will have an average *I-RE* of 80 and this is below the standard *I-RE*. All this must be considered when making shipments to customers that have rigid specifications for quality.

TABLE   17:5

DATA ON BATCHES OF CARBON SUBMITTED FOR BLENDING

Batch	M-RE	I-RE
#033	80	60
#037	120	100
Reference standard	100	100

Control of the Activation Environment

Thus far we have considered synthetic test solutions solely as a means for classifying finished product, and for providing trouble-free shipments. But synthetic test solutions can be of great utility in establishing and maintaining a proper activation environment. Activation of carbon is an art; the significance of this concept is illustrated in an analogy; namely, pyrometers, flowmeters, heaters, burners, and other tools for activation are analogous to orchestral instruments in that both need performers and a guiding score to make them effective.

The score for conducting the activation of carbon is found in experiences, both planned and unplanned. The unplanned experiences are gained by close attention to the influence of factory operating conditions on the adsorptive power, but such information is often fortuitous and fragmentary. More nourishing data can be gathered by a systematic research in which deliberate changes are made in the activation environment. The effects of such changes on the carbon being processed are measured with test solutions. When adequate data have been assembled, it becomes possible to discern the types of adsorptive power created by specific conditions of activation. This know-how can be useful when a batch of production fails to meet specifications on one or more test solutions. In such a situation, a comparison with data taken from the record of research will often identify and locate the faulty operating conditions and point the way to corrective measures.

The foregoing approach can have great utility to activation on an industrial scale, but it would be unfortunate to give the impression that the interpretation of the data is easily mastered. Problems arise from varied factors. First of all, in order to examine all diverse properties of the carbon being processed, the study may include so many test solutions that the total data, on initial inspection, may appear more confusing than informative. Some situations are further complicated by the fact that a particular type of adsorptive power may be created by more than one set of activating conditions. Therefore when some properties of the finished carbon are off normal, this may be not a consequence of a deviation in any single factor but rather a resultant of several factors. Still further, activated carbon has a "memory" and the prior history, e.g., the manner of carbonization, can influence the way in which a char responds to subsequent activation.

It is not surprising that even the veteran operator will not always be able to pin-point the source of a difficulty. But, given adequate and appropriate test solutions, he can narrow the field of view and shorten the time needed to restore normal operation and provide production of standard quality. That all this can be done has been demonstrated in the many years of satisfactory experience of long established manufacturing operations.

10. SYNTHETIC TEST SOLUTIONS

General Aspects

To qualify for use, synthetic test solutions should contain substances having adsorptive characteristics that reflect those of the impurities in the application under consideration. There are practical advantages in selecting substances of which the concentrations are readily measured, *e.g.,* dyes.

Because many carbons contain small amounts of acid or alkaline constituents, it is advisable to buffer a synthetic test solution with non-adsorbable agents such as mono- and di-sodium phosphate. Buffering is especially necessary when using colors that are altered by *p*H.

Dye Test

Solutions of various dyes provide indirect tests. Concentrations which are convenient for use are shown in Table 17:6, which also

TABLE   17:6

DYE SOLUTIONS FOR SYNTHETIC TESTS

Solvent	Solute	Concentration g/liter	Other conditions	Suggested carbon dosage g/100ml
Water	Methylene Blue	0.80		0.15–0.50
Water	Malachite Green	0.91		0.10–0.50
Water	Alizarin Red	0.73	Buffered to pH 6.2	0.10–0.50
Water	Ponceau Red	1.09	Buffered to pH 7.0	0.20–1.00
Ethanol	Methylene Blue	0.50		0.40–3.50
Ethanol	Malachite Green	0.43		0.60–8.00

gives the probable range of carbon dosages to be used. Test solutions containing a mixture of suitable dyes are of value in revealing versatile adsorptive properties.

Procedure

1. Into 200-milliliter Erlenmeyer flasks introduce 100-milliliter portions of the solution together with weights of carbon to give an adsorption range of 60–99% removal of the solute originally present.
2. Agitate the contents of the flasks by a suitable device for 1 hour at approximately 25°C; (for control tests, the time can be shortened to 15 minutes. The rate of adsorption is very rapid for the first 10 minutes and then decreases sharply. After 1 hour, further changes are slight).

3. Then filter the carbon-dye mixture through a 15-centimeter No. 5 Whatman filter paper (or equivalent) in a gravity funnel, allowing all the solution to filter.
4. Stir the filtrates to ensure homogeneity before determining the concentration of color. Complete freedom from turbidity is necessary in order to obtain an accurate measurement of the dissolved color.

The data are calculated as described in Chapter 4.

The adsorption characteristics of many dyes used in synthetic test solutions will vary from one batch to another. Therefore, when tests are to be correlated over a period of time, it is important to procure an ample supply of a selected batch of each dye, sufficient to last for the duration of the study. When several laboratories participate in a joint study, all should use reagents from an identical supply.

Molasses Test

Batches of molasses obtained from different sources will vary considerably in depth of color and in the ease with which the color bodies are adsorbed. Because of this, it is not practical to state the concentrations of molasses to be used in preparing the stock solution. Depending upon the type of molasses, somewhere between 20 and 60 grams of molasses per liter of solution will be suitable.

The appropriate weight of molasses, together with 15 grams of disodium phosphate, is dissolved in 500 milliliters of water and sufficient phosphoric acid is added to give a pH of 6.5. The mixture is diluted to 1 liter. If turbid, the solution should be filtered before use through a 1/4-inch thick layer of filter aid on a cloth mat placed in a Buechner funnel.

Dilute test solutions of molasses undergo chemical and biological changes that alter the adsorbability. Preservatives are of questionable utility in this. If kept in a refrigerator, a dilute molasses solution will usually be stable for 4 to 8 hours. When a new supply of molasses is procured, the stability of a diluted solution should be determined by making decolorizing tests with a standard carbon at hourly intervals.

Many supplies of molasses are acceptable but there is appreciable variation in the characteristics of adsorptive behavior. In general, a darker molasses will show a broader spectrum of adsorptive power.

Each batch of molasses has eccentricities of behavior and whenever a new batch is obtained, time is needed to learn the characteristics. Therefore, when a suitable new batch is located, one should procure a quantity sufficient to last several years—or at least for the period during which tests are to be correlated. The reserve supply should be kept in cold storage.

Two methods are commonly followed for conducting the molasses test.

Method A

1. Measure a 50-milliliter portion of the solution into a 150 milliliter beaker, add 0.5 gram of carbon, and stir until wetted down.
2. Place the beaker on a heating unit and bring contents to a boil
3. Filter through a 15–centimeter No. 5 Whatman filter paper (or equivalent) in a gravity funnel and measure the color of the filtrate.
4. Repeat steps 1 to 3 using other weights of carbon suitable to give an adsorption range between 70 and 90% removal of the original color.

Method B

1. Place separate portions of molasses solution plus appropriate carbon dosages in bottles and shake them in a steam chest at 100°C for 50 minutes.
2. Turn off the steam and agitate the bottles for 45 minutes longer.
3. Filter the solutions and measure the color of the filtrates.

The adsorption may be calculated by using the Freundlich isotherm. Often the removal of color is expressed as a percentage and plotted as shown in Figure 17:4. By inspection the weight of carbon is determined that will reduce the color to an acceptable point. This weight is compared with the amount of a standard carbon that is required to remove the same amount of color.

Care is required in selecting the comparison point to be used in this or in any other indirect test. Comparisons should be made within a range at which a variation in carbon dosage produces a readily measurable change in the amount adsorbed. For instance,

an inspection of Figure 17:4 shows that, in this case, it is undesirable to make comparisons when the removal of color exceeds
95%. Above 95%, a large increase in carbon dosage makes only
a small change in the amount of color adsorbed. Conversely,
a small error in measuring residual colors could cause a large
error in calculating the relative value of two carbons.

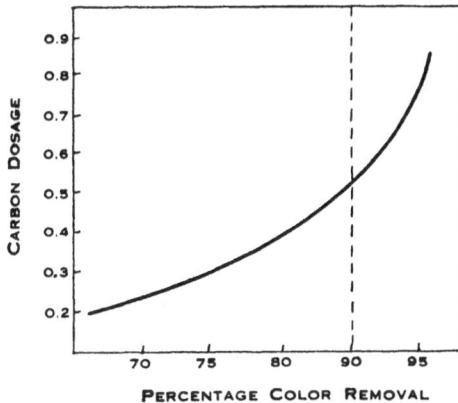

Fig. 17:4   Adsorption of molasses color
Carbon dosage expressed as g/50 ml. solution

It is noted to be in Table 17:7 that the relative efficiency of different types of carbon can depend on which molasses testing procedure, A or B is used.

Caramel Test

As in the case of molasses, the color of different samples of
caramel varies in intensity. Usually 4 to 6 grams of caramel per
liter of water will give a satisfactory solution. The method of
buffering the solution and the procedure for conducting the test
parallel those used for molasses. During the early development of
carbons for the treatment of liquids, this test was extensively

TABLE 17:7

TESTS ON MOLASSES

Grade carbon	M-RE	
	Method A	Method B
L	80	70
M	90	95
R	70	70
S	70	100
T	100	90
U	95	95
W	100	100
Y	100	75
Standard	100	100

used. Since then, it has been found that the test has very little relation to the removal of impurities from most industrial solutions and is seldom used now.

Iodine Test

For the test with iodine, a stock solution is prepared, containing 2.7 grams of iodine and 4.1 grams of potassium iodide per liter. The stock solution should be kept in dark bottles and stored in a cool place. The strength should be determined before use. When the iodine concentration falls below 2.65 grams per liter, an entirely fresh solution should be prepared.

Procedure

1. Into a 250-milliliter Erlenmeyer flask, introduce 0.5 gram of the carbon being tested and 10 milliliters of 5% hydrochloric acid.
2. Swirl the flask until the carbon is wetted, then add 100 milliliters of stock iodine solution and shake for 5 minutes.
3. Filter through a 15-centimeter No. 5 Whatman paper or equivalent in a gravity funnel, allowing all of the solution to filter.
4. Stir the filtrate to ensure homogeneity and titrate an aliquot portion with 0.1 N sodium thiosulfate, using starch as an indicator.

Repeat the test using appropriate carbon dosages in the range 0.2 to 0.4 g per 100 ml iodine solution.

Baylis Phenol Test

This test with phenol was developed to evaluate carbons for the removal of tastes and odors from potable water supplies. The amount of carbon required to reduce the phenol concentration in a liter of water from 0.10 to 0.01 parts per million is known as the *phenol number*.

Successful use of this test requires much practice and close adherence to the details of the procedure. The reader is referred to the original publication.[1,11]

Permanganate Test

For the test with permanganate, add 0.4 gram of carbon to 25 milliliters of 0.5 N $KMnO_4$ in a 150-milliliter beaker, stir for 30 seconds and allow the mixture to stand an additional 9.5 minutes. Filter the permanganate-carbon mixture through a Gooch crucible fitted with an asbestos fiber mat. Transfer the last traces of the permanganate-carbon mixture from the beaker, using 50 milliliters of distilled water in 10-milliliter portions. Wash the crucible with 50 milliliters of distilled water in 10-milliliter portions. Combine the filtrate and washings and determine iodometrically the amount of remaining permanganate. Commercial carbons will remove from 25 to over 50% of the permanganate.

Other Tests

Other indirect tests have been developed which yield significant information to certain industries. For the greater part, the pattern of their procedure is similar to the tests that have been described.

In some industrial products, the impurities are in such small concentrations that they present analytical difficulties in the measurement of adsorptive power. In some of such cases, it is practical to conduct the adsorption on mother liquors in which the high concentration of color and other impurities simplifies the analytical measurement of the adsorptive power.

Another method is to add artificial color to a test portion of the product being purified and use this to measure the adsorptive power of carbons. This method gives data of less significance than is provided by the use of mother liquors.

Multiple Tests

When several tests are used to control the quality of activated carbon production and shipments, each teŝt should be chosen to catch deviations and mutations not detected by other control tests in use. The selection of tests can be based on a laboratory study in which a number of batches of the same carbon are tested with various synthetic test solutions.

Abridged data on such a study appear in Table 17:8. Inspection of the table discloses that the adsorption of aniline blue closely parallels the *M-RE*; hence if the *M-RE* is being used, nothing will be gained by adding the test with Aniline Blue.

The *I-RE* should be included because it gives a different response from the *M-RE* on several batches. However, there is no need to include the phenol test because this almost duplicates the figures furnished by the *I-RE*. As the Ponceau test shows departures from both the *M-RE* and the *I-RE*, this could be included as a third test for customers that require more rigid quality control.

It is to be emphasized that the data given for Carbon *W* in Table 17:8 and the deductions as to suitable control tests apply only to that particular grade of carbon. Another grade, particularly one made by a different process, would furnish other assortments of adsorptive power. A corollary is that specifications for the purchase of a certain grade of carbon should not be used automatically for the purchase of a different grade. Furthermore, if and when a switch to another carbon is being considered the decision should rest on a comparative performance with the solution to be purified. Carbons must never be substituted for one another on the basis of synthetic test solutions in an industrial application.

11. SPECIFICATIONS

The experimental conditions to be established for tests conducted on industrial process liquors are governed by economic and technological factors. With synthetic tests, *e.g.*, iodine, dye, molasses solutions, analytical considerations influence the selection of experimental conditions for conducting the tests. Precise standardization is essential on all details such as time of contact,

TABLE 17:8

RELATIVE SPECIFIC ADSORPTIVE CAPACITY OF DIFFERENT
BATCHES OF A CARBON PRODUCED DURING AN EXPERIMENTAL
DEVELOPMENT PROGRAM.

Batch No. Carbon W	Molasses M-RE	Aniline Blue RE	Ponceau R RE	Phenol RE	Iodine RE
30	100	100	70	100	100
33	95	100	100	90	95
37	95	90	90	80	85
42	75	75	100	100	95
48	100	95	80	90	90
50	100	100	100	95	100
54	95	90	75	70	75
59	60	65	90	100	95
62	70	70	100	80	85
65	100	100	95	95	100
70	80	75	100	100	95
73	100	100	95	60	65
Standard	100	100	100	100	100

temperature, method of filtration, measuring instrument. All participating laboratories should use reagents from the same standard supply. Even with rigorously defined procedures, however, different laboratories may not always duplicate one another's data. Such contingencies can be resolved by agreement on a standard sample of carbon that will serve as the final gauge.

Agreement is also necessary as to the amount of tolerance to be allowed, and this in turn will depend on the magnitude of the experimental errors inherent in the test procedure, plus provision for deviations in quality that normally occur in the manufacture of activated carbon. For this the manufacturers experience is the best guide in the case of established tests. With new and untried tests, it is well to base specifications on experimental data covering at least a dozen different batches of the grade of carbon selected.

New specifications must be developed whenever a change is made from one type of carbon to the purchase of a different brand or grade.

Standard Carbon

The sample of standard carbon should be from a representative batch of the grade of carbon to be used. Sufficient quantity should be reserved for all tests likely to be required during the period in which that grade of carbon is to be used. The supply should be stored in glass-stoppered bottles.

12.   PROCEDURES FOR GRANULAR ACTIVATED CARBONS[13,14,15,16,17]

It would seem that the laboratory evaluation of granular carbons should be conducted under conditions that would parallel those that are employed in plant operation, namely, percolation columns. For reasons that will become clear as the discussion proceeds there are limitations to doing this. In practice, the evaluation of granular carbons involves a combination study of batch-contact and use of columns.

The preliminary studies—to learn proper pH, temperature, and correlation with other steps in processing—all these can be effectively made by batch-contact methods. For these, the granular carbons are pulverized and used in a particle size comparable with that of powdered carbons. Work so conducted requires less liquid, and can be completed in far less time than is required for column studies. This last advantage is important when working with process liquors that are perishable, or in which the quality does not remain constant but varies from time to time.

When a batch-contact process is to be used in plant operation, the laboratory batch test reproduces in miniature practically all the features needed to lay out the operation on plant scale; and one can move almost directly from the laboratory to the drafting table. Batch tests contribute helpful information for a column set-up. However, in the case of the application of percolation system to a new type of purification, batch tests leave undetermined some essential data.

In a single stage batch contact treatment, the carbon when removed from the system is in equilibrium with the liquid in its final purity. Even with countercurrent batch-contact, the carbon is in equilibrium with a partially purified liquid. Hence, in both cases, when the carbon is removed from the system, each gram carries away less impurity than in a percolation operation in which the carbon is not removed from the system until it is in equilibrium with the original liquid; and consequently its adsorptive capacity is then saturated.

Therefore, batch-contact tests do not directly disclose the exact amount of impurity carried away by the carbon, when it leaves a percolation system in equilibrium with the original unpurified liquid. The amount can be approximated by extrapolation of the isotherm but should be confirmed by experimental percolation studies.

Experimental studies are also needed to furnish data of conditions appropriate to the characteristics of the product to be purified. Among these are: velocity of liquid flow; dimensions of mass transfer zone; required residence time; suitable size of carbon granules.

Published descriptions are available of various types of laboratory and bench type percolation units.

The data obtained from such a study can be scaled up to design a plant operational unit. However, it is to be emphasized that in making any plans for the design of a plant operation, it is essential to obtain the services of engineers qualified by experience.

13. RECLAMATION OF SUBSTANCES BY ADSORPTION-DESORPTION

The procedures described thus far involve processes in which the impurities are adsorbed and the desired product remains in solution. In some applications, purification is accomplished by selectively adsorbing and then desorbing the desired product.

Adsorption-desorption is preferably studied by carrying out the entire process, that is, the material is first adsorbed by the carbon, which is then separated and eluted with various agents.

The adsorption stage should be conducted to obtain maximum removal of the desired substance from the solution. However at that point the concentration of adsorbate on the carbon is small. Moreover, initial portions that are adsorbed are often held irreversibly and cannot be desorbed.

Therefore, in order to accomplish adequate removal of adsorbate from the solution, and also store sufficient amounts of extractable adsorbate on the carbon, it is necessary to employ countercurrent batch adsorption; or better still, percolate the liquor through a bed of carbon.

The procedure is illustrated by an experimental study to recover dye from a waste solution. Isotherms were determined of several granular carbons and one that appeared promising was selected for further study.

The waste dye solution was percolated through a bed of the carbon until the color of the effluent showed that the carbon was saturated with adsorbed dye.

One half of the carbon bed was dried and then divided into separate 5 gram portions, each of which was extracted with a different eluting agent. The other half of the carbon bed was also divided into separate portions and eluted without the intervening drying step.

The results of such a preliminary test will often suffice to indicate whether additional study should be undertaken.

The eluting power of different solvents can be approximated by a modification of this procedure. Some of the substance to be processed is dissolved in each of the solvents to be studied. Each solution is then treated with carbon, and the solvent from which the poorest adsorption is obtained can be expected to be the best eluting agent.

It is to be recognized that the data provided by such simplified procedures have limited significance. Their main value is that eluants and conditions of no value can be eliminated from further consideration. Conclusions reached from such simplified procedures—as to the best eluants and conditions—should be confirmed by a supplementary test using the full adsorption-desorption procedure on the substance to be processed. This is especially necessary when the preliminary studies employ a synthetically prepared solution; these studies do not fully reflect the behavior of a substance when present in industrial or biological preparations containing coadsorbates.

Also to be considered is the fact that the value of an eluting agent may depend on the ability to selectively extract the desired substance and leave most of the impurities on the carbon. Obviously, such information is secured only from studies using the liquid to be processed.

14. FLEXIBILITY IN TESTING PROCEDURES

Testing and evaluation procedures, when established as part of the manufacturing routine, must never be regarded as immutable. The true function of synthetic test solutions must be kept in mind at all times. Synthetic tests have no inborn worth; instead they acquire utility to the extent that they mirror the needs of the markets being served. As new processes and products enter the market place, some of them will call for activated carbons with new and different forms of usefulness. And some of these new needs will in turn call for appropriate new test procedures.

Experiences with adsorption-desorption processes indicate a need to investigate additional tests in which an organic solvent is substituted for water.

Case History of the Development of Testing Procedures

The way in which testing methods keep in touch with changing market patterns is illustrated by excerpts taken from the history of the original process for making activated carbon in America. The source material for that process was black-ash residue, a waste product from the manufacture of soda pulp. Although this early activated carbon had much less adsorptive power than brands of today, it was found useful for decolorizing coconut oil, phosphoric acid, and other industrial products. Complaints were received from customers stating that the quality was uneven, and as these complaints became more frequent it was necessary to institute quality controls.

For this program, the caramel test as described in some scientific literature of the time was adopted. Events proved this to be a bad choice because the adsorption of caramel had very little relation to the utility of carbon for decolorizing the products being purified: Two batches of carbon would be equal for adsorption of caramel, but be very unequal for decolorizing coconut oil or phosphoric acid. As the caramel test did not decrease the customers' complaints, the next step was to pre-test each batch of carbon on the various products for which a market then existed. This eliminated complaints but created a burdensome testing program, particularly when the market later expanded to include an increasing number of products.

A search to simplify the testing program resulted in the invention of a test solution in which an oil-soluble dye, Sudan III, was dissolved in kerosene. The use of this test restored good customer relations, and the test became a useful production guide that led to improved manufacturing methods and an elevation of the quality of the activated carbon. A steady improvement in the adsorptive properties led to a decision to enter the sugar-refining field; but it was found that the Sudan III test did not measure the decolorizing power for sugar syrups. Following this, the molasses test which had long been used for evaluating bone char was adopted. This test proved helpful to develop the carbons for sugar refining, and also led to the development of carbons suitable for many other products. The molasses test (M-RE) has continued in use, and provides a broader spectrum of decolorizing power than any other single test thus far devised.

Shortly after 1920, a project was undertaken to recover iodine from petroleum brines; this required a carbon with a good M-RE and also adequate adsorptive power for iodine. For this, the iodine test (I-RE) was included in the study of activation conditions.

A new type of situation arose with the use of carbon for potable water supplies. It was soon recognized that decolorizing properties as measured by the M-RE would be needlessly costly because they have no direct value for removing tastes and odors from water supplies. A search for a test that would reveal the power of a carbon for removing taste and odor resulted in the phenol test. This made it practicable to produce carbons that not only were more effective for control of taste and odor but also could be manufactured at a much lower cost. All of this contributed much to great subsequent expansion in the use of carbon for potable water supplies.

Thus far the story has been a recital of successful experiences, but for future guidance it is well to mention one of the less fortunate ones. Activated carbon was used with apparent success in the initial manufacture of streptomycin, but at times the operation was marred by low recovery of streptomycin. This was traced to several factors, one of which was an uneven quality of the carbon delivered. Unfortunately, the nature of the unevenness could not be identified or measured by any of the synthetic tests solutions then in use; and before suitable tests could be developed the manufacturers of streptomycin had turned to other methods of processing. The experience was unfortunate, not only because a large and valuable market was lost but also because the episode appeared to cause a waning of interest of research groups in exploring the potential use of activated carbon for the production of other antibiotics.

Future Testing Procedures

The question arises: Is it possible to anticipate and be prepared for such situations? Obviously we cannot know in advance the identity of the processes and products that will appear in the future, nor of the specific properties they will require of carbon. But we do know that many forms of usefulness have their roots in adsorptive powers that can be detected and measured by appropriate synthetic tests. Consequently, experiences such as that with streptomycin suggest a need to invent new tests that will measure properties not revealed by existing tests. Such new tests could unfold previously unrecognized properties of activated carbon.

Research should continue to learn ways and means by which new types of adsorptive power can be created, which should include both new methods of activation, and modifications in the environment of established operations. If and when a distinctive new grade of carbon is developed, batches of experimental size can be prepared for receptive research groups. Should any such experimental carbon be found especially useful for an application, old or new, information would be at hand to know proper methods of manufacture and quality control, and also to evaluate the economic potential. Interestingly enough, activation processes, that are more difficult to keep on the beam in normal operation, for that very reason contain a flexibility that may be useful for meeting the varied needs of diverse markets.

REFERENCES

1. Deitz, V. R., *Bibliography of Solid Adsorbents,* National Bureau of Standards, Washington, D.C., 1944: References to the scientific literature during the years 1900–1942.

Supplementary volume issued in 1956 with abstracts for years 1943 to 1953.

2. Mantell, C. L., *Industrial Carbon,* D. Van Nostrand Co., New York, 1946.

3. Sanders, M. T., *Ind. Eng. Chem.,* 15:784 (1923); *ibid.,* 20:791 (1928).

4. Helbig, W. A., in *Colloid Chemistry* (J. Alexander, Editor), Vol. 6, p. 814; Reinhold Publishing Corp., New York, 1946.

5. Spaulding, C. H., *Am. J. Public Health,* 21:1038 (1931); *J. Amer. Water Works Assoc.,* 24:1111 (1932); 34:877 (1942).

6. Braidech, M. M., *Chem. Abstracts,* 32:9351 (1938).

7. Laughlin, H., *J. Am. Water Works Assoc.,* 32:1191 (1940).

8. Hulbert, R., and Feben, D., *J. Am. Water Works Assoc.,* 33, 1945 (1941).

9. Hassler, W. W., *Taste and Odor Control J.,* 8 Nos. 7, 9, 11, 13— contains 891 classified references.

10. Gleysteen, L. F., and Scheffler, G. H., *Proceedings of the Fourth Conference on Carbon,* p. 48; Pergamon Press, New York, 1960.

11. Baylis, J. R., *Water Works and Sewerage,* 80:220 (1933).

12. Braidech, M. M., Billings, L. C., Glicreas, F. W., Kershaw, N., Scott, R. D., and Spaulding, G. R., *J. Am. Water Works Assoc.,* 30:1133 (1938).

13. Literature from Pittsburgh Chemical Company, Pittsburgh, Pa.

14. Communication, Jonathan C. Cooper of Pittsburgh Chemical Company.

15. Communication, Franklyn M. Willliams of Pittsburgh Chemical Company.

16. Communication, R. S. Joyce of Pittsburgh Chemical Company.

17. Proceedings Technical Session Bone Char Research: I, 1949; to 1959 Edited by Victor R. Deitz; published by Bone Char Research, c/o Revere Sugar Refinery, Charlestown, Mass.

18

General Properties of Activated Carbons

In addition to a knowledge of the adsorptive behavior of an active carbon, it is often desirable to have information as to other properties that can influence the utility and value of a carbon. Table 18:1

CHARACTERISTICS OF ACTIVATED CARBON THAT MAY REQUIRE
CONSIDERATION FOR SOME PRODUCTS AND PROCESSES

Density
Particle size distribution
Porosity   Content of
Surface area   Sulfur
Filtration rate   Sulfides
Wettability   Sulfates
Dustiness   Phosphates
Ignition temperature   Chlorides
Electrical conductivity   Iron
Oil retention   Copper
Moisture   Zinc
pH   Calcium
Total ash   Magnesium
Resistance to attrition   Silica
Hardness
Water-extractable inorganics
Acid-soluble inorganics

1. MOISTURE

Active carbon is generally priced on a moisture free basis, although occasionally some moisture content is stipulated, *e.g.*, 3, 8, 10%. Unless packaged in airtight containers, some activated carbons when stored under humid conditions, will adsorb considerable moisture over a period of months. They may adsorb as much as 25 to 30% moisture and still appear dry. For many purposes, this moisture content does not affect the adsorptive power, but obviously it dilutes the carbon. Therefore, an additional weight of moist carbon is needed to provide the required dry weight.

Active carbon, purchased on a dry-weight basis, should be evaluated on that same basis in the laboratory. This can be done by drying the carbon at 110°C. for 24 hours and then keeping the sample in a desiccator over a suitable dehydrating agent. An objection to this procedure exists in the case of those carbons which lose appreciable adsorptive power when dried. In such cases, test portions used for measuring adsorptive power are weighed out as is, and calculated on a dry basis. The moisture determination is made on a separate portion.

2. INORGANIC CONSTITUENTS

As activated carbons contain inorganic constituents derived from the source materials and from activating agents added during manufacture, the total amount of inorganic constituents will vary from one grade of carbon to another. Permissible limits depend on a number of factors: the chemical nature and concentration of the individual inorganic constituents, the carbon dosage to be used, and the standards of quality required for the product to be treated. Some applications require a thoroughly washed carbon, whereas unwashed carbons are acceptable for many products. Even when large dosages of 4 or 5% carbon are used, the amount of inorganic constituents eluted from the average unwashed carbon is relatively small. In processes that include a subsequent crystallization or distillation, the extracted inorganic constituents do not reach the finished product.

Ash

To determine the content of ash, a weighed quantity (2 grams of powdered carbon, or 10 to 20 grams granular carbon) is placed in a porcelain crucible and heated in air in a muffle furnace until the carbon has been completely burned. The temperature should be kept below 600°C to minimize volatilization of inorganic constituents, and also to leave the ash in a suitable condition for further examination.

The inorganic constituents in a carbon are usually reported as being in the form in which they appear when the carbon is ashed. This can be misleading because the analysis of an ash does not necessarily show the form in which the inorganics exist in the carbon. Table 18:2 shows the amount of soluble matter found in the ash of a carbon was more than could be eluted from the carbon with water.

TABLE 18:2

AMOUNTS OF WATER-SOLUBLE INORGANIC
SUBSTANCES IN CARBON

Carbon Grade	Found in ash from 100 g carbon	Eluted from 100 g original carbon
EE	0.75 g	0.50 g
FF	0.80	0.41
GG	1.52	1.55
MM	2.50	1.20
NN	3.50	2.05

The form in which inorganic constituents are present is influenced by the method of preparing the carbon. A sample of carbon *KK* from an activating furnace was divided into two portions: one portion was cooled in the presence of air; another portion was cooled in complete absence of air. The amount of water-extractable inorganic matter differed in the two samples, whereas the watersoluble substances in the ash were identical for both samples.

Ion Exchangers, Inorganic Constituents

As previously mentioned, the amount of inorganic substances that can be leached from some carbons is less than the water-soluble matter found in the ash of the carbon. The deviation varies, being negligible in some carbons and appreciable in others.

In some types of carbon, a normally soluble inorganic compound is bonded to the carbon in the form of an ion exchanger. This behavior is illustrated by the data in Table 18:3. Carbon

EFFECT OF VARIOUS TREATMENTS ON
ASH CONTENT AND pH

Carbon	Na_2SO_4 in the ash from 100 g carbon	pH water extract of carbon
Washed with water	2.2	8.0
Washed with HCl	0.9	1.5
Washed with HCl, then neutralized with NaOH	1.9	7.2
Washed with HCl, then heated to 800° C	1.0	7.6

JJ was thoroughly washed with water and then when ashed was found to contain 2.2 grams of sodium sulfate per 100 grams of original carbon, none of which could be leached from the carbon by washing with water. A portion of carbon *JJ* was extracted with hydrochloric acid and, based on an analysis of the ash, the content of sodium sulfate had decreased to 0.9%.

The acid extraction, however, lowered the pH of carbon *JJ* from 8.0 to pH 1.5, and the pH was not altered to any appreciable extent by washing with water. The carbon could be made neutral by heating to an elevated temperature to drive off the acid, or by adding a base. When the adjustment was made by adding a base such as sodium hydroxide, the carbon regained much of the original sodium sulfate content and this could not be removed by washing with water. This suggests that the apparent extraction of inorganic matter from carbon *JJ* was in reality an ion-exchange in which sodium ions left the carbon to be replaced by hydrogen ions from the acid. The net effect is that carbon *JJ* after extraction with acid contains sulfuric acid instead of sodium sulfate.*

Although such ion-exchange does not occur with all types of carbon, the possibility must be kept in mind. Inasmuch as a process liquor may contain exchangeable ions, it follows that in measuring the amount of inorganic constituents that can be eluted from a carbon, the test should be based on using the liquor to be processed, not water.

*We refer to the ion-exchange as though a sulfate ion were bonded to the carbon, but we do not know the actual identity. The designation *sulfate* is applied only because that happens to be the form in which the sodium appears in the ash in this instance.

Miscellaneous Inorganic Constituents

This section will consider only those inorganic constituents commonly reported.[1,2,3]

Iron:

The concentration of iron ranges from traces to over 1 per cent. In carbons prepared by steam-activation, iron may be present in a magnetic state, and some of it can be removed by a magnet. Acid-washed carbon can contain iron in both ferric and ferrous forms. Iron is seldom released from a carbon except in solutions having a low pH. Actually most types of carbon can remove much iron from a solution having a pH of 4 or higher.

Some of the iron in carbon can be removed by washing with acid or by electrolysis, but complete removal is seldom if ever accomplished, because iron salts are so strongly held by carbon and only a small portion is released with each extraction. Therefore a washed carbon can release some iron to acids such as citric, tartaric, hydrochloric. For applications in which the release or iron is objectionable it may be necessary to use a carbon prepared from source materials that are low in iron, or else take supplementary measures to remove the dissolved iron from the carbon treated product.

Sulfur:

Depending on the type of carbon, total sulfur will range from traces to over 2 per cent. It exists in varied forms—free sulfur, sulfides, sulfates, thiosulfates. In some types of carbon, much of the sulfur is bonded in forms not yet identified. Sulfur can be present in forms that act as ion-exchangers. In some carbons much of the sulfur can be volatilized by heating in a stream of nitrogen, hydrogen, or steam at temperatures over 800°C. It is reported that certain tightly bonded sulfur compounds can be removed by adding 5% of phosphoric acid to the carbon and then heating at 900°C.

Phosphates:

The content of phosphates will range from zero to over 3 per cent. Phosphates have been reported as aiding the adsorption of some impurities. Little or no information is available on other phosphorus compounds that may be present in carbon.

Calcium:

The content of calcium compounds ranges from zero to over 1 per cent. Calcium has often been reported as a cause of haze, but in a number of instances the claim is unfounded.

Chlorides:

The content of chlorides ranges from zero to over 0.5 per cent. Chlorides are not acceptable in a number of applications.

Sodium:

Compounds of sodium range from traces to over 3 per cent. In some carbons much of the sodium content is not extractable with water.

Copper:

The content of copper will range from zero to 20 ppm. As it is usually tightly held, it is seldom released to solution. Instead, many types of carbon can adsorb traces of copper.

Silica:

Even when present in concentration as high as 25 per cent, silica is not released except in strongly alkaline solutions.

3. pH OF CARBON

What is called the pH of carbon may be defined as the pH of a suspension of a carbon in distilled water. The numerical value of the pH will be affected by the experimental conditions, *e.g.*, time and temperature of extraction and the carbon-water ratio. Consequently, the process conditions, under which a carbon is used, should be considered when selecting a method for measuring the pH.

In a method suitable for most general purposes, a suspension of 2 grams of carbon in 50 milliliters of distilled water (pH 7.0) is heated to 90°C and then cooled to 20°C. The pH of the suspension is determined electrometrically.

In some procedures, the suspension is allowed to settle and the pH of the supernatant water is determined. However, there is evidence that the pH of the suspension and that of the supernatant liquid are not identical. A similar objection exists to determining the pH of a carbon filtrate.

With industrial aqueous solutions, the main influence of pH in carbon arises from changes produced on the pH of the liquid being treated. The extent of such change depends on characteristics of the individual system. A wide range of pH in carbon can be tolerated when small carbon dosages are applied to buffered solutions, whereas the permissible range of pH can be very narrow where a large carbon dosage is applied to liquids sensitive to the influence of pH.

In many cases the user can offset any adverse pH effect by adding an appropriate amount of acid or base to offset the pH of the carbon. To avoid the necessity for such adjustment, the desired pH of the carbon should be specified on the purchase order. In a few applications, a pH as high as 10 is acceptable, and even desirable, whereas other applications may call for a pH as low as 3.0. For the great majority of applications a neutral range of pH 6.0–8.0 is safest for all-round service. Over-all aspects must be kept in mind; thus, with corn sugar, a low pH will give better decolorization but it will impair the removal of iron.

For reasons not well understood, many non-aqueous liquids, e.g., edible oils, are sensitive to the pH of the carbon and when this is so, it is important to use a carbon with the proper pH: Usually a neutral pH is satisfactory but each case should be individually studied.

The pH of most commercial carbons is due to inorganic ingredients originating in the source materials or added during manufacture. In the case of carbons that are washed after activation, the pH is conveniently adjusted during the washing cycle, after which the carbon is dried. Drying is a major cost in the washing of a carbon. Where a carbon can be used in the wet state, the cost of drying can be eliminated by washing at the site of use. But, except for granular carbons, this seldom is acceptable to the customer.

After activation, most carbons are alkaline, and experience has shown that for many applications, the steps of washing and drying can be eliminated by blending a powdered carbon with an appropriate quantity of acid.

Sulfuric acid is often used for blending, preferably in a dilute form of approximately 30% acid. Moisture is necessary to ionize the acid so that it will react with and neutralize the alkalinity of the carbon during the blending stage. If concentrated acid is used, the neutralization will not take place and the surface of the carbon will contain patches of both unreacted acid and un-neutralized alkalinity. The effect is more apt to be adverse when the carbon is applied to non-aqueous liquids.

Phosphoric acid has a disadvantage of high cost; moreover greater quantities are required to provide a given adjustment in pH. In some applications, however, phosphoric-adjusted carbons furnish definite benefits.

Hydrochloric acid can be used for some applications, but it is not acceptable in a number of processes because of objections to chlorides.

Alkaline Carbons

Carbon containing mineral alkali can often be used as is or after washing with tap water. If a neutral pH is desired, the usual practice is to adjust the carbon to pH 7 and then wash with neutral water. Salts formed by neutralizing an alkali can usually be completely removed by washing with water.

4. STRUCTURAL CHARACTERISTICS

Shape and Form

Activated carbons are available in diverse forms:
  a) Symmetrical pellets
  b) Irregular shaped granules
  c) Powder
  d) Specialties such as pre-formed shapes; wool; slurry for coating paper, cloth, metal, or other supporting media

The *pellets* are available in several sizes, *e.g.*, $\frac{1}{8}$ and $\frac{3}{16}$ inch in diameter, and are recommended where low resistance to fluid flow is necessary.

The *granular carbons* are available in many sizes from 4×6 mesh to as small as 20×50 mesh. A small size provides more rapid rate of adsorption; a larger size is used where low resistance to flow of fluid is needed, as, for example, in the treatment of viscous liquids.

Powdered carbons for batch-contact processes are pulverized as fine as practicable to provide a great number of particles to scatter throughout the liquid being treated (Table 18:4). Fine grinding also makes the internal surface more readily accessible to molecules of solute.

The particle size of powdered carbons must not be too small because the rate of filtration is thus impaired. The rate of filtration is influenced by particle size but even more by the distribution of particle size. When particles of widely different sizes are present as a mixture, the smaller particles become lodged in the interstices between the larger particles and impede the flow of liquid. It can happen that a batch of finely ground particles of fairly uniform size will give faster filtration than another batch that is more coarsely ground, on the average, but contains an assortment of very unevenly sized particles.

TABLE 18:4

BULK DENSITY AND SCREEN ANALYSIS OF
POWDERED ACTIVATED CARBON

Carbon Code	Bulk density lb/cu ft	Screen analysis	
		on 100 mesh, %	through 200 mesh, %
E	31	6	68
F	28	21	61
G	23	3	85
H	15	30	42
I	18	18	54
J	32	7	73
K	20	0	100

The preparation of proper distribution of particle size requires appropriate pulverizing equipment. Other factors include: structural characteristics of the source material and the activation process employed. The presence of much moisture in the carbon during pulverizing can have an adverse effect on the particle size distribution.

Carbons for Water Purification

Carbons intended for water purification are pulverized more finely than those intended for industrial purposes.

For purifying water, powdered carbon is used in dosages of a few parts per million, and at such low dosages each unit weight of carbon should contain very many particles that can scatter throughout the water being treated and come rapidly into contact with all molecules carrying taste and odor.

The scattering capacity is disclosed by the opacity produced by a given concentration of carbon. A convenient method of measurement is to make a suspension of 0.70 gram of carbon per liter of water and measure the transmission of light in a suitable colorimeter. The reading should be made immediately after the suspension has been properly formed.

The opacity is influenced by the method of preparing the suspension. Therefore when carbons are being compared, the conditions for preparing all suspensions should be identical.

Distilled water should be used in preparing the suspensions. In water containing inorganic ingredients, the particles of carbon may coalesce to form flocs that settle rapidly. Flocculation is not rapid in distilled water unless the concentration of carbon exceeds a critical value. The critical concentration depends on the carbon being tested.

Density

The apparent density of a carbon is the weight of a unit volume of it including the pores and the voids between the particles. The apparent density is ascertained by measuring the volume of a weighed sample in a graduate after tamping till no further shrinkage in volume occurs. The data can be expressed as density (water = 1), or as pounds per cubic foot.

Surface Area

Measurements of surface area are seldom made by users of activated carbon, and suppliers are fully familiar with the procedure. An adequate presentation of the procedure is outside the scope of this text. The interested reader is referred to the many papers published on the subjects.[1]

Pore Size Distribution

The comments under surface area apply also to the determination of pore size distribution.[1]

Hardness, Resistance to Attrition

During processing operations using granular carbons, the carbon granules continually rub against one another, and thereby cause a wearing away of the particles that results in what is known as attrition loss. Because processing with granular carbon is a recycling operation, it follows that the attrition loss is cumulative. Therefore it is important to select a granular carbon that has the strength to resist abrasion and thereby reduce the attrition loss to a minimum.

Studies have shown that the type of strength that is needed can best be measured by a test developed by the National Bureau of Standards.[3,4] In that test, the carbon granules are caused to rub against one another, thus simulating the action that occurs in plant operation.

The strength of the granules is disclosed by measuring the average particle size at the beginning and end of the test period, and also by the amount of dust that is formed.

5. SOME OTHER PROPERTIES OF ACTIVATED CARBONS

Oil Retention

The power of carbon to retain oil is measured in various ways. In one laboratory method, 10 g carbon are mixed with 200 g oil at 80°C, then filtered on a Buechner funnel placed in a drying oven in which an inert atmosphere (nitrogen) is maintained. Even under the best experimental conditions, the retention of oil so measured is higher than in plant operation in which all loosely held oil is recovered by blowing the press with steam.

Retention of oil can also be measured by extracting a dried portion of the carbon cake from the filter press. This method, however, can seldom be applied directly to carbon unless a suitable laboratory press is available. Plant operations generally use carbon in a mixture with clay.

The retention of oil by carbon is somewhat greater than by clay. This adverse feature of carbon is often offset when the carbon is used in a mixture because a carbon-clay mixture is often more effective than clay alone for decolorizing an oil.

Filtration Rate

Rates of filtration are measured for several purposes:
1. To determine the filter capacity required for plant operation.
2. To determine the relative filtration rates of different carbons.
3. To check shipments of carbon against a standard sample.

The capacity of the filter required for a new application can be based on flow rates obtained in the laboratory, using a suitable unit such as the bomb filter. In many cases, a test with a Buechner funnel gives significant information. In any such study, it is well to consult with various manufacturers of filters as their knowledge and experience can be very helpful.

In evaluating the relative filtration rate of different carbons, it is important to consider that the filtration rate is a function of mechanical resistance that not only results from the shape and size of the carbon particles, but also depends on the filterability of the liquid. Some carbons are so effective for removing lyophilic colloids, gums, and resins that they give a better over-all filtration performance than does a carbon that has better mechanical structure, but lacks ability to remove colloidal substances. Consequently, comparisons of filtration of different carbons should be based on the solution to be purified, and not on water, as is sometimes done. A further precaution is to employ the dosage of each carbon needed to give the desired final purity. Thus, if 1 part of carbon *A* accomplishes the same desired purity as 2 parts of carbon *B*, then these respective dosages should be used in studying the filtration. The final thickness of the carbon cake should be sufficient to permit comparison with that obtained in the industrial operation.

Sometimes it is desirable to check the uniformity of shipments of the same brand of a commercial carbon. In such a case, any difference in rate of filtration would be apt to arise from variations in size and structure of particles. A test for uniformity may be carried out as follows: Make a slurry of 20 grams of carbon in 200 milliliters of distilled water at room temperature, and filter it through a cloth mat on a Buechner funnel. When the first dry spot appears on the surface of the carbon cake, add 250 milliliters of distilled water at room temperature. Keep the vacuum at 15 inches of mercury and observe the time elapsing until the first dry spot appears on the surface of the carbon cake.

Sedimentation

Although sedimentation rates in water are of little value for measuring the size of carbon particles, they have practical importance in evaluating a carbon for water purification. Here carbon is applied to a relatively quiescent fluid in which it is desirable for the particles to remain suspended for sufficient time to come into contact with most of the adsorbable taste and odor molecules. Sedimentation tests should employ a dosage corresponding to that to be used in the plant operation. Incidentally, the volume occupied by a given weight of settled carbon is not the same in all liquids.

Temperature of Ignition

Reported ignition points for carbons range from a low of less than 300°C for some specimens to over 600°C for others. The temperature at which a carbon will ignite in air depends on the experimental conditions. Consequently, the different methods that have been devised will not show an identical ignition temperature for a particular carbon. Because of this, comparisons of the ignition temperature of different carbons are of significance only when determined under identical experimental conditions.

It is reported that the temperature at which a carbon will ignite depends, to some extent, on the temperature at which the prior carbonization and activation were conducted. Some non-carbon constituents have an influence, thus sodium carbonate lowers the ignition point, whereas phosphoric acid elevates the ignition point.

Electrical Conductivity

The electrical conductivity is not the same for all types of activated carbons, and appears to depend in part on method of preparation.

Corrosion

Activated carbon may augment the corrosive action of low pH liquids; therefore when selecting materials for processing equipment, tests should be conducted with carbon present.

REFERENCES

1. Deitz, V. R., *Bibliography of Solid Adsorbents* United States Cane Sugar Refiners and Bone Char Manufacturers and the National Bureau of Standards, Washington, D.C. 1944 Supplementary volume issued in 1956.
2. Manufacturers and Suppliers can furnish helpful information
3. Rinehart, T. M., Scheffler, G. H., Helbig W. A., *Symposium on Activated Carbon.* Atlas Chemical Industries
    Fornwalt, H. J. and Hutchins, R. A. *Purifying Liquids with Activated Carbon,* Atlas Chemical Industries
4. Carpenter, F. G., Proceedings of the Fifth Technical Session on Bone Char, Bone Char Research Project, 1959, U.S. National Bureau of Standards.

PART VII

APPENDIX

19

Final Gleanings

During the preparation of this text, many items were uncovered which for various reasons could not be woven into prior chapters. Therefore such items that are of general interest have been selected for inclusion in this chapter.

Measurement of Adsorption

In studies of vapor phase, the amount of adsorbed substance can be measured by a direct weighing. This direct approach is not feasible in liquid systems and the adsorption is calculated from measurements of changes in concentration. The figure thus furnished does not show the actual weight of the adsorbed substance. As A. M. Williams[12] pointed out many years ago, calculations based on changes in concentration do not take into account the amount of solvent adsorbed; hence such calculations give the *apparent excess in adsorption of a solute,* and this is less than the true gravimetric quantity.

This aspect has little or no significance to most industrial users, most of whom use carbon to correct an undesired characteristic. Their interest is in knowing the extent of the improvement. If in a given situation, the removal of 90% of objectionable color provides a finished product of satisfactory quality the operator seldom has any need to know the weight of color removed.

In some instances, however, knowledge of gravimetric quantities can be helpful, *e.g.*, when a substance is to be reclaimed by subsequent elution. But practically all such applications deal with small initial concentrations and in such cases there is little deviation between the calculated and the actual gravimetric quantities.

The deviation becomes appreciable only when the solute is present in a large concentration in a very adsorbable solvent; then the apparent adsorption can be misleading as to the true situation.

Bartell and Sloan[13] conducted adsorption experiments on mixtures of ethanol and ethyl carbonate. They found that ethanol is preferentially adsorbed when the original concentration of ethanol is less than that of the ethyl carbonate. When ethanol is present in greater concentrations, however, ethyl carbonate is preferentially adsorbed. Still further, they found that there is no apparent adsorption of either ingredient when the two are present in approximately equal molar concentrations. This is explained on the basis that at that point both ingredients are adsorbed to the same extent, leaving the concentration unchanged.

Freundlich Isotherm*

Here we will consider certain aspects of the Freundlich isotherm not covered in Chapter 7. As there stated, the Freundlich isotherm expresses a relationship between the amount of substance that is adsorbed and that which remains unadsorbed.[1,2,8,9,14,15,16,17]

Many adsorption isotherms, when plotted logarithmically, fail to give a straight line, particularly when the adsorption is studied over a wide range of concentration. They are said to fail to obey the Freundlich equation. Such failure does not cancel the utility of an isotherm because the data usually are interpreted by an inspection of the graphs. There can be no doubt that it is easier to make comparative studies when dealing with straight-line isotherms, but one should not make a fetish of them. After all, the Freundlich equation is only a statement of behavior that has been observed in many but not all cases of adsorption. Deviations often reveal significant information on the adsorption being studied, as is illustrated in the following examples.

As previously mentioned, in adsorption from solution we do not measure the true weight of solute adsorbed, but rather the extent to which the solute is adsorbed in excess of the solvent. The excess adsorption of the solute diminishes at higher concentrations if the solvent is very adsorbable—this can lead to behavior illustrated by isotherm A in Figure 19:1.

---

*The Darcograph, supplied by Atlas Chemical Industries. Wilmington Del., is helpful in studies using the adsorption isotherm.

In many industrial solutions, the adsorption is measured by properties, *e.g.*, color, which may be due to a number of ingredients that are not equally adsorbable. This may result in a discontinuity exhibited by isotherm *B* in Figure 19:1. Helbig[2] suggests that this phenomenon can be traced to the fact that the color below the break is associated with an ingredient that is not appreciably adsorbed by the carbon.

When the various ingredients are not too different in adsorbability, the isotherm may show a gradual curvature instead of a sharp break (*C* in Figure 19:1). Isotherms of this type can also occur when carbon contains extractable material such as

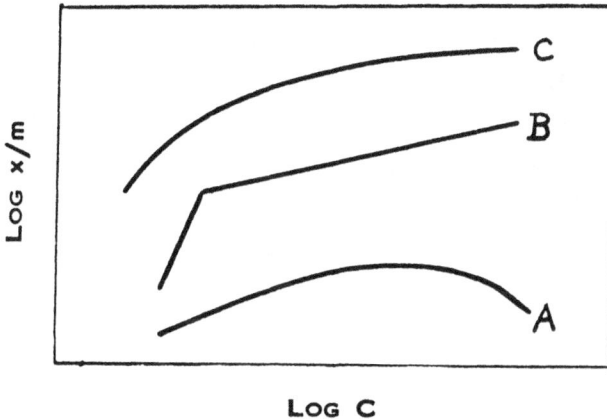

Fig. 19:1   Isotherms to illustrate deviations from Freundlich equation

iron which may form color bodies with certain solutes. As a result, the color removed by the carbon will be partially offset, an effect that is more pronounced at higher carbon dosages.

Helbig[2] calls attention to deviations that may be caused by incorrect measurement of the original concentration. If an erroneously high value is obtained for the original color, this will cause the isotherm to curve upward in the region of low carbon dosages; conversely, it will curve downward if the original color value is erroneously low.

Theoretical Equations

The Freundlich equation reveals the operation of mass action in adsorption phenomena; efforts to find greater theoretical significance have not been very successful. This has led to a study of other equations to interpret adsorption phenomena.

The development of theoretical equations can be approached from a thermodynamic, statistical, or kinetic angle. The thermodynamic approach is presented in the equations of Polányi, Berényi, and others. The equation of Langmuir represents the kinetic approach. The thermodynamic treatment has a great advantage that it does not require a knowledge of the intimate mechanism of the process and, in particular, involves no assumptions with regard to the attractive forces between molecules. However, no relation for the rate of adsorption can be obtained from thermodynamics alone. Kinetic treatment permits the deduction of the rate of an adsorption reaction, but when an attempt is made to allow for the interaction of molecules, the mathematics become so involved as to be impossible of solution.

No attempt is made in this text to discuss the various theoretical equations.

Isosteres

When carbon, containing adsorbed gas, is heated to a higher temperature, some of the adsorbed gas escapes unless a compensating increase is made in the pressure. Conversely, when the temperature falls, additional gas will be adsorbed unless the pressure is reduced. An isostere is formed by plotting the pressure required to maintain constant adsorption at different temperatures.[7]

When the vapor pressure of a liquid at two different temperatures is known, it is possible to calculate the heat of condensation by means of the Clausius-Clapeyron equation. A similar procedure is employed to calculate the differential heat of adsorption from the isosteres.

Data for plotting isosteres can be derived from isotherms of separate adsorptions, each adsorption being conducted at a different suitable temperature.

Hydrophilic and Hydrophobic Characteristics

Attraction and repulsion of water can be roughly gauged by shaking a solid in powdered form with a mixture of water and an immiscible organic liquid, *e.g.,* water and benzene. Substances such as clay, which go into the aqueous phase, are described as *hydrophilic* (water-liking) and substances such as mercuric sulfide, which go into the organic liquid, are termed *hydrophobic* (water-hating).

Activated carbons are hydrophobic; some are more hydrophobic than others. The difference can be shown by an experiment in which a suspension of a carbon in water is shaken with an immiscible organic liquid. A very hydrophobic carbon will leave the water and go completely into the organic liquid. In contrast, a carbon that is less hydrophobic, that is, slightly hydrophilic, will gather at the interface between the two liquids. The reaction is more evident when the aqueous phase contains an alkali.

Orientation at Interfaces

Five types of interfaces are possible:

liquid || gas

liquid || liquid

liquid || solid

solid || gas

solid || solid

Much is known as to the nature of the liquid || liquid, and the liquid || gas interfaces.[1] Careful analysis of a vast quantity of experimental data by many workers places our understanding of such boundaries on a firm foundation. We have fewer data leading to an understanding of interfaces that include a solid, and our concepts of such interfaces rest on the premise of a similarity to liquid || liquid, and liquid || gas interfaces.

Liquid || Gas Interface

Let us begin by considering behavior at the liquid || gas interface. If propane vapor is passed over the surface of a layer of liquid ether, many molecules of propane will enter the liquid and dissolve—this being in line with the well-known principle that *like dissolves like.*

If the experiment is repeated, but with water substituted for the ether, the propane molecules—being unlike water—will not dissolve, but instead will rebound into the vapor space. The next step in the study is to convert propane into propyl alcohol by introducing an hydroxyl group, and then pass the alcohol vapors over the surface of the water. When the alcohol vapor molecules impinge on the surface of the water they will be captured by the pull of the water for the hydroxyl head* (*like for like*), and many molecules of the vapor will be dragged into the body of the water. Inasmuch as the C_3H_7 tails resist immersion, many of the captured propyl alcohol molecules will remain at the surface. Molecules remaining at the surface will be oriented, the heads dipping into the water and the tails extending into the vapor space (Figure 19:2).

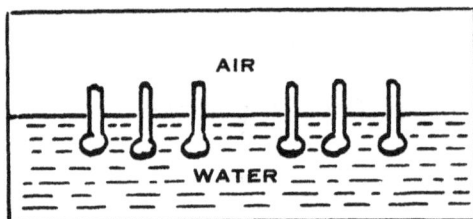

Fig. 19:2   Orientation of propyl alcohol molecules at air/water interface

Incidentally, evidence of orientation is shown in a simple laboratory experiment in which a thin layer of melted beeswax is allowed to cool slowly on the surface of water; when solidified, the beeswax will be wettable on the underside and non-wettable on the top side. Orientation constitutes an important feature of adsorption which helps us to understand the specificity so often found.

Liquid ‖ Liquid Interface

A liquid ‖ liquid interface can be established in diverse ways.

*In the subsequent discussion we shall employ the conventional terminology, referring to the hydrocarbon portion of an organic molecule as the *tail* and to the hydroxyl or other polar part as the *head*.

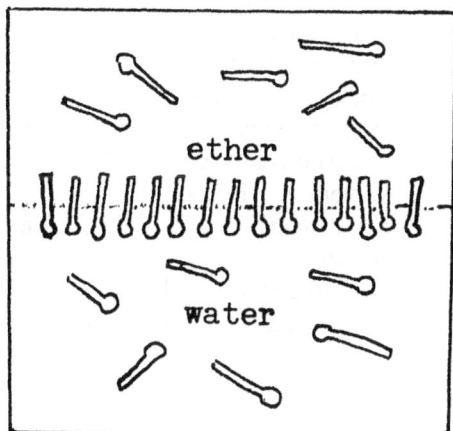

Fig. 19:3   Orientation at ether/water interface

For the purpose of this discussion we shall consider an interface
that forms when a layer of liquid ether is placed over an aqueous
solution of propyl alcohol. Some molecules of alcohol will remain
in the body of the aqueous solution, others will migrate into the
ether layer, pulled there by the solubilizing power of ether for the
C_3H_7 tail (*like dissolving like*). But many alcohol molecules will
be anchored into the interface where each end of the alcohol
molecule is drawn into the solvent in which it is more soluble—
the heads dipping into the water and the tails being pulled into the
ether (Figure 19:3) As both of the pulls cooperate and reinforce
the anchorage into the interface, the concentration of propyl
alcohol molecules will be greater in the interfacial region than in
either bulk phase—water or ether.

Interfaces Involving Activated Carbon

Let us now seek to interpret adsorption by carbon in the light
of deductions drawn from the liquid ‖ gas and the liquid ‖ liquid
interfaces. If activated carbon is placed in contact with propyl
alcohol (as either vapor or liquid) much of the alcohol will be
taken up. Assuming that the attractive power of a solid surface

is akin to the solubilizing power of a liquid then, as like dissolves like, we may deduce that the propyl alcohol is held as shown in Figure 19:4 with the tails joined to the carbon, and the heads directed toward the liquid or vapor space. Many data support the view that the orientation is as shown in Figure 19:4 but this does not preclude the existence of adsorption in which polar groups are bonded to the carbon surface. As the orientation will be influenced by the solvent used, benzene, which has a solubilizing attraction for the hydrocarbon tails, will promote an orientation unlike that with water.

Thus far we have considered simple structures. As we progress to complex structures, views as to their orientation are at best only a speculation.

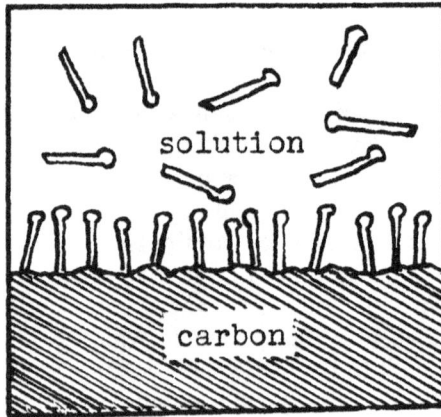

Fig. 19:4   Orientation of propyl alcohol at carbon/aqueous solution interface

Phenomena of Adsorption and Solution

Certain similarities observed between adsorption and solution have led some to believe that both may be different aspects of the same fundamental phenomenon.[1] The amount of a gas that is adsorbed is a function of the pressure, and this is also true of the amount of a gas that dissolves in a liquid. The way in which a solute distributes itself quantitatively between the adsorbent and solution recalls a similar distribution that occurs when a solute is placed in a mixture of two immiscible solvents—for instance, acetic acid in water and benzene. With solutions in which the solute has the same molecular structure in both solvents, the equation for the distribution may be written:

$$C_a = kC_b$$

where C_a is the concentration in one solvent and C_b the concentration in the other solvent. When, because of association or dissociation, the solute does not have the same molecular form in both solvents the equation is modified:

$$C_a = kC_b{}^n \quad \text{or} \quad C_a{}^{1/n} = kC_b$$

The exponent is a function of the difference in the molecular weight of the solute in the two solvents. If this interpretation is placed on the exponent in the Freundlich equation, it can lead to deductions that are inadmissible. For example the adsorption of hydrogen by some carbons has been found to be proportional to the cube root of the concentration. According to the distribution equation, this would mean that adsorbed hydrogen has molecular weight that is less than that of the hydrogen atom.[8] Others point out that the known laws of solution are derived from the behavior of dilute solutions, whereas many adsorptions conceivably involve high concentrations at the interface. Moreover, it has been pointed out that the quantitative relations, existing in a system in which all components are mobile, might require modification in systems in which one components is mechanically rigid and therefore, unable to diffuse.

Rate of Adsorption

The rate of adsorption may be retarded by the presence of previously adsorbed gases, some portions of which are weakly held and are readily displaced; but the last traces are often held with such tenacity as to require months for complete displacement. This condition can be corrected in laboratory studies by degassing. To do this, the carbon is heated to a high temperature in a vacuum. McBain[8] found that degassing enables equilibrium in adsorption to be reached rapidly, after which, no further change occurs in years. In the absence of degassing, a slow adsorption often continues for a period of weeks or months.

The subsequent slow adsorption has also been ascribed to a gradual interpenetration of the adsorbed molecules into the carbon structure to form a solid solution. Such behavior has been studied in connection with the adsorption of iodine. The data of different investigators are not in full agreement, but there appears to be a rapid initial adsorption which is complete in less than an hour, after which a very slow sorption of iodine continues for years. That the slow stage consists of a penetration of the iodine molecules into the carbon structure is indicated by work of Davis[18] who found that iodine is less readily desorbed after long contact with carbon. In some adsorptions, such as that of oxygen on carbon, the slow stage seems to involve a chemical reaction.

The time required to establish an adsorptive bond, after contact is made, is not the same for all adsorbents or for all adsorbates. Unequal rates of adsorption have been attributed to differences in the pore size of dissimilar carbons; larger molecules would have greater difficulty in penetrating into a carbon having smaller pores. The inconsistent behavior of different carbons has also been attributed to dissimilar qualities of adsorptive powers.

Adsorption is primarily a surface phenomenon, but it is often accompanied by a deeper penetration of the vapor or solute into the body of the solid adsorbent. This deeper penetration is akin to the formation of a solid solution and is termed *absorption*. Since it is seldom practicable to distinguish clearly between absorption and adsorption, McBain[8] suggested the use of the broader term *sorption* to cover both phenomena.

Traube's Rule

In an aqueous solution, an increase in the molecular size in a homologous series usually results in a greater number of molecules being adsorbed.[1] This behavior is known as *Traube's Rule*.[9]

Some carbons, however, do not conform to the rule,[1] in that they adsorb smaller molecules in preference to larger ones. Such failure to obey Traube's Rule is more apt to occur with a carbon that has been given only a brief period of activation. In the early stages of activation, the pores may not be sufficiently enlarged to admit the larger size molecules. This belief finds support in the fact that during activation, the volume of a carbon particle does not diminish in proportion to the decrease in weight. This suggests that erosion occurs within the pores and that they are thus enlarged. The separating walls would simultaneously become thinner —a change that could explain why carbon particles become softer and more friable after longer activation.

Whether or not a carbon obeys Traube's Rule can also depend on the temperature used in activation. Dubinin[19] found that carbons activated at 500°C adsorb organic acids according to Traube's Rule, whereas carbons activated at 800°C show a reversal; that is, the lower members of a homologous series are better adsorbed. One explanation is that activation at 800°C produces smaller pores than it does at 500°C. Prolonged activation at 800°C, however, promotes the adsorptive power of a carbon for larger molecules.

Deviations from Traube's Rule are minimized when the carbon remains in contact with the solution for a longer period.[20] Deviations are also minimized when the carbon is finely pulverized.[21]

TABLE   19:1

DENSITIES OF CHARCOAL*

Liquid	Density g/cc
Water	1.821
Benzene	1.994
Carbon tetrachloride	1.860
Petroleum ether	2.083

* J. L. Culbertson and A. Dunbar, *J. Am. Chem. Soc.* **59** : 306 (1937); reprinted by permission of the American Chemical Society.

True Density

It might be expected that the true density of carbon, that is, the density of the carbon skeleton, could be determined by the classical method of Archimedes—placing a known mass in a pycnometer and displacing the air by a liquid. With carbon, this method presents difficulties, for it has been found that the observed density depends on the liquid employed (Table 19:1).

Two interpretations of this behavior have been offered.[22] One is that the film of liquid at the carbon surface is compressed by adsorption forces and, therefore, has a density greater than that of the bulk liquid. According to this interpretation, densities so obtained are fictitiously high depending on the compressibility of the film of the liquid used. Against this view are the data of Howard and Hulett[23] who found a high density of 2.1 for cocoanut carbon in helium at 25°C; and at this temperature, helium is not appreciably adsorbed.

This leads to the other interpretation which explains the variations in density as caused by unequal ability of different adsorbates to enter into the smallest pores. This concept finds support in work by Cude and Hulett[24] who found that the density gradually drifts to higher values during prolonged immersion in the liquid. This drift may continue for months before a constant value is reached. When a system is placed initially under great pressures (500 to 8,600 atmospheres), the final density is attained within a few hours. Thus, in water, at a pressure up to 8,600 atmospheres, a sample of carbon reached a density of 1.902 in a relatively short time; whereas, when immersed in water at atmospheric pressure for 4 months, the observed density was only 1.86.

In an analysis of methods for determining true densities, Coolidge[25] points out that highly porous adsorbents approach molecular dispersity. Thus, their *true density* is in principle as elusive as the true density of a solute in an ordinary solution.

Effect of pH on Specific Carbons

The fact that the optimum pH for adsorption is not necessarily the same for all carbon indicates a direct but as yet unknown effect on the carbon. In a study of caramel adsorption, Hauge and Willaman[26] found the order of efficiency of various carbons to be A > C > D > E > F at pH 7.0 whereas at pH 4.0 the order was F > A > E > D > C. Any direct effect of pH on the properties of carbon are minor, however, compared to the much stronger influence of hydrogen and hydroxyl ions on the properties of many solutes.

Influence of Specific Ions on Adsorption

The influence of a salt on adsorption is reported to be specific in some instances. Kruyt[27] found considerable specificity in the influence of the anion. The adsorption of Methylene Blue was promoted by anions in the order Cl, SO_4, PO_4 Br, NO_3; whereas for Crystal Violet, the order was PO_4, SO_4, Cl, Br, NO_3; and for auramine, Cl, SO_4, Br, NO_3, PO_4. Kozakevich[28] found Br ions more effective than Cl ions for promoting adsorption of benzoic acid from ethanol, whereas the reverse was true with methanol or water as the solvent.

Adsorptive Characteristics of Water Vapor

For many experimental purposes, activated carbon can be dried by keeping it in a desiccator for several days. The adsorption of water vapor differs from that of most gases. At low humidities, the adsorption of water is very small but it increases rapidly at a relative humidity of 0.3 to 0.4. McBain[29] suggests that when a sufficient number of water molecules simultaneously impinge on the surface of a carbon, they are held by mutual polarization which augments the attractive power of the carbon.

Moisturized Carbon for Dust Control

In past years a frequent objection to the use of powdered activated carbon was the dust and dirt involved. In more recent years this problem has been avoided in many applications by improved methods of handling. Those methods, however, are not always feasible for those that would use carbon intermittently or in small quantities. In considering such situations it could be of interest to mention some laboratory studies which indicate that dust and dirt can often be eliminated by using powdered carbon in moisturized form.

One of the problems in handling powdered carbon arises from the difficulty of immersing it in the liquid to be treated. The difficulty is caused by a film of adsorbed gas that is displaced when a liquid wets the carbon. When displaced, the gas tends to carry some of the powdered carbon and thus create dust. The situation is aggravated if the gas is released suddenly, as when the carbon is mixed with a hot liquid, especially a hot oil or wax. Because of this, carbon should be mixed with a liquid at the lowest practicable temperature.

Laboratory studies show that when the adsorbed gas is predisplaced by moisturizing, the rate of wetting is accelerated without creating dust. Another benefit arises from the fact that moist carbon particles adhere to one another; consequently, moisturized carbon does not raise dust when handled. The amount of moisture needed will range from 30% to over 50%, depending on the type of carbon.

Moisture should be blended with the carbon after the grinding operation because pulverizing a damp carbon can have an adverse effect on the particle size distribution, and cause the carbon to have poor filtration characteristics. Moisturized carbon should be packed in a type of container that will ensure retention of the moisture during storage.

Moisture does not impair the adsorptive capacity; carbons have been stored in water for several years without any impairment. Moisturized carbon is suitable for all aqueous solutions and also for some organic liquids.

Inasmuch as moisturized carbon would go mainly to users of small quantities, the added freight cost would be negligible for those wishing to ensure the benefits of minimum dust and dirt.

Changes in Adsorptive Capacity with Time

Questions are sometimes asked as to the effect of age when activated carbon is to be stored for long periods of time. In most cases, this need cause no concern because experience has shown that most commercial carbons retain almost all their initial capacity over a period of years. However carbons that lose adsorptive capacity when dried may deteriorate with age.

Scientific data on this question are meager and those available are somewhat conflicting.[30] An examination of the data indicates that the influence of storage on the keeping qualities of a carbon depends on several factors. An important factor is the method employed for preparing the carbon. Zaverina and Dubinin[31] found that carbons activated between 300° and 500°C lost no adsorptive power for iodine during a storage period of 5 years, whereas carbons prepared at higher temperatures showed a loss of adsorptive power for iodine. King[32] observed other properties on which the effect of time depended upon the temperature at which the carbon had been activated. Carbon activated at higher temperatures showed a decrease in pH after several months of storage whereas no change occurred in carbons prepared at low temperatures. When changes do occur, they seem to be associated with the effects of oxidation. Moist air causes a decay in certain catalytic properties.[33]

The extent and direction of a change in carbon with age may depend upon which properties are examined. Bente and Walton[34] found that exposure of activated carbon to moist oxygen caused a decay in the power to catalyze the oxidation of potassium urate; but no change resulted in a similar property for hydroquinone. Zaverina and Dubinin[31] examined a carbon in which the adsorptive power for benzene vapor was fully retained when the carbon was stored, whereas the adsorptive power for acids decreased.

Instances of increased adsorptive power with age are reported. Zaverina found age enhanced the adsorptive power for sodium hydroxide; and Lawson[35] noted that some chars become more hygroscopic on standing. Bolam and Phillips[36] found that a decreased adsorptive power for acids was accompanied by an almost equivalent increase in adsorptive power for silver nitrate.

The effect of age may also depend upon which period of the history of the carbon is examined. In carbons prepared by McBain and Sessions,[37] the adsorptive power for iodine decreased during the first two years, after which very little decay occurred.

Activated Carbon—an Organic Polymer

Garten and Weiss[38] point out that as activated carbon contains from 2% to 25% of oxygen and considerable quantities of hydrogen, the term *carbon* is a misnomer. They suggest that activated carbon be viewed as a complex organic polymer having a large internal surface and electrical conductivity. From certain points of view, activated carbon could be regarded as a giant molecule in which resonance is possible between chemical structures situated at a distance but connected by a conducting system of bonds.

Surface Oxides

Support is found for the general outlines of Shilov's theory of surface oxides, but there is some question that they can be defined as rigidly as he suggested.[38,39] Moreover, although many data support the concept that surface oxides participate in the adsorption of strongly ionized acids and bases at well as in the hydrolytic adsorption of inorganic salts, the phenomenon is more complex than was originally assumed. The adsorption of benzoic acid and of aliphatic acids above butyric does not seem to be limited to the oxide that adsorbs dilute hydrochloric acid; in fact these organic acids can be adsorbed by carbons that adsorb strong bases such as sodium hydroxide. Even in the case of hydrochloric acid the influence of the oxide is less apparent in strong concentrations of acid. Similar inconsistencies are found with bases, the adsorptive behavior of the relatively strong tetraethylamine hydroxide is unlike that of either potassium or ammonium hydroxide. Evidently, the mechanism by which many acids and bases are adsorbed by carbon is more complex than a simple reaction involving hydrogen and hydroxyl ions.

To what extent surface oxides influence adsorptions other than those of strong acids and bases is not clear.[1] Carbons prepared below 500°C are more hydrophilic, that is, they adsorb water vapor more readily than do carbons prepared at higher temperatures. Moreover, when a carbon is reheated at a different temperature it tends to develop the water adsorbing properties characteristic of the final temperature given.

The adsorption of polar compounds may be influenced by surface oxides.

The temperature at which certain other adsorptive powers develop often corresponds to a temperature favorable to the formation of one of the surface oxides. This, however, does not establish a direct connection between a specific adsorptive power and the surface oxide. Let us consider the adsorption of Malachite Green. For this, certain temperatures of activation are favorable but only when the carbon has been given suitable previous treatment. Moreover, adsorptive power for Malachite Green when once developed is not readily lost by reheating at a different temperature. In fact, some sequences of temperature enhance the adsorption. From various studies, one might reasonably conclude that the temperature of activation influences the development of specific adsorptive powers apart from those related to surface oxides.

Knowledge of the existence of surface oxides on carbon dates back to de Saussure and many studies have since been made to determine their chemical nature.[1] Rhead and Wheeler[40] studied the fixation of oxygen by carbon some years before a relation was sought to adsorptive power. They considered the *surface oxide complex* to be a primary product of the interaction between carbon and oxygen, and one that is capable of being subsequently decomposed into the ultimate gaseous products of combustion—carbon monoxide and dioxide. They were puzzled by the behavior of this carbon-oxygen complex for, although it resembles a chemical compound, yet it cannot be wholly decomposed into carbon monoxide and dioxide at any one temperature. Instead, the reaction proceeds by steps through a wide range of temperature. The gases come off in rushes at characteristic temperatures.

Rhead and Wheeler[40] considered the possibility that at each temperature the complex might partly decompose, splitting off gaseous oxides and leaving a residual solid oxide containing a smaller ratio of oxygen which is capable of undergoing further decomposition at a still higher temperature. They questioned this hypothesis on finding that each rise of a few degrees in temperature causes a limited evolution of gaseous oxides which would imply a tremendous number of stages in the degradation process.

Shah[41] found that the gaseous oxides liberated when carbon is heated are not readsorbed when the temperature is lowered, and this indicates that the oxides are held in a form different from that in which they are evolved. The ratio of carbon monoxide to dioxide in the liberated gases not only depends on the temperature at which the gases are driven off, but also is a function of the temperature at which the oxygen was originally fixed. This indicates that various types of oxides can form.

Strickland-Constable[42] offers a constructive interpretation of the behavior of carbon when subjected to increasing temperature, based on analogous behavior of organic compounds containing oxygen. Such organic compounds decompose to yield carbon monoxide and dioxide at widely different temperatures; the decomposition temperature is about 500°C for acetone and 400°C for diethyl ether. The covalent attachment of oxygen to a heterogeneous carbon surface could provide opportunities for a wide assortment of different groupings; and groups similar to those found in ketones, alcohols, acids may all be present simultaneously. Each form would have characteristic properties, including a specific decomposition temperature. The over-all effect would permit a progressive liberation of gaseous oxides over a wide temperature range. The studies of Frumkin[1] and his coworkers are of interest. He heated an ash-free charcoal in a vacuum to 1000°C and found that this carbon adsorbed neither mineral acids nor alkalis provided air was completely excluded. After contact with air, the carbon adsorbed acids but not alkalis. In another study, carbon prepared at a high temperature in an atmosphere of hydrogen was able to adsorb alkali when hydrogen was present. This carbon did not adsorb acid from dilute solutions although it did so from solutions stronger than 0.1 normal.

These and other studies led Frumkin to believe that carbon acts like a gas electrode. Adsorbed oxygen, in the presence of water, is transformed into hydroxyl ions which are held at the carbon surface by electrovalent forces and can be exchanged for other anions.

On the other hand adsorbed hydrogen in the presence of water forms hydrogen ions which can be exchanged for other cations. The reaction is reversed by making the carbon suspension definitely acid, in which case, hydrogen is evolved as a gas.

Steenberg[39] in a comprehensive and informative review of many aspects of surface oxides has contributed much to our understanding of their characteristics. He classifies activated carbons into H and L types. The L carbons are those that preferentially adsorb alkali, and the H types are those that adsorb mineral acids and but little alkali. Many associated properties are described and discussed.

In recent years, infrared internal-reflection spectometry has been employed to examine activated carbon surfaces to obtain spectrometric evidence of the nature of the functional surface groups. Studies have also been conducted on functional surface groups containing other elements such as sulfur and the halogens. References to that work are included in the bibliography in Chapter 20, section 19.

Catalytic Influence of Surface Oxides

Carbons prepared at 850°C to 950°C appear more effective for some oxidations, e.g., hydroquinone, sodium thiosulfate, and for decomposition of hydrogen peroxide; whereas carbons prepared at lower temperatures, in the range of 450°C to 600°C are found more effective for oxidation of sodium arsenite, and potassium nitrite.[1] It should be mentioned that certain of these specific effects are not supported by all data.

King[43] found that a carbon prepared at 900°C. when later heated at 500°C, lost the properties of a higher-temperature carbon and acquired those characteristic of the lower-temperature carbon. If this carbon were again reheated to 850°C, however, the catalytic properties peculiar to the higher temperature would be regained. The influence of the temperature of activation on catalytic powers recalls a similar phenomenon observed in the effect on properites attributed to surface oxides. Because of this, it has been suggested that specific catalytic effects depend on the type of surface oxide formed on the carbon. According to this theory, certain reactions can be promoted by the surface oxide formed at low temperatures, whereas the surface oxide formed at higher temperatures catalyzes other reactions.

Obviously, the evidence is circumstantial, and the catalytic properties could be due to other changes occurring in the carbon at each temperature. Studies by Rideal and Wright[44,45] indicate that the oxidizing action is not due directly to oxygen held in the form of stable oxides.

The significance of surface oxides in oxidation reactions is placed in doubt by studies that trace oxidations to active centers containing other non-carbon elements. The oxidation is often enhanced when a carbon is prepared from nitrogeneous raw materials; similarly the presence of iron accelerates many oxidations.

pH and Surface Oxides

The pH of a carbon, as measured in a suspension in distilled water, arises mainly from associated components incident to the activation process or added subsequently to give a desired pH. There is evidence, however, that some pH effects can arise from constitutive properties of the carbon. In studies of a carbon with a low ash, King[46] found that the pH depended on the temperature at which the carbon had previously been ignited; the pH increased as the temperature was raised from 25° to 900°C. Above 900°C, the pH again fell.

Wiegand,[47] in a study of colloidal carbons, found that the pH of an aqueous suspension ranged between 3 and 11, depending on the temperature to which the carbon had been heated previously; the pH increased with an elevation in temperature. With ash-free carbons, the pH effect is found with the suspension rather than in the supernatant liquid. This led Wiegand to suggest a mechanism occurring at or near the surface of the carbon. He considered that surface oxides, present in low-temperature carbons, adsorb water hydrolytically in such a manner that the hydroxyl group is held close to the surface, and the hydrogen is held loosely as ions. Although the hydrogen ions cannot wander far from the surface, they have sufficient freedom to affect the potential of a glass electrode placed in a carbon suspension. Carbons heated to high temperatures lose some surface oxides, and then the water is adsorbed so that hydrogen ions are held to the surface and the loosely held hydroxyl ions give an alkaline reaction.

Peptization

Carbon may be peptized in distilled water. Here the solubilizing effect is produced by the ions in the double layer on the carbon. Osmosis causes the loosely held ions in the double layer to diffuse throughout the liquid and, in turn, they drag the oppositely charged ions together with the attached carbon particles. Peptization from this cause may be corrected by increasing the concentration of carbon, or by adding suitable electrolytes, $e.g.$, HCl, $CaCl_2$, $Al_2(SO_4)_3$, in sufficient concentrations. It is to be noted, however, that certain electrolytes, such as NaOH and also some dyes, tend to stabilize a carbon suspension.

New Ventures in the Manufacture of Activated Carbon

In undertaking a new venture to manufacture activated carbon, a first step is to determine the kind of adsorptive powers to be developed, and this in turn requires a knowledge of the markets to be reached.

One approach would be to obtain process liquors from representative industries and use them to measure the effectiveness of the chars produced during the various stages of the development. This can be very time-consuming and, more important, the data will be inconclusive unless based on adequate knowledge of what each particular industry requires of activated carbon.

In another approach, we can select one or more of the established commercial carbons and determine the adsorptive powers as measured by various synthetic tests. Then we study how to provide similar properties in the *new* carbon. This does not mean a slavish endeavor to provide an exact replica of the established carbon. In the first place, it is seldom possible to reproduce such properties in an exact copy. Even if it can be done, it is certainly desirable to explore possible variations in adsorptive power that may enable the new carbon to enter fields not previously reached

In countries where all supplies of activated carbon are imported, there is natural incentive to construct manufacturing facilities to handle domestic needs. Plans for such an undertaking should include a preliminary market survey to determine the quality of the carbon being imported. If it is found that several types of carbon are needed, it may be desirable to design manufacturing facilities that will produce several types of carbon, otherwise it would still be necessary to import certain grades of carbon.

An approach to Future Research

Among the major projects for future research on activated carbon are ways and means to provide greater markets. Sales outlets will expand in proportion to the ability of suppliers to surmount existing hurdles and to furnish new forms of usefulness. For this, a first step is to be aware of problems and needs that exist. These can become known through intelligent reporting by sales and service representatives. In this, management will do well to keep clearly in mind that the things we least like to hear often can do us the most good. And so when gripes, complaints, and requests for unusual forms of utility are relayed by sales representatives, they should be welcomed as furnishing an indication of what needs to be done.

REFERENCES

1. Deitz, V. R., *Bibliography of Solid Adsorbents,* United States Cane Sugar Refineries and Bone Char Manufacturers and the National Bureau of Standards, Washington, D.C., 1944. Supplementary volume published in 1956.

2. Helbig, W. A., in *Colloid Chemistry* (J. Alexander, Editor), Vol. 6, Reinhold Publishing Corp., New York, 1946.
3. Mantell, C. L., *Adsorption*, Mc-Graw-Hill Book Co., New York 1944, *Industrial Carbon*, D. van Nostrand, New York, 1946.
4. Kausch, O., *Die aktive Kohle*, W. Knapp. Halle, 1928, Supplement 1932.
5. Aehnelt, W. R., *Entfärbungs-und Klärmittel*, T. Steinkopff, Dresden u. Leipzig, 1943.
6. Bailleul, G., Herbert, W., and Reisemann, E., *Active Carbon*, Stuttgart, 1937.
7. Brunauer, S., *Physical Adsorption*, Princeton University Press, Princeton, 1943.
8. McBain, J. W., *The Sorption of Gases by Solids*, George Routledge and Sons, London, 1932.
9. Freundlich, H., *Colloid & Capillary Chemistry*, E. P. Dutton & Co. Inc., New York, and Methuen & Co., Ltd., London, 1926.
10. Adam, N. K., *Physics and Chemistry of Surfaces*, Oxford University Press, 3rd Edition, London, 1941.
11. Adamson, A. W., *Physical Chemistry of Surfaces*, Interscience Publishers, Inc., New York, 1960.
12. Williams, A. M., *Medd. K. Vetenskapsakad. Nobelinst.* No. 27, 23 pp. (1913); *Chem. Abst.*, 8:1630.
13. Bartell, F. E., and Sloan, C. K., *J. Am. Chem. Soc.*, 51:1643 (1929).
14. Sanders, M. T., *Ind. Eng. Chem.*, 15:784 (1923); 20:791 (1928).
15. Freundlich, H., and Losev, G., *Z. Physik. Chem.*, 59:284 (1907).
16. Freundlich, H., *Z. physik. Chem.*, 57:385 (1907).
17. Ostwald, Wo., and Izaguirre, R. de., *Kolloid-Z.*, 30:279 (1922).
18. Davis, O. C. M., *J. Chem. Soc.*, 91:1666 (1907).
19. Dubinin, M. M., *Z. physik. Chem.*, A140:81 (1929); A150: 145 (1930).
20. Dubinin, M. M., *Z. physik. Chem.*, A155:116 (1931).
21. Bruns, B., *Kolloid-Z.*, 54:33 (1931).
22. Harkins, W. D., and Ewing, D. T., *J. Am. Chem. Soc.*, 43:1787 (1921).

Morrison, J. A., and McIntosh, R., *Canadian J. Research* 24B, 137 (1946).

23. Howard, H. C., and Hulett, G. A., *J. Phys. Chem.,* 28:1082 (1924).
24. Cude, H. E., and Hulett, G. A., *J. Am. Chem. Soc.,* 42:391 (1920).
25. Coolidge, A. S., *J. Am. Chem. Soc.,* 56:554 (1934).
26. Hauge, S. M., and Williaman, J. J., *Ind. Eng. Chem.,* 19:943 (1927).
27. Kruyt, H. R., and van der Made, J. E. M., *Proc. Acad. Sci. Amsterdam,* 20:636 (1918).
28. Kozakevich, P. P., and Izmailov, N. A., *Z. physik. Chem.,* A150:295 (1930).
29. McBain, J. W., Porter, J. L., and Sessions, R. F., *J. Am. Chem. Soc.,* 55:2294 (1938).
30. Firth, J. B., *J. Soc. Chem. Ind.,* 42:242 (1923).
    Lamb, A. B., Wilson, R. E., and Chaney, N. K., *J. Ind. Eng. Chem.,* 11:420 (1919).
31. Zaverina, E., and Dubinin, M., *J. Phys. Chem. Russia,* 12:397 (1938).
32. King, A., *J. Chem. Soc.,* 889 (1935).
33. Larsen, E. C., and Walton, J. H., *J. Phys. Chem.,* 44:70 (1940).
34. Bente, P. F., and Walton, J. H., *J. Phys. Chem.,* 47:133 (1943).
35. Lawson, C. G., *Trans. Faraday Soc.,* 32:473 (1936).
36. Bolam, T. R., and Phillips, W. A., *Trans. Faraday Soc.,* 31:1443 (1935).
37. McBain, J. W., and Sessions, R. F., *J. Am. Chem. Soc.,* 56:1 (1934).
38. Garten, V. A., and Weiss, D. E., *Reviews of Pure and Applied Chemistry,* 7:69, June 1957.
39. Steenberg, B., *Adsorption and Exchange of Ions on Activated Charcoal,* Almquist and Wiksells, Upsala, 1944.
40. Rhead, T. F. E., and Wheeler, R. V., *J. Chem. Soc.,* 103:461, 1210 (1913).
41. Shah, M. S., *J. Chem. Soc.,* 2661, 2676 (1929).
42. Strickland-Constable, R. F., *Trans. Faraday Soc.,* 34:1074, 1374 (1938).
43. King, A., *J. Chem. Soc.,* 1688 (1936).

44. Rideal, E. K., and Wright, W. M., *J. Chem. Soc.*, 1347 (1925); 1813, 3182 (1926).
45. Wright, W. M., *Proc. Cambridge Phil. Soc.*, 23:187 (1926).
46. King A., *J. Chem. Soc.* 22 (1934); 889 (1935).
47. Wiegand, W. B., *Ind. Eng. Chem.*, 29:953 (1937).

INDEX